Dimitrios Kolymbas

Tunelling and Tunnel Mechanics

Dimitrios Kolymbas

Tunelling and Tunnel Mechanics
A Rational Approach to Tunnelling

With 332 Figures

Professor Dr. Dimitrios Kolymbas
Universität Innsbruck
Fakultät für Bauingenieurwesen und Architektur
Institut für Geotechnik und Tunnelbau
Technikerstr. 13/III
6020 Innsbruck
Austria
dimitrios.kolymbas@uibk.ac.at

Library of Congress Control Number: 2005926885

ISBN-10   3-540-25196-0 Springer Berlin Heidelberg New York
ISBN-13   978-3-540-25196-5 Springer Berlin Heidelberg New York

This work is subject to copyright. All rights are reserved, whether the whole or part of the material is concerned, specifically the rights of translation, reprinting, reuse of illustrations, recitation, broadcasting, reproduction on microfilm or in other ways, and storage in data banks. Duplication of this publication or parts thereof is permitted only under the provisions of the German Copyright Law of September 9, 1965, in its current version, and permission for use must always be obtained from Springer-Verlag. Violations are liable to prosecution under German Copyright Law.

Springer is a part of Springer Science+Business Media
springeronline.com

© Springer-Verlag Berlin Heidelberg 2005
Printed in Germany

The use of general descriptive names, registered names, trademarks, etc. in this publication does not imply, even in the absence of a specific statement, that such names are exempt from the relevant protective laws and regulations and therefore free for general use.

Typesetting: Dataconversion by author
Final processing by PTP-Berlin Protago-T$_E$X-Production GmbH, Germany
Cover-Design: medionet AG, Berlin
Printed on acid-free paper   62/3141/Yu – 5 4 3 2 1 0

Dedicated to Ingrid

Preface

Tunnelling is an exciting and rapidly evolving technology. Pioneering processes are commonplace and innovative thinking continues to rewrite the rules. In civil engineering, tunnelling is one of the few areas where new horizons are constantly being discovered.
But for the profession to reach its full potential, tunnelling needs to be more accessible to those talented engineers in search of new challenges and keen to make lasting contributions to society.
In the eyes of too many, tunnelling is still seen as the exclusive domain of too few: a mysterious art form, accessible only to those who have already spent countless years perfecting their approach, a skill whose secrets remain suppressed.
Over the following pages I hope to show that tunnelling need not be a closed book. I have omitted methods and definitions that depended more on historical precedent than modern scientific evaluation. Instead of confusing the reader with countless details and definitions that are in any case open to change, I have focused on the underlying concepts that make tunnelling easier to grasp.
So while this book is designed to provide a concise, up-to-date and useful frame of reference to all those newly qualified and engaged in the field, I hope that it will also serve to reveal to those talented engineers who thought they had found their niche above ground the very real opportunities and unanswered questions that await them underground.
As rock mechanics is less developed and less known than soil mechanics, the principles of this young discipline are included in this book. I also attempted to integrate theoretical and practical viewpoints, since I consider both of them indispensable to engineering. The often encountered attitude of practitioners to blame theory (and vice versa) is regrettable. The theory addressed here refers to the mechanics of soil/rock behaviour and its interaction with the various support structures. Some chapters of the book aim at the description of planning and construction processes and, thus, need no reference to theory. Some other chapters contain both, practical and theoretical aspects. Some

others again are of theoretical nature and refer also to open questions and yet unsolved problems. They aim not only at informing the reader about the present state of knowledge but also at highlighting the possibilities of theoretical approach and stimulating research. In many cases the link between practical and theoretical aspects is established by cross-references. Lengthy theoretical derivations are presented in appendices. Those readers who are merely interested in basic concepts and practical applications can skip the theoretical parts.

Cut and cover tunnelling is not included in this book, as the underlying approaches, such as diaphragm walls etc., are described in textbooks of geotechnical engineering. The reader should take into account that all quantitative statements, referring to e.g. strength of materials, costs, durations etc., are understood as mere examples, as they are subject to technical progress.

As in every book, the distribution of emphasis is subjective and mirrors the scientific profile of the author. My aim is to inform the reader about concepts relevant for tunnelling rather than putting together all available information on the topic. The exhaustive completeness encountered in codes and standards has been avoided, as it would render the text too lengthy and falls beyond the scope of this book.

I wish to thank my co-workers Dr. Andreas Laudahn, Dr. Pornpot Tanseng and Dipl.-Ing. Markus Mähr for their valuable help in preparing the manuscript. Professors Gerd Gudehus, Konrad Kuntsche and Christos Vrettos as well as Dipl.-Ing. Sigmund Fraccaro contributed with many valuable suggestions. I owe particular thanks to Yannis Bakoyannis for his many and invaluable tips. In the final stage of the concluding and finishing the book the work piled up to such an extent that this book would not have seen the light of day without the dedicated, considerate and skilful aid of my co-worker Dipl.-Ing. Ansgar Kirsch. Also the contribution of Michaela Major and Alexander Schuh to the drawings is thankfully acknowledged.

Innsbruck,
April 2005

Dimitrios Kolymbas

Contents

Part I Design

1 Introduction 3
   1.1 Benefits from tunnelling 3
   1.2 Statistical review 3
   1.3 Notations in tunnelling 6
   1.4 Cross sections 7
        1.4.1 Road tunnels 15
        1.4.2 Rail tunnels 17
   1.5 Alignment 18
   1.6 Underground water conduits 20
   1.7 Standards and Recommendations 22
   1.8 Costs 24
   1.9 Planning and contracting 25
        1.9.1 Cost and time management 28
        1.9.2 Experts 29

2 Installations in tunnels 31
   2.1 Installations for traffic control 31
   2.2 Installations for telecommunication 32
   2.3 Ventilation 32
        2.3.1 Ventilation during construction 32
        2.3.2 Design of construction ventilation 35
        2.3.3 Ventilation of road tunnels 37
        2.3.4 Control of ventilation 41
   2.4 Fire protection 42
        2.4.1 Fire-resistant concrete 45
        2.4.2 Fire detectors and extinguishers 46
        2.4.3 Example: Refurbishment of the Montblanc tunnel . 48
   2.5 Illumination of road tunnels 49
   2.6 Drainage 52

X   Contents

	2.7	Examples for the equipment of modern road tunnels	53
	2.8	Rating of safety in road tunnels	55

3 Investigation and description of the ground 57
- 3.1 Geotechnical investigations 57
    - 3.1.1 Preliminary investigation 58
    - 3.1.2 Main site investigation 59
    - 3.1.3 Investigation during and after construction 59
- 3.2 Site investigation 60
    - 3.2.1 Exploration drilling 62
- 3.3 Geophysical exploration 64
- 3.4 Joints 64
- 3.5 Weathering 69
- 3.6 Rock rating and classification 69
    - 3.6.1 RMR-System 70
    - 3.6.2 Q-system 71
- 3.7 Reports 72

4 Heading 75
- 4.1 Full face and partial face excavation 75
- 4.2 Excavation 80
- 4.3 Drill & blast 84
    - 4.3.1 Drilling of blastholes 85
    - 4.3.2 Charging 86
    - 4.3.3 Tamping 86
    - 4.3.4 Ignition 86
    - 4.3.5 Distribution of charges and consecution of ignition . 87
    - 4.3.6 Explosives 88
    - 4.3.7 Explosive consumption 89
    - 4.3.8 Safety provisions 89
    - 4.3.9 Ventilation 90
    - 4.3.10 Backup 90
    - 4.3.11 Shocks and Vibrations 90
- 4.4 Shield heading 92
    - 4.4.1 Shield heading in groundwater 103
    - 4.4.2 Tunnelling with box- or pipe-jacking 109
    - 4.4.3 Microtunnels 110
    - 4.4.4 Speed of advance 111
    - 4.4.5 Drive-in and drive-out operations 112
    - 4.4.6 Problems with shield heading 112
- 4.5 Comparison of TBM with conventional heading 114
- 4.6 Rock excavation 115
    - 4.6.1 Drilling of boreholes 116
    - 4.6.2 Rock excavation with disc cutters 118

		4.6.3	Abrasion	124
		4.6.4	Drilling: history review	125
	4.7	Profiling		126
	4.8	Mucking		127

5 Support ... 131
5.1 Basic idea of support ... 131
5.2 Shotcrete ... 132
5.2.1 Steel fibre reinforced shotcrete (SFRS) ... 136
5.2.2 Quality assessment of shotcrete ... 137
5.3 Steel meshes ... 137
5.4 Rock reinforcement ... 137
5.4.1 Connection with the adjacent rock ... 138
5.4.2 Tensioning ... 142
5.4.3 Testing ... 143
5.4.4 Application ... 143
5.5 Timbering ... 144
5.6 Support arches ... 146
5.7 Forepoling ... 147
5.8 Face support ... 149
5.9 Sealing ... 149
5.10 Recommendations for support ... 150
5.11 Temporary and permanent linings ... 152
5.12 Permanent lining ... 153
5.12.1 Reinforcement of the permanent lining ... 154
5.12.2 Quality assessment of the lining ... 155
5.13 Single-shell (monocoque) lining ... 156

6 Grouting and freezing ... 159
6.1 Low pressure grouting ... 159
6.2 Soil fracturing, compensation grouting ... 161
6.3 Jet grouting ... 163
6.4 Grouts ... 163
6.5 Rock grouting ... 166
6.6 Advance grouting ... 168
6.7 Soil freezing ... 168
6.7.1 Frost heaves ... 169
6.8 Propagation of frost ... 170

7 The New Austrian Tunnelling Method ... 171
7.1 HSE Review ... 173

Contents

8 Management of groundwater 177
- 8.1 Flow within rock 177
  - 8.1.1 Porosity of rock 177
  - 8.1.2 Pore pressure 178
  - 8.1.3 Permeability of rock 179
- 8.2 Inflow in the construction phase 181
- 8.3 To drain or to seal? 183
- 8.4 Drainage 183
- 8.5 Water ingress into a drained circular tunnel 187
  - 8.5.1 Seepage force 189
- 8.6 Influence of drainage 190
- 8.7 Sealing (waterproofing) 191
- 8.8 Geosynthetics in tunnelling 195

9 Application of compressed air 197
- 9.1 Health problems 199
- 9.2 Influence on shotcrete 201
- 9.3 Blow-outs 201

10 Subaqueous tunnels 203
- 10.1 Towing and lowering method 204
- 10.2 Caissons 206

11 Shafts 211
- 11.1 Driving of shafts 211
- 11.2 Earth pressure on shafts 214

12 Safety during construction 217
- 12.1 Health hazards 217
- 12.2 Electrical installations in tunnelling 221
  - 12.2.1 Hazards due to failure of vital installations 221
  - 12.2.2 Special provisions 221
  - 12.2.3 Energy supply of excavation machines 222
  - 12.2.4 Illumination during construction 223
- 12.3 Controls 223
- 12.4 Risk management 224
- 12.5 Emergency plan and rescue concept 225
- 12.6 Quantification of safety 226
- 12.7 Collapses 226
  - 12.7.1 Heathrow collapse 229

Part II Tunnelling Mechanics

13 Behaviour of soil and rock ... 235
- 13.1 Soil and rock ... 235
- 13.2 General notes on material behaviour ... 235
- 13.3 Elasticity ... 237
- 13.4 Plasticity ... 239
- 13.5 Strength ... 240
  - 13.5.1 Strength of soil ... 240
  - 13.5.2 Strength of rock ... 243
  - 13.5.3 Brittle and ductile behaviour ... 246
- 13.6 Post-peak deformation ... 248
  - 13.6.1 Point load test ... 249
  - 13.6.2 Griffith's theory ... 250
  - 13.6.3 Acoustic emission ... 251
  - 13.6.4 Friction of joints ... 251
- 13.7 Anisotropy ... 252
- 13.8 Rate dependence and viscosity of soil and rock ... 253
- 13.9 Size effect ... 256
  - 13.9.1 Size effect in rock ... 256
  - 13.9.2 Size effect in soil ... 258
  - 13.9.3 Rodionov's theory ... 258
- 13.10 Discrete models ... 260
- 13.11 Rock mass strength ... 261
- 13.12 Swelling ... 265
- 13.13 Field tests ... 267

14 Stress and deformation fields around a deep circular tunnel 271
- 14.1 Rationale of analytical solutions ... 271
- 14.2 Some fundamentals ... 271
- 14.3 Geostatic primary stress ... 276
- 14.4 Hydrostatic primary stress ... 279
- 14.5 Plastification ... 281
  - 14.5.1 Consideration of cohesion ... 284
- 14.6 Ground reaction line ... 285
- 14.7 Pressuremeter, theoretical background ... 288
- 14.8 Support reaction line ... 289
- 14.9 Rigid block deformation mechanism for tunnels and shafts ... 290
- 14.10 Squeezing ... 293
  - 14.10.1 Squeezing as a time-dependent phenomenon ... 293
  - 14.10.2 Neglecting time-dependence ... 295
  - 14.10.3 Interaction with support ... 297
  - 14.10.4 Squeezing in anisotropic rock ... 301
- 14.11 Softening of the ground ... 301

15 Supporting action of anchors/bolts ... 305
- 15.1 Impact of pattern bolting ... 306
  - 15.1.1 Ground stiffening by pre-stressed anchors ... 307
  - 15.1.2 Pre-stressed anchors in cohesive soils ... 309
  - 15.1.3 Stiffening effect of pattern bolting ... 311

16 Some approximate solutions for shallow tunnels ... 313
- 16.1 Janssen's silo equation ... 313
- 16.2 Trapdoor ... 316
- 16.3 Support pressures at crown and invert ... 319
- 16.4 Forces acting upon and within the lining ... 325
- 16.5 Estimations based on the bound theorems ... 327
  - 16.5.1 Lower bound of the support pressure ... 327
  - 16.5.2 Upper bound of the support pressure ... 328

17 Stability of the excavation face ... 329
- 17.1 Approximate solution for ground with own weight ... 329
- 17.2 Numerical results ... 330
- 17.3 Stability according to the bound theorems ... 330
- 17.4 Stand-up time of the excavation face ... 333

18 Earthquake effects on tunnels ... 335
- 18.1 General remarks ... 335
- 18.2 Imposed deformation ... 336

19 Settlement of the surface ... 339
- 19.1 Estimation of settlement ... 339
- 19.2 Reversal of settlements with grouting ... 344
- 19.3 Risk of building damage due to tunnelling ... 344

20 Stability problems in tunnelling ... 349
- 20.1 Rockburst ... 349
- 20.2 Buckling of buried pipes ... 349
  - 20.2.1 Buckling of pipes loaded by fluid ... 350
  - 20.2.2 Buckling of elastically embedded pipes ... 352

21 Monitoring ... 353
- 21.1 Levelling ... 354
- 21.2 Monitoring of displacements and convergence ... 354
- 21.3 Extensometers and inclinometers ... 355
- 21.4 Monitoring stresses within the lining ... 357
- 21.5 Measurement of primary stress ... 360
  - 21.5.1 Hydraulic fracturing ... 360
  - 21.5.2 Unloading and compensation methods ... 362
- 21.6 Cross sections for monitoring ... 363

22 Numerical analysis of tunnels 365
22.1 General remarks 365
22.1.1 Initial and boundary conditions.................... 366
22.1.2 Coping with non-linearity 368
22.1.3 Constitutive equation 370
22.2 Method of subgrade reaction 371
22.3 Difficulties related to the design of shotcrete lining 373

Part III Appendices

A Physics of detonation 379
A.1 Detonation... 379
A.2 Underground explosions 381
A.3 Interaction of charges 382

B Support of soil with a pressurised fluid 385

C A simple analytical approximation for frost propagation.... 387

D Rigorous solution for the steady water inflow to a circular tunnel.. 393

E Aerodynamic pressure rise in tunnels 397

F Multiphase model of reinforced ground.................... 399

G Deformation of a tunnel due to seismic waves 403

H A rational approach to swelling 405

Glossary.. 409
9.1 English - German 410
9.2 German - English 419

Index 429

Part I

Design

1

Introduction

1.1 Benefits from tunnelling

The necessity for tunnels and the benefits they bring cannot be overestimated. Tunnels improve connections and shorten lifelines. Moving traffic underground, they improve the quality of life above ground and may have enormous economic impact. The utilisation of underground space for storage, power and water treatment plants, civil defense and other activities is often a must in view of limited space, safe operation, environmental protection and energy saving. Of course, the construction of tunnels is risky and expensive and requires a high level of technical skill.

1.2 Statistical review

Tunnels are continuously constructed and the variety of owners and purposes is very large. As a consequence, any statistics can only be an approximation of the present situation. The following tables give some information about well-known tunnels.[1]

---

[1] Overviews of tunnel history can be found in: K. Széchy, Tunnelbau, Springer-Verlag, Wien, 1969; Z.T. Bieniawski, Rock Mechanics Design in Mining and Tunnelling, Balkema, 1984, M. Sintzel und G. Girmscheid: Vom Urner Loch zum NEAT-Projekt: das Abenteuer der Alpendurchquerung, *Spektrum der Wissenschaft*, August 2000; A. Muir-Wood, Tunnelling: Management by design, Spon, London, 2000.

The oldest tunnels	
Eupalinos tunnel (Samos, 500 BC)	1 km
Urner Loch (1707, first alpine tunnel in Switzerland)	64 m
Mont-Cenis (France - Italy, 1857-1870, also called Fréjus tunnel)	12 km
St. Gotthard rail tunnel (Switzerland, 1872-1878)	15 km
The longest tunnels	
Seikan (Japan, 1981-1984)	54 km
Euro-tunnel (France - England, 1986-1993)	50 km
Simplon I (Switzerland - Italy, 1898-1906)	20 km
Grand-Apennin (Italy, 1921-1930)	19 km
New Gotthard (Switzerland, 1969-1980)	16 km
The longest tunnels in Austria	
Arlberg (1974-1977) *road*	14.0 km
Arlberg (1880-1884) *rail*	10.3 km
Plabutsch	9.9 km
Gleinalm (1973-1977)	8.3 km
Tauern (1903-1909) *rail*	8.5 km
Karawanken (1902-1906)	8.0 km
Landeck	7.0 km
Pfänder	6.7 km
Tauern (1970-1975) *road*	6.4 km
Bosruck	5.5 km
Katschberg (1971-1974)	5.4 km
Felbertauern (1963-1966)	5.3 km
Large metro systems	
New York	221 km
London	414 km
Paris	165 km
Moscow	254 km

The tunnelling activity in **Austria** is shown in the following table:[2]

	in operation	in construction	planned
road tunnels	343 km	41 km	119 km
rail tunnels ÖBB	150 km	50 km	275 km
metro	15 km		

---

[2] As for end of 2004. The length of tunnels is here given as total length of tubes. E.g. the 55 km long Brenner base tunnel will comprise 160 km of tunnel tubes.

The Landeck road bypass tunnel (7 km long) has been accomplished in 2000, and the construction of the 5.8 km long twin tube road tunnel of Strengen started in 2001 (completion scheduled for 2005).

The tunnelling activity in **Germany** is shown in the following table:[3]

	in construction	planned
road tunnels	37 km	130 km
rail tunnels DB	25 km	247 km
urban rail	32 km	84 km

The corresponding figures for **Switzerland** are:[4]

	in operation	in construction
road tunnels	322 km	86 km
rail tunnels	392 km	101 km

Swiss rail tunnels			
Tunnel	Year	Length (km)	Max. cover (m)
Gotthard	1873-1882	14.9	1,800
Simplon 1	1898-1906	19.8	2,150
Lötschberg	1906-1913	14.5	1,600
Simplon 2	1912-1921	19.8	2,150
Furka	1973-1982	15.4	1,500
Vereina	1991-1999	19.0	1,500
Swiss road tunnels			
Gotthard	1970-1980	16.9	1,500
Seelisberg	1970-1980	2×9.3	1,300
Alp Transit tunnels			
Gotthard	1993-	2×56.9	2,300
Lötschberg	1994-	2×41.9	2,300

With 57 km the Gotthard base tunnel will become the longest tunnel in the world, to be completed probably in 2015.

In the **Netherlands** several major projects are completed, currently running or at the planning stage:[5]

- 2nd Heinenoord tunnel, beneath the Old Maas (2×945 m)

---

[3] As for beginning of 2004. Source: STUVA (www.stuva.de)

[4] As for beginning of 2003. Source: Tunnelling Switzerland, Swiss Tunnelling Society, Bertelsmann 2001

[5] E. Gurkan, W. Fritsche, Transport Tunnels in the Netherlands built by Shield Tunnelling, *Tunnel* 8/2001, 6-17

1 Introduction

- Botlek railway tunnel, beneath the Old Maas (2×1,835 m), part of the 160 km long Betuweroute that connects Rotterdam with Germany for freight trains.
- Westerschelde tunnel: Two 6.6 km road tubes.
- Sophia tunnel: Two 4.24 km long tubes for the Betuweroute.
- Tunnel beneath the Pannerdesch Canal: Two 1.62 km long tubes for the Betuweroute.
- Groene Hart tunnel: 7.16 km long tunnel for high-speed trains.

The new high-speed train route Bologna-Firenze in **Italy** includes ca 78 km of twin-track tunnel with a cross-sectional area of 140 m^2, representing 92 % of the entire route. The longest tunnels are Firenzuola (14.3 km), Raticosa (10.4 km) and Pianoro (9.3 km).

In **Spain**, the 28.4 km long twin tube rail tunnel Guadarrama is being driven with 4 double shield (two of Herrenknecht and two of Wirth) through hard rock (gneiss and granite) including weak zones.

For the Japanese high-speed rail Shinkansen many tunnels were constructed, the longest being Daishimizu with 22.2 km. As of April 2000 **Japan** had road tunnels with a total length of 2.575 km.[6]

Optimistic estimations for the **PR China** predict the annual construction of 180 km pressure tunnels, 40 km mining tunnels and 300 km traffic tunnels.

1.3 Notations in tunnelling

Considering the cross and longitudinal sections of tunnels shown in Fig. 1.1 and 1.2, the various locations are denoted by the indicated names. The word 'chainage' is used to identify a point along the axis of a tunnel defined by its distance from a fixed reference point.

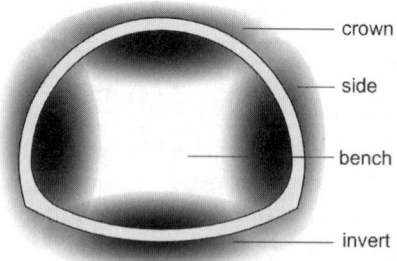

Fig. 1.1. Parts of a tunnel cross section

---

[6]H. Mashimo, State of the road tunnel safety technology in Japan, *Tunnelling and Underground Space Technology*, 17 (2002), 145-152

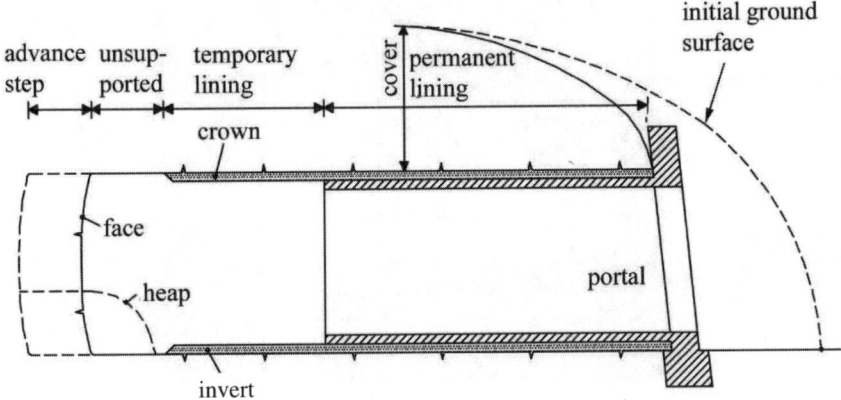

Fig. 1.2. Longitudinal sections of heading

1.4 Cross sections

The shape of a tunnel cross section is also called profile. Various profiles are conceivable, e.g. rectangular ones. The most widespread ones, however, are circular (e.g. Fig. 1.9) and (mostly oblate) mouth profiles (Fig. 1.4). The choice of the profile aims at accommodating the performance requirements of the tunnel. Moreover it tries to minimise bending moments in the lining (which is often academic, since the loads cannot be exactly assessed) as well as costs for excavation and lining. Further aspects for the choice of the profile are: ventilation, maintenance, risk management and avoidance of claustrophobia of users.

A mouth profile is composed of circular sections. The ratio of adjacent curvature radiuses should not exceed 5 ($r_1/r_2 < 5$). The minimum radius should not be smaller than 1.5 m. Note that in the case of weak rock the lower part of the lining also receives load from the adjacent ground. Therefore, a curved profile is advisable from a statical point of view also in the invert (see also Sec. 16.3).

The following relations refer to geometrical properties of mouth profiles. With the initial parameters r_1, r_2, r_3 it is obtained:

$$\sin \beta = \frac{r_1 - r_2}{r_3 - r_2}$$

$$c = \sqrt{r_3^2 - 2r_2(r_3 - r_1) - r_1^2}$$

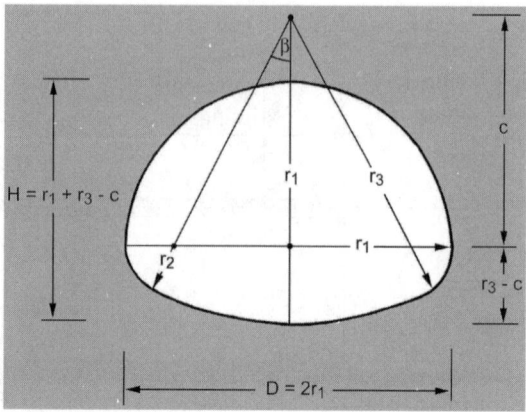

Fig. 1.3. Geometry of a mouth profile

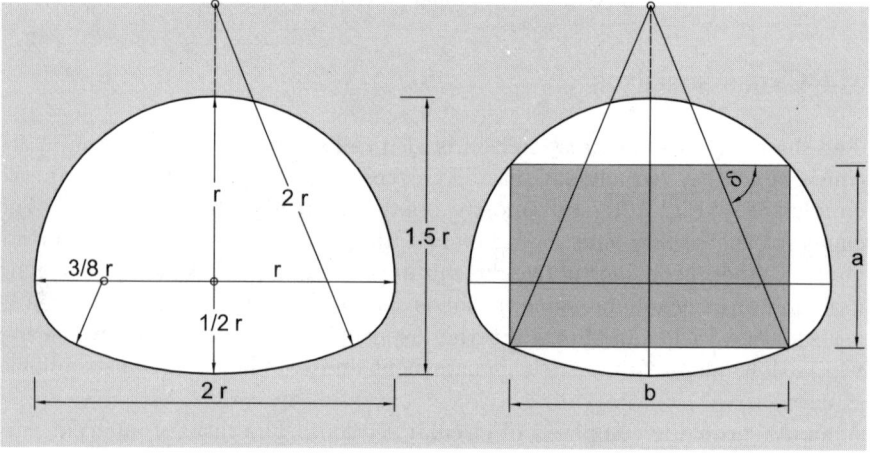

Fig. 1.4. Example of a mouth profile

cross section area $A = \dfrac{\pi}{2}r_1^2 + \left(\dfrac{\pi}{2} - \beta\right)r_2^2 + \beta r_3^2 - (r_1 - r_2)c$;   (β in rad!)

$$\text{height } H = r_1 + r_3 - c$$
$$\text{span } D = 2r_1$$

A typical problem of tunnel design is to fit a rectangle into a mouth profile (Fig. 1.4), i.e. to choose r_1, r_2, r_3 in such a way that the mouth profile comprises the rectangle. One possible procedure is as follows:

1. choose r_1
2. evaluate $\cos \delta = \dfrac{b}{2\sqrt{3}}$

3. evaluate $r_3 = \dfrac{1}{\sin \delta} \left(a + c - \sqrt{r_1^2 - \dfrac{b^2}{4}} \right)$

4. choose r_2, e.g. $r_2 = r_3/5$.

It is assumed that the lower edges of the rectangle are on the circle with radius r_3. Often, the size of a tunnel is given by its cross section area. Typical values are:

Typical tunnel cross sections	Area (m^2)
sewer	10
hydropower tunnels	10 - 30
motorway (one lane)	75
rail (one track)	50
metro (one track)	35
high speed rail (one track)	50
high speed rail (two tracks)	80 - 100

10   1 Introduction

In the following, some examples of tunnel cross sections are given.

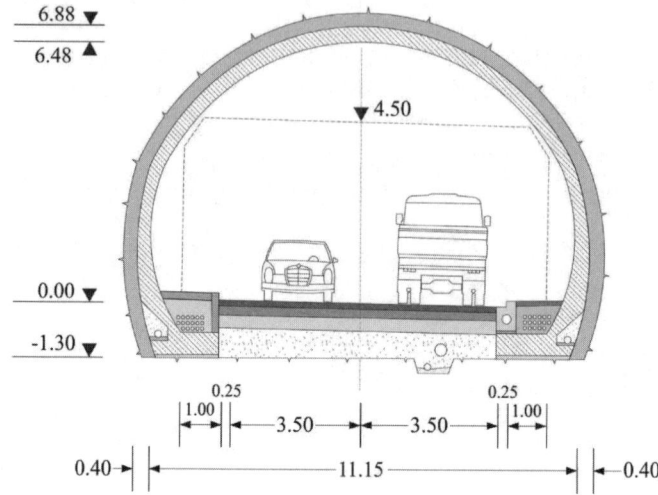

Fig. 1.5. Hochrheinautobahn – Bürgerwaldtunnel, A 98 Waldshut-Tiengen

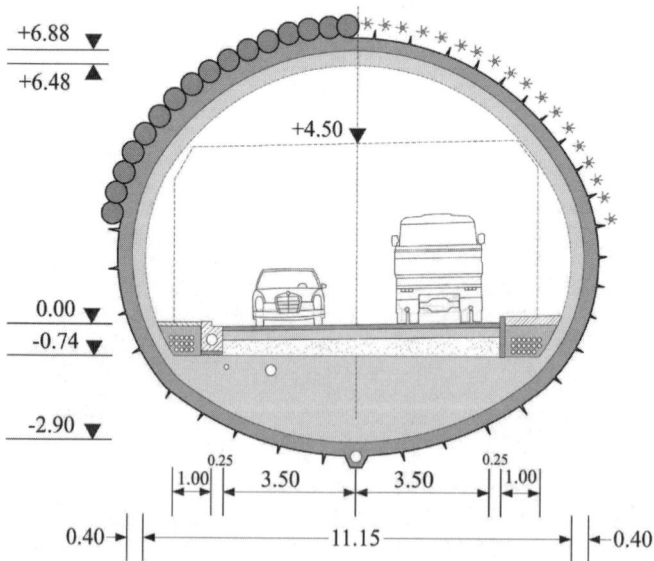

Fig. 1.6. Hochrheinautobahn – Bürgerwaldtunnel, A 98 Waldshut-Tiengen

1.4 Cross sections 11

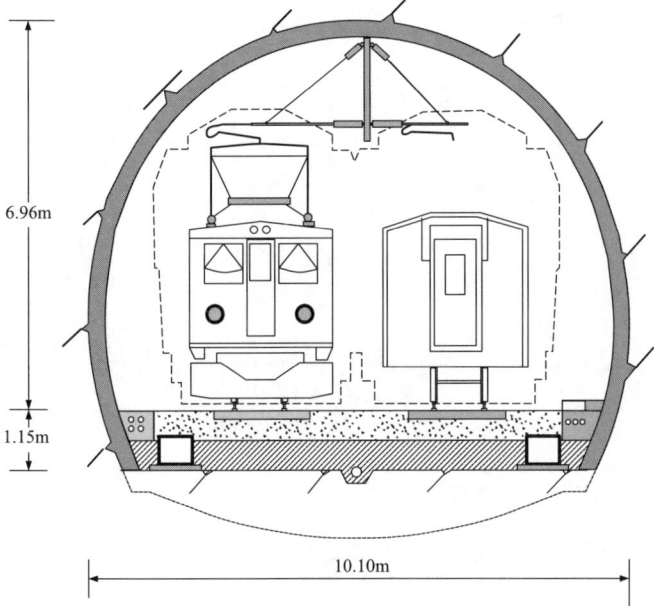

Fig. 1.7. Vereina Tunnel South, Rätische Bahn; two tracks

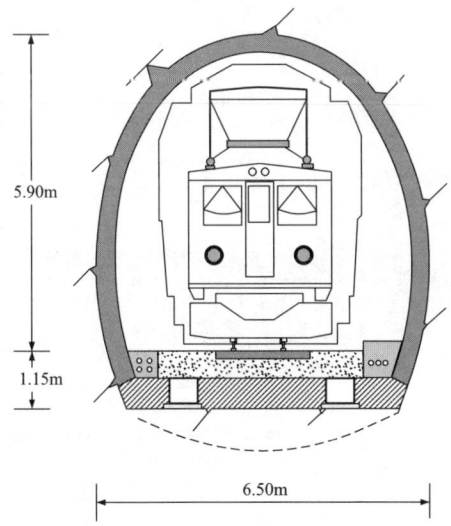

Fig. 1.8. Vereina Tunnel South, Rätische Bahn; one track

12   1 Introduction

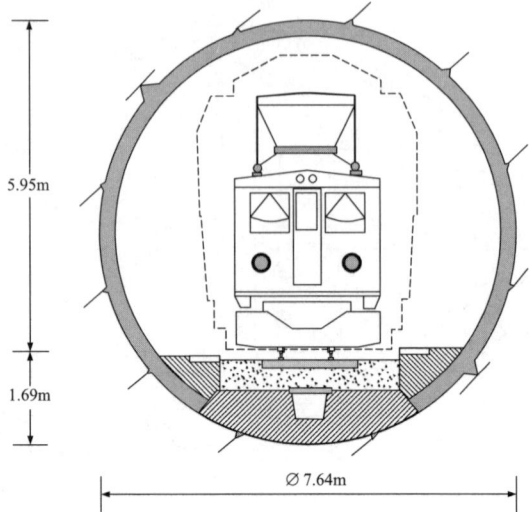

Fig. 1.9. Zugwaldtunnel Rätische Bahn; one track

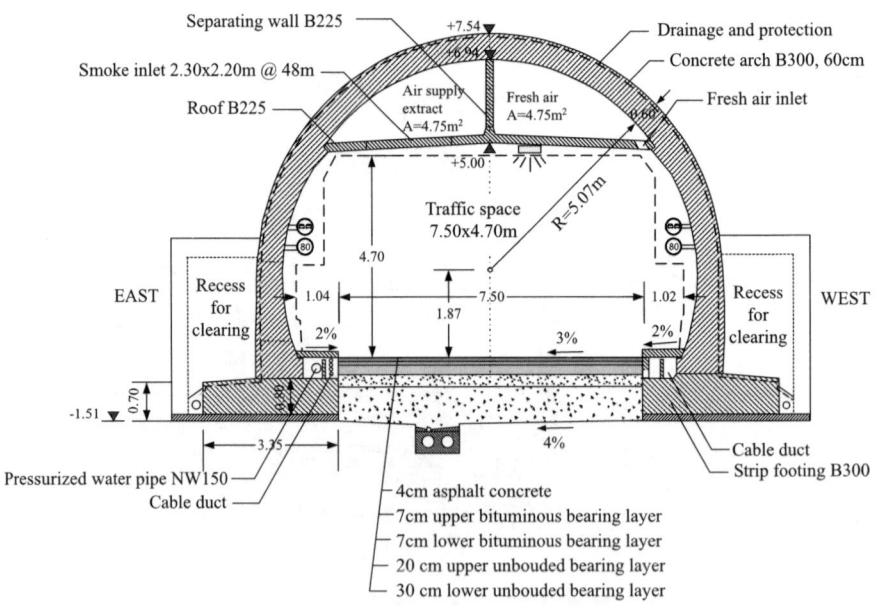

Fig. 1.10. Southern bypass Landeck tunnel

1.4 Cross sections 13

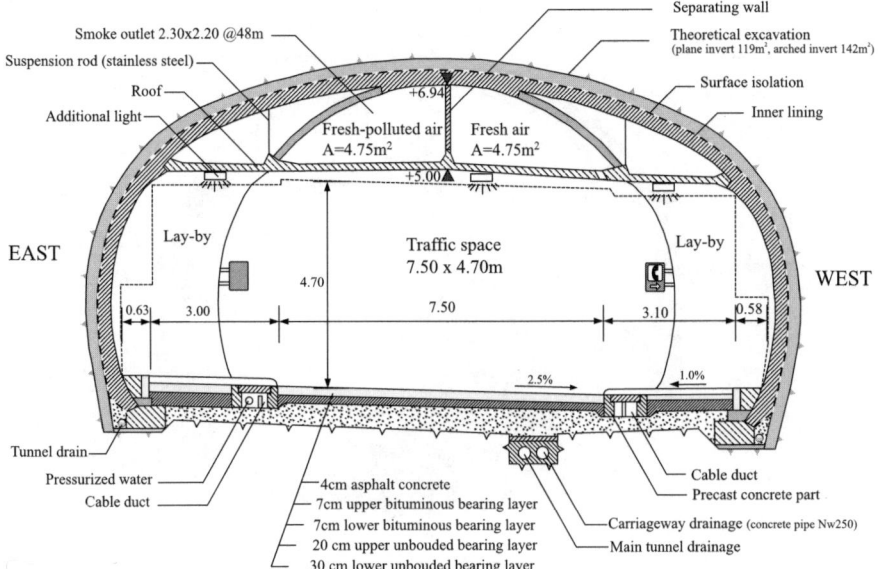

Fig. 1.11. Southern bypass Landeck tunnel

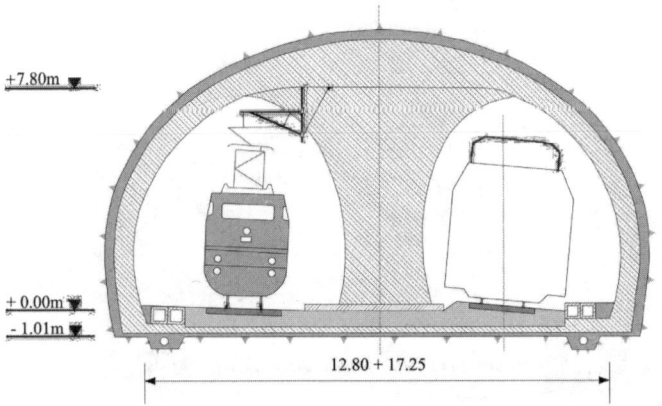

Fig. 1.12. Markstein/Nebenwegtunnel

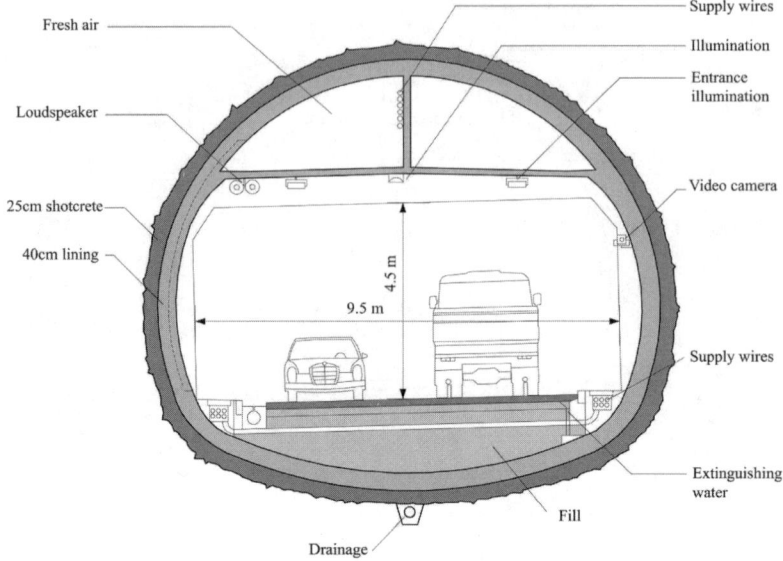

Fig. 1.13. B14 Bypass Heslach

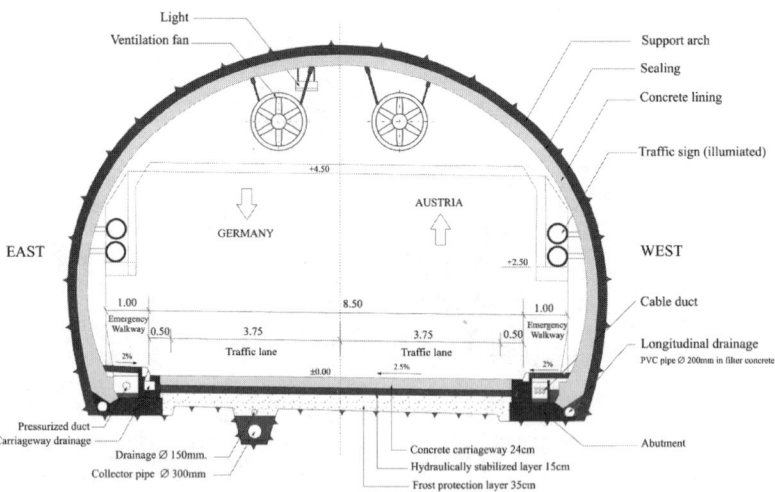

Fig. 1.14. Füssen tunnel

1.4.1 Road tunnels

The several parts of road tunnels are named as shown in Fig 1.15.[7] The regulations for their design vary across the several countries, the figures given in this section originate mainly from the PIARC recommendations. The design of the tunnel should take into account the expected traffic capacity, safety measures (emergency exits, evacuation tunnels, lay-by's, vehicle turn around points), provisions for breakdowns[8], horizontal and vertical alignment (see Section 1.5), widths of the several elements shown in Fig. 1.15, and vertical clearances, in accordance to the regulations prevailing for a considered tunnel. Here are given only some general hints:

Capacity: The traffic capacity depends on several factors, among them the width of the lanes, the percentage of heavy vehicles, the longitudinal inclination etc. The traffic capacity in tunnels appears to be slightly higher than in open sections of a highway. This is possibly due to increased concentration of drivers in tunnels. According to the French regulations, the

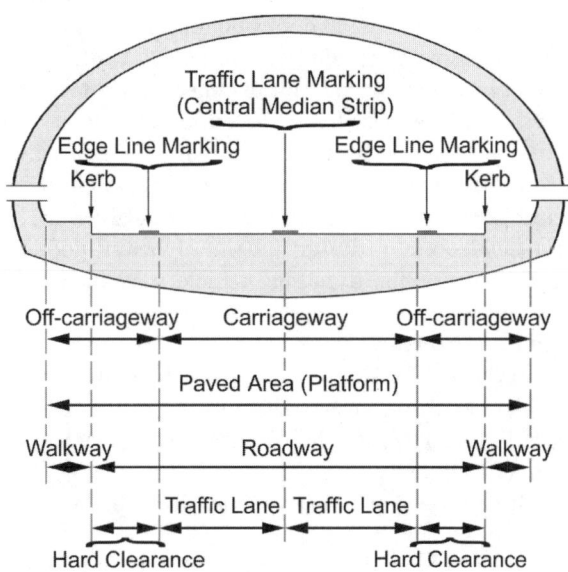

Fig. 1.15. Partition of road tunnels according to PIARC

---

[7] World Road Congresses; Technical Committee on Road Tunnels PIARC, Cross Section Design for Bi-Directional Road Tunnels, 2004

[8] A breakdown is defined as a vehicle stopping inside the tunnel for any reason except traffic congestion

traffic capacity of bi-directional urban tunnels is 2200 pcph (passenger cars per hour), for bi-directional mountain tunnels it is 2350 pcph. The speed in bi-directional tunnels should be limited to 80-90 km/h.

Widths: For traffic lanes, the recommended width is 3.50 m (in North America: 3.65 m). The width of traffic lane markings should be $\geq$ 15 cm. For safety reasons, the width of the roadway should be $\geq$ 8.50 m, so as to enable a heavy goods vehicle to overtake another vehicle that is stopped without completely interrupting traffic in the opposite direction. The hard clearance is intended to:
- increase lane capacity
- provide a redress space for drivers crossing the edge line
- provide space for breakdowns
- provide an emergency lane to give access to rescue services
- facilitate maintenance activities

and should, therefore, have a width of 2.50 - 3.00 m. The width of the hard clearances can be reduced to a minimum of 0.50 m if their functions as emergency and breakdown lane are taken over by a central median strip having a width between 1.0 and 2.5 m.

Walkways:[9] Walkways are used by staff, and in case of incidents, by pedestrians. They also serve to enable door opening (from stopped cars and also emergency escape doors etc.). The recommended width is 0.75 m. Regular pedestrian and bicycle traffic should, in general, not be allowed in tunnels. The walkway should be raised 7 to 15 cm above the carriageway with a vertical kerb[10]. Alternatively, walkways can be demarcated by roll-over kerbs. In this case, they can also be used by breaking down cars.

Vertical clearances: It should be distinguished between
- Minimum headroom = design height of heavy goods vehicles, plus necessary allowance for dynamic vehicle movements. Recommended value: 4.20 m, and
- Maintained headroom [11].

Additional allowances for signs, luminaries, fans etc. vary between 0.20 and 0.40 m. The headroom over walkways should be 2.30 m (if the walkways are accessible also to cars via roll-over kerbs, then the aforementioned headrooms should be kept).

Climbing lane: It the speed of heavy vehicles drops below 50 km/h, a climbing lane of 3.00 m width should be planned.

Lay-by's: Lay-by's are planned to accommodate breakdowns. Their spacing should decrease with increasing traffic density. It should be noted, however, that it is unprobable that a breakdown occurs exactly where a lay-by is to be found. A variant of lay-by's are turn-around points.

---

[10] U.S.: curb

[11] U.S.: clear height = minimum headroom + confort margin, i.e. headroom, which shall be preserved at all times, e.g. after resurfacing, to ensure safe passage of permitted traffic. Recommended value: 4.50 m (RVS: 4.70 m)

For a daily traffic volume[12] exceeding 10,000 vehicles per lane, a twin tube tunnel with uni-directional traffic should be considered. Two tubes are much safer and also preferable for maintenance reasons. In case of fire, a twin tube with cross walks and cross drives offers better chances for escape and also better access for rescue services. Note that the supplementary construction of the second tube faces a reduced geological risk, as information from the first heading can be used.

1.4.2 Rail tunnels

Often a choice between the economically preferable one tube with two tracks and the safer twin tubes is necessary.[13] Required cross sections have been specified by the rail companies. The German Rail (DB), e.g., prescribes a radius of 4.9 m for circular cross sections for one-track high speed train tunnels. For two-track tunnels cross sections of 92 to 101 m^2 are required. With improved electric wire suspension this area can be reduced to ca 85 m^2.

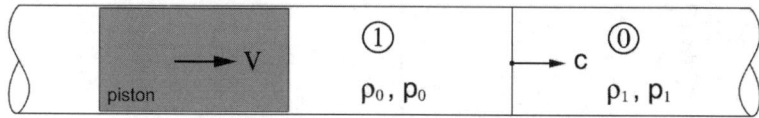

Fig. 1.16. Piston moving within a tube

The aerodynamic air pressure rise due to the entrance of a high speed train into a tunnel may cause discomfort of the passengers (unless the cars are pressure-tight) and constitutes a load acting upon the lining. The prevailing aerodynamic effect can be understood if we consider a piston entering with a high speed into a tube (Fig. 1.16). Neglecting the gap between piston and tube we observe that the piston moving with the velocity V imposes the same velocity upon the air in front of it. This gives rise to a compaction shock that moves ahead with the velocity c, which reads (cf. Appendix E):

---

[12]also called Annual Average Daily Traffic (AADT) = Total traffic flow in one year divided by 365 days. Expressed in vehicles per day.

[13]see R. Grüter und W. Schuck: Tunnelbautechnische Grundsatzentscheidungen für den Bau neuer Schnellfahrstrecken für die Deutsche Bahn A.G., Vorträge der STUVA-Tagung 1995 in Stuttgart, 38-44. See also E. Knoll: Sicherheit im Bahn Tunnel – Der Schlern Tunnel in Südtirol – Eine neue Sicherheitsphilosophie. *Österreichische Ingenieur- und Architektenzeitschrift*, 142. Jg., Heft 11-12/1997, 789-800

$$c = \frac{1}{2}\left(V + \sqrt{V^2 + 4\frac{p_0}{\rho_0}\kappa}\right) . \tag{1.1}$$

In reality, air can escape backwards through the gap between train and tunnel wall. Thus, the pressure in the air cushion pushed ahead of the train rises gradually, as the length of the gap increases. When the bow shock front reaches the portal, then a rarefaction shock travels backwards to the piston/train and causes there a sudden pressure drop. Realistic figures can only be obtained numerically and are shown in Fig. 1.17.

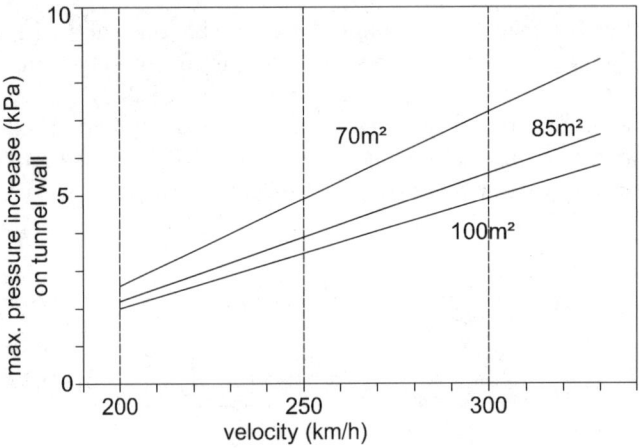

Fig. 1.17. Aerodynamic pressure rise on the tunnel wall due to entrance of high speed trains

With respect to double tracks in one tube, one should take into account that a high speed train should not encounter a freight train inside a tunnel, because it could be endangered by falling items.

1.5 Alignment

For the choice of the proper alignment, several aspects must be taken into account: geotechnical, traffic, hydrological and risk management issues. The main geotechnical aspect is to avoid bad rock and adverse groundwater. The choice of alignment depends also on the excavation method (drill & blast or TBM). To minimise the disturbance of the environment, aspects of vibration (e.g. due to blasting), noise and ventilation ((i) the disposal of polluted air should not impair abutters, (ii) ventilation shafts should not be too long) should be considered. In road tunnels, straight alignments longer than 1,500 m should be avoided, as they could distract the driver. Furthermore, to avoid excessive concentration on one point, the last few meters of a tunnel should have a gentle curve in plan view.

A dilemma is the choice between a high level tunnel and a base tunnel (Fig. 1.18). On the one hand a high level tunnel is much shorter and has reduced geological risk (because of the reduced cover). On the other hand the operation is more expensive because of increased power consumption and increased wear of the waggons. Velocity is reduced and traffic interruptions or delays during winter must be factored in. A base tunnel is much longer and, therefore, much more expensive and difficult to construct. But it offers many operational advantages. Consider e.g. the Engelberg tunnel on the Stuttgart-Heilbronn motorway. The old high level tunnel had a length of ca 300 m and access ramps with a 6% slope. As a result, heavy trucks had to slow down, which lead to traffic jams. The new base tunnel (completed in 1999) has a length of 2 × 2.5 km and a maximum slope of 0.9%. Each tube has three drive lanes and one emergency lane. Thus, the traffic capacity has increased dramatically.

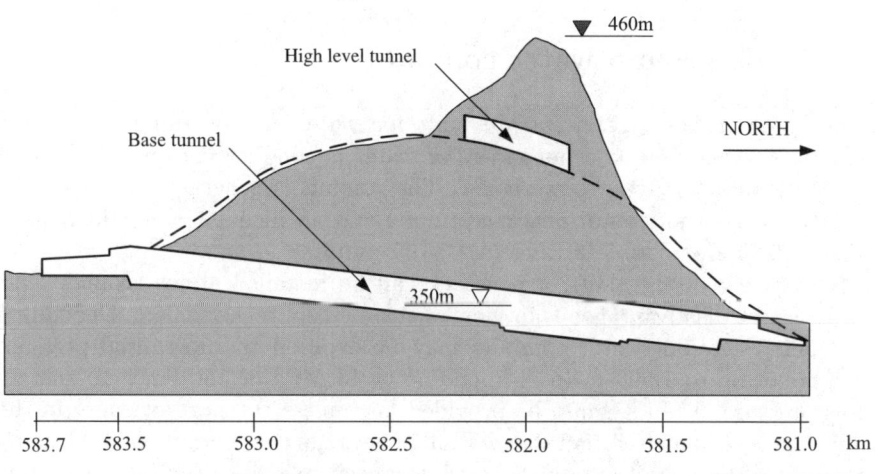

Fig. 1.18. Engelberg tunnel

Further aspects for the choice of alignment are:

Depth: For tunnelled undercrossings, a sufficient cover is needed to avoid surface settlements and daylight collapses. It requires, however, longer ramps.

Longitudinal slope s: There are the following limitations:

20   1 Introduction

$s > 0.2 - 0.5$ % for water drainage
$s <$ 2 %   for rail tunnels
$s <$ 15 %   for road tunnels (due to exhaust gases). Inclinations of more than 3.5% make ventilation more difficult, as it has to overcome the chimney effect. The latter becomes particularly important in case of fire. Therefore, considerably lower limits for longitudinal inclination are recommended: $s < 4\%$ in bi-directional read tunnels. If the tunnel length exceeds 400 m, then the inclination s should not exceed 2 %.

With an increasing slope the production of exhaust gases also increases and, therefore, ventilation costs rise.

Portals: Lengthy cuts in slope debris should be avoided as well as large jumps in illumination (to avoid high illumination costs in the tunnel). Thus, tunnel entrances in east-west direction should be avoided and the distance between two consecutive tunnels should not be too small.

1.6 Underground water conduits

Underground water conduits are built for water supply and water power plants. For hydroelectric purposes the water is conducted from an elevated reservoir to a low lying powerhouse. The conduit is generally subdivided in a more or less horizontal headrace tunnel and an inclined or vertical shaft. Fig. 1.19 shows possible arrangements of conduits.

Alternatively to the shaft, a penstock can be installed above ground. This is, however, disadvantageous in view of slope creep, rock falls etc. Depending on their elevation, water conduits may be exposed to substantial pressure heights of up to 1,500 m (corresponding to 15 MPa or 150 bar). To sustain this pressure, a lining must be provided for, unless the pore pressure in the adjacent rock is higher than the internal pressure in the conduit. An additional reason for the installation of lining is the need to reduce the hydrodynamic wall friction, which can be considerable for mined tunnels. However, the walls of TBM-headed conduits are sufficiently smooth. Usually, headrace tunnels are either unlined or lined with a concrete shell, which can be reinforced or even pre-stressed. Pre-stressing occurs either with tendons or with grouting of the gap between lining and rock. The concrete wall may be covered with a waterproofing membrane.

The shafts (also called 'pressure shafts') are usually equipped with a steel lining.[14,15] Fracturing of the rock (cf. Sect. 21.5.1) is avoided if the minimum

---

[14] Attention should be paid, when that the shaft is emptied for inspection and maintenance, because a high pore pressure may act externally upon the lining.

[15] In Norway, many shafts with pressures up to 1.5 MPa (water pressure heights of up to 150 m) are unlined.

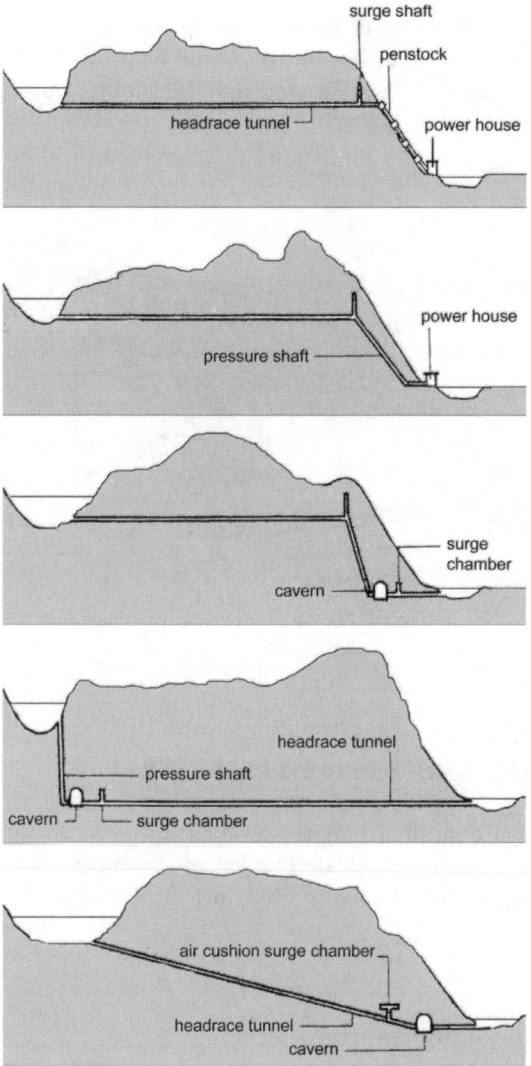

Fig. 1.19. Possible arrangements of water conduits for hydroelectric plants

principal component of the primary stress is larger than the pressure inside the shaft. For safety reasons, the rock tensile strength is not taken into account. The pressure increase due to surging is usually not taken into account because of its short duration. To reduce the pressure rise due to fast closure of the outlet valve, surge shafts are usually provided for. An alternative is to provide unlined surge chambers with volumes of up to 90,000 m^3 filled with compressed air (with pressures of up to 78 bar). Tests with pressurized air or water are used to check the permeability of the surrounding rock.

1 Introduction

A difficult heading operation is to tap a lake, i.e. break through into an existing lake to create a water intake. The intake should be protected against landslides and erosion. The final plug is blasted and the debris falls into a rock trap (Fig. 1.20).

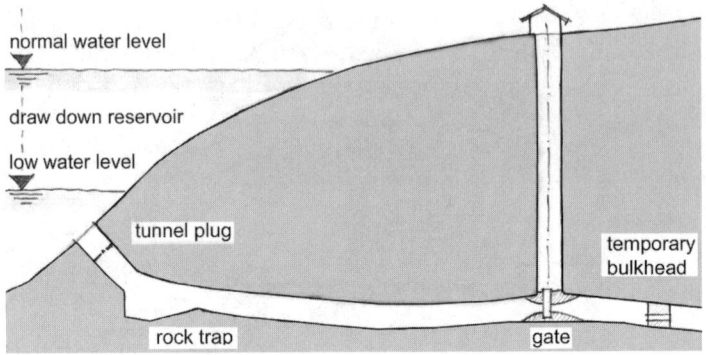

Fig. 1.20. Layout for a lake tap

1.7 Standards and Recommendations

The ever increasing number and volume of standards, codes and recommendations can hardly be overlooked. The list of standards given below is by no means complete. It should merely give an idea about some of the presently existing ones.

AUSTRIA
Codes and guidelines for tunnels:

- ÖNORM B 2203-1(2001): Driving of mined tunnels - contract standards. Part 1: cyclic excavation
- ÖNORM B 2203-2 (in preparation): Driving of mined tunnels - contract standards. Part 2: continuous excavation

Guidelines of the Research Association for Roads and Traffic:
(Richtlinien der Forschungsgemeinschaft Straße und Verkehr)[16]

- RVS 9.231 (in preparation): tunnel alignment
- RVS 9.232 (1994): tunnel cross sections
- RVS 9.234 (2001): installations in tunnels
- RVS 9.24 (1992): general and geotechnical preparatory work

- RVS 9.261 (2001): basics of ventilation
- RVS 9.262 (1997): fresh air demand
- RVS 9.27 (1981): illumination
- RVS 9.281 (2002): operation and safety constructions
- RVS 9.282 (2002): operation and safety equipment
- RVS 9.286 (1987, in preparation): equipment for radio communication
- RVS 9.31 (1993, 1994): static design, cut and cover
- RVS 9.32 (1993): static design, tunnels mined in soft underground covered with buildings
- RVS 9.34 (1995): concrete for permanent lining
- RVS 9.35 (2002): concrete cover of reinforcement
- RVS 9.4 (1982, revision in progress): maintenance and operation
- RVS 13.73 (1995): controls of structures
- RVS 13.74 (1999): controls of operational and safety installations

Guidelines for tunnel drainage (Richtlinie Ausbildung von Tunnelentwässerung, Österreichische Vereinigung für Beton- und Bautechnik) June 2003

GERMANY
Codes and guidelines:

- German rail (DB), DS 853: design, construction and maintenance of rail tunnels
- ETB: Guidelines of the working group "Tunnelling", Ernst und Sohn, Berlin 1995
- RABT (2003): Guidelines for the equipment and operation of road tunnels (Richtlinien für die Ausstattung und den Betrieb von Straßentunneln. Forschungsgesellschaft für Straßen und Verkehrswesen.)

SWITZERLAND
Codes and guidelines for tunnels:

- SIA 196: underground ventilation
- SIA 197/1: design of rail tunnels
- SIA 197/2: design of road tunnels
- SIA 199: exploration of underground
- SIA 198: underground works
- SIA 260: basics for the design of structures
- SIA 267: geotechnical engineering
- SIA 272: waterproofing of underground structures

---

[16] www.fsv.at

EUROPEAN DOCUMENTS

Directive 2004/54/EC of the European parliament and of the council on minimum safety requirements for tunnels in the Trans-European Road Network of 29 April 2004[17]

1.8 Costs

Costs for tunnel construction do not only depend on technical features, such as ground quality and the current rates, but also on other factors such as:[18]

- project culture (cooperation)
- laws, standards etc.
- legal procedures
- tendering
- contract
- risk management.

Table 1.1 gives an impression of the variability of costs of recent metro extensions. [19]

City	Duration (years)	Length (km)	Stations	Costs (m $)	Costs/km (m $)
London	9	16	11	6,000	375
Athens	12	18	21	2,800	156
Paris	8	7	7	1,090	155
Lisbon	8	12,1	20	1,430	118
Madrid	4	56	37	1,706	30

Table 1.1. Costs of recent metro extensions [20]

The reasons behind the success of Madrid are good and extremely lean management (enabling decisions in 24 hours), own project management (solution of disputes before they arise) and the appropriate selection of construction methods (full face methods forbidden, open faces no greater than 5 m^2, selection of extremely powerful EPB tunnelling machines).

---

[17] http://europa.eu.int./eur-lex/en/archive/2004/l_20120040607en.html

[18] NEAT Project Control: Gotthard and Lötschberg Base Tunnels in International Comparison, *Tunnel* 1/2002, 48-50. This publication contains also a table indicating the construction costs of some recent rail projects with a large proportion of tunnels.

[19] The duration comprises design and construction. The given costs include rolling stock.

[20] Brochure of the Community of Madrid

Table 1.2 gives approximate costs for tunnel construction according to HOEK.[21] It should be noticed, however, that these costs do not include concrete lining, tunnel fittings or tunnels driven by TBMs. d is the tunnel span in meter. Costs are given for 6 m $\leq d \leq$ 16 m.

Case	Costs in US $/m
Estimated minimum costs worldwide	1,000 + 600 (d - 6 m)
Good ground requiring minimum support	3,000 + 800 (d - 6 m)
Average tunnelling costs	5,000 + 1,000 (d - 6 m)
Poor ground requiring extensive support	7,000 + 1,200 (d - 6 m)
Faulted ground with severe squeezing	9,000 + 1,400 (d - 6 m)

Table 1.2. Approximate costs for 1 m tunnel excavation and support

1.9 Planning and contracting

Planning of tunnels is staged and addresses many aspects such as:

- financing
- site investigation
- water rights
- disappropriation
- waste disposal
- tendering
- design
- construction
- operation
- maintenance
- management of accidents.

Tunnelling is risky in the sense that the difficulty (and, thus, the costs) of construction due to adverse ground conditions may be higher or lower than anticipated (i.e. predicted by the consultant on the basis of the exploration results). The risk, which is not only due to unforeseen ('changed') geologic conditions but also due to contamination, third-party impacts (such as unknown utilities, settlement induced damage, delays for property procurement and permit acquisition) and design flaws can be taken either by the owner[22] or by the contractor. The first way may cause considerable costs to the owner and demotivates the contractor from working with efficiency. In the second

---

[21] E. Hoek, Big tunnels in bad rock (36th Terzaghi Lecture), ASCE, *Journal of Geotechnical and Geoenvironmental Engineering*, September 2001, 726-740

[22] Also termed 'promoter', 'client', 'employer'

case the owner is burdened by costs either due to increased prices in the bid or due to claims. Therefore, the risk should be shared by both parties. This can be achieved if additional works, due to unexpected ground conditions, are reimbursed to the contractor, on the one hand, and his efficiency is rewarded, on the other hand.

Appropriate design is expected to manage the risk as far as possible. Therefore, the two approaches to risk allocation mentioned above are mirrored in the two main types of contract between Owner and Contractor (who can be a Joint Venture Partnership): [23]

Design-Build:[24] The Contractor is responsible not only for the construction but also for the design (and, sometimes, also for financing and operation). The owner's engineer develops a preliminary design that incorporates the essential project requirements, owner's preferences for design and configurations and sufficient assessment of ground conditions and third-party impacts. The Contractor has increased opportunities to apply innovative solutions and his design can interact closely with the construction process. Partly overlapping design and construction ('design as you go') speeds up the total time for the project. These advantages are counterbalanced by a series of flaws that render Design-Build contracts not so attractive, especially in tunnelling.[25] The disadvantages of this type of contracts are:
- The Owner looses (at least partly) control of the process
- The Contractor is mainly interested in the least expensive construction approach but not in the long-term performance. Furthermore, unifying Designer and Contractor reduces the traditional checks, balances and monitoring, which help to avoid flaws. The amount of design and construction documentation is minimised.
- The description of underground conditions, provided by the Owner, is not adapted to a particular method of construction.
- Since the Contractor is charged not only with the construction but also with the design, his subjection to unanticipated geologic conditions, contaminations and third party claims is increased. Therefore, the Contractor tries to shed much of the risk buck to the Owner by including, as part of his proposal, a long list of assumptions, guidelines, concerns, exclusions and understandings. This makes the comparison of the various proposals very difficult.
- The preparation of a Design-Build proposal is much more costly for the Contractor than in the case of traditional Design-Bid-Build. To

---

[23] R.A. Robinson, Application of Design-build Contracts to Tunnel Construction. In: Rational Tunnelling, D. Kolymbas (ed.), Advances in Geotechnical Engineering and Tunnelling, Logos, Berlin, 2003

[25] The American Consulting Engineers Council "strongly believes that the use of the traditional design and separate bid/built project delivery system is in the best interest of the owner as well as protecting the health, safety and welfare of the general public".

encourage a larger number of proposing teams, some owners include a stipend or honorarium payment, ranging from 0.05% to 0.3% of the estimated total design and construction cost, for the loosing bidders.
- Design-Build contracting is rather new and, therefore, considerable experimentation is still required to determine how best to use risk sharing principles.

Owner-Design (or Design-Bid-Build): This is the traditional (or conventional) type of contract. Here, the Owner is responsible for the design. He contracts with designers to develop feasibility studies, environmental impact statements, preliminary design, and final design. In several iteration steps, the design undergoes reviews and many of the problems and issues of a project are revealed during the various stages of design.

Apart from the above stated both general types of contracts, the following contract types can also be considered:[26]

Admeasurement contracts are based on bills of quantities and incorporate the principle of payment by remeasurement of completed work at initially tendered or subsequently negotiated rates. Thus considerable changes to the work can be made.

Cost-reimbursable contracts: The contractor is paid the costs incurred in carrying out the work. Where costs are difficult to assess, simplifying assumptions may be made and rates may be established. In these contracts, management, overheads and profit are usually paid on a fee basis.

Lump sums contracts: A single price is given for all the work or complete sections of the work. Payment may be on completion of all the work or on the completion of key events.

Target contracts are based on an estimate of a probable cost for the work which is adjusted for changes in the work and escalation of cost. The contractor's actual costs are monitored as a cost-reimbursable contract, and any difference between actual cost and the target cost is shared in a specified way. There may be a separate fee for the management overheads and profit elements. Time target as well as cost targets may be set.

Risk should by no means lead to an antagonism between the involved parties. This would deprive them of the benefits of synergy and cause additional costs. MUIR-WOOD[27] writes:

> The most dominant philosophical basis of bad management is the notion of the zero-sum game, a defect which springs from the dominance of lawyers ... Where lawyers earn more from the failure of projects than do the most skilled engineers from success, clearly there are fundamental systemic faults.

---

[26] CIRIA Report 79 (May 1978), Tunnelling—improved contract practices
[27] Tunnelling: Management by design. Spon, London, 2000

1 Introduction

Dispute Review Boards (DRB), consisting of independent specialists, can help to resolve controversies without costly litigations. A DRB typically consists of three experts — one appointed by the Owner, one by the Contractor and the third appointed on the recommendation of the first two experts.

To be successful, a DRB should develop a good understanding of the project and its participants. This is best accomplished through regular on-site meetings and site visits, whether or not there is a dispute to be heard. Hearings can be made more efficient if the parties provide a short written statement of position in advance.[28] A DRB is most effective when a well defined set of contract documents and a Geotechnical Baseline Report (see Section 3.7) are available.

Opting for the cheapest offer often results in financial loss. Quality may cost more, but makes a much better long term option than having to commit to significant repair and refurbishment work in the short term. In the future, assignment criteria in the construction industry will increasingly have to relate to initial quality, longevity, low maintenance and operational costs.

1.9.1 Cost and time management

Underground projects have often been completed with substantial additional costs and delays. Therefore, efficient cost and time management[29] is a necessity.[30] Let us consider, e.g., two recently completed Swiss tunnels: The costs for the 15.4 km long Furka tunnel overrun the initially alloted 75 million Swiss Francs by additional 150 millions, whereas the 19 km long Vereina tunnel went into operation in 1999, half a year ahead of schedule and with costs less than estimated.

Reasons for overshooting of budgeted costs can be:

- political pressure, e.g. towards low costs
- favourable or even optimistic interpretation of geology
- overtaxing of the involved persons and companies
- lack of cost controlling.

The costs are estimated in two stages:
In the pre-project status the costs are usually determined on the basis of values by experience, the accuracy of the estimation is 25%.

---

[28] R.J. Smith, Dispute review boards - When and how to benefit from a successful contracting practice. In: Proceed. 'World Tunnel Congress/STUVA-Tagung' 95, STUVA Volume 36, 219-213

[29] also termed 'cost and time (or schedule) controlling'

[30] F. Amberg, Br. Röthlisberger, Cost and schedule management for major tunnel projects with reference to the Vereina tunnel and the Gotthard base tunnel. In: Rational Tunnelling, D. Kolymbas (ed.), Advances in Geotechnical Engineering and Tunnelling, Logos, Berlin, 2003, 277-336

In the construction phase, which forms the basis for tendering, the costs are estimated with detailed calculations, that take into account excavation, support, lining, installations etc., on the basis of current material prices, wages and machinery costs. The accuracy is better than 10%.

Regulations for inflation must be adjusted to the considered construction activities.

Controlling should have a clear price basis at the starting point and is achieved with cycles of feedback, executed in short time intervals:

- Daily recording of quantities on site. These have to be assessed by mutual agreement of the involved parties.
- Weekly cost control
- Quarterly balance-sheets with forecasts.

Time (schedule) controlling is also based on feedback loops. Deadlines are defined for all phases of a project, and early recognition of variances should be ensured. Tools for scheduling are bar charts and line charts (path-time charts).[31] Network planning methods, such as PERT, are costly and have not been established in the construction sector. The evaluation of progress occurs on the basis of daily journals, where all activities are compiled.

1.9.2 Experts

In tunnelling projects many types of experts are involved:[32]

Construction supervisors have to ensure that the construction procedures and quality are as specified in the final approved designs. They are recruited from the construction industry and they are seldom required to have a very high level of theoretical expertise.

Construction experts are asked to advise on special topics such as concrete quality design, design of blasting, operation of TBM.

Designers: Usually the design is carried out by teams. Routine calculations can be performed by relatively inexperienced graduates directed by an experienced designer. The latter should have the authority to bring in technical experts to resolve critical technical issues.

Design checkers carry out a comprehensive check on those designs which are critical in terms of technical difficulty or cost.

Panels of Experts [33] provide technical advice on the most complex technical, contractual and schedule issues. They also act as technical auditors for funding agencies. Panels of Experts are usually very small, typically two or three persons having the highest level of technical skill and experience. They should be completely independent of the involved organisations and

---

[31] See e.g. the software TILOS, www.astadev.de
[32] E. Hoek, The Role of Experts in Tunnelling Projects, 2001, www.rockscience.com
[33] also termed Consulting Boards

they should have sufficient seniority and maturity to give sound unbiased and balanced opinions. A special case of panel is the Disputes Review Panel (or Board).

Expert witnesses: When the dispute cannot be resolved by the DRB, it moves to arbitration or to the courts. There, the Owner and the Contractor are represented by expert witnesses. These should be not only firm in technical point of view, but they should also present their evidence clearly and convincingly under the unfamiliar and sometimes confrontational examination by lawyers.

The still prevailing lack of rational approaches to tunnelling renders the acquisition of expertise in this field a process of trial and error. As remarked by HOEK, sometimes experts are considered as persons who have gained their experience at someone else's expense. With increasing scientific approach, this view will loose ground.

2

Installations in tunnels

The significance of installations in tunnels can be easily illustrated when looking at the cost. For the 9.2 km long Plabutsch western tube (road tunnel) in Austria, the following sums have been spent:

	costs in m €
Planning, controlling	11.0
Main construction	91.0
Geotechnical measurements	2.3
Electrical equipment	18.0
Ventilation	5.8
Water supply for fire fighting	1.0

2.1 Installations for traffic control

Installations for traffic control comprise:
- Road signs
- Traffic lights at tunnels with emergency call provisions (to be placed at the portals, U-turn (turn-around) niches and at trafficable cross-overs).
- Traffic guide equipment (floor labelling, side reflectors)
- Traffic census. Peak of maximum allowable traffic (with respect to ventilation capacity) is indicated in the tunnel control-room.
- Height control to catch oversized vehicles before they enter the tunnel (measured e.g. with photo-sensors)
- Video-monitoring of those tunnels longer than 1,500 m or with a high traffic density. The entire tunnel length as well as the areas in front of the portals should be monitored.
- Modern sensors provide warning of traffic slow down (e.g. due to a fire).

The frequency of accidents in road tunnels is reduced by ca 50% compared with open roads.[1] The reasons are:

- Speed limits in tunnels are, in general, accepted by the users.
- Snow, ice, wind, rain and fog are rarely encountered in tunnels.

One should, however, take into account the consequences of accidents in tunnels which are much more severe than on open roads.

2.2 Installations for telecommunication

Equipment for emergency calls: These have to be provided in tunnels of more than 500 m length with a spacing of 150 m. Portals and U-turn-niches should also be equipped with emergency call facilities. Telephone boxes should be provided with glass doors that can be opened toward the tunnel.

Service telephones: These telephones have to be provided in tunnels of more that 1,000 m length at every service station and also in every control room. They are dispensable if radio communication facilities are provided.

Radio communication: In those tunnels longer than 1,000 m or with a high traffic density radio communication should be provided for fire brigade, police and road administration as well as for traffic announcements via radio. Radio re-broadcasting equipment and loudspeakers serve the information of the public.

2.3 Ventilation

With reference to tunnel ventilation, two different systems of ventilation have to be distinguished: ventilation during construction (i.e. during the heading of the tunnel) and service ventilation (i.e. during the operation of the tunnel). Expenditures on the latter amount up to ca 30 % of the total construction costs.

2.3.1 Ventilation during construction

Ventilation during construction has the following aims:

Supply with oxygen: The O_2-content of air should not fall below 20 vol. %. Below 18 vol. %[2] breathing is not possible and protection masks should be used. The lack of oxygen is due to:
- Combustion motors

---

[1] According to G. Brux, Safety in road tunnels, *Tunnel*, 6/2001, 52-61, in the year 2000 up to 10 accidents per km tunnel.

[2] This can be checked by the fact that a match cannot be lit.

- Breathing
- Oxidation of wood, coal etc.
- Ground water dissolves more O_2 than N_2

Cleaning-up of air: The following pollutants have to be removed
- Dust from rock excavation and shotcreting
- Combustion motors
- Blasting fumes
- Gas egression from rock
- Radon decay products.

With respect to air pollution in working environments, maximum allowable concentrations apply (MAC).[3] These refer to exposures of 8 hours per day and up to 42 hours per week. In addition, there are short time limits (*Kurzzeitgrenzwerte*, KZG), which should not be exceeded in short exposures, and also instantaneous values (*Momentanwerte*, R), which should never be exceeded.[4]

Pollutant	MAK
CO_2	$5000\,cm^3/m^3$
CO	$30\,cm^3/m^3$
NO_x	$5\,cm^3/m^3$
SO_2	$5\,cm^3/m^3$
H_2S	$10\,cm^3/m^3$
fine dust	$4\,mg/m^3$
fine quartz	$0.125\,mg/m^3$
respiratory asbestos fibres	$10,000\,fibres/m^3$

Fine SiO_2 dust with a particle diameter $\varnothing < 5\,\mu m$ is deposited in the lungs and can cause the lethal illness 'silicose' by hardening the lung tissue. Radioactivity due to radon and its decay products should not exceed 1,000 to 3,000 Bequerel per m^3 air.

Methane concentrations between 4 and 14 vol. % may cause explosions.

Ventilation also serves cooling purposes. One should take into account that e.g. in the Simplon tunnel the temperature due to geothermy was about 55°C and in the Lötschberg tunnel 34°C. Apart from geothermy, the hydratation of concrete can be an additional source of heat. In Switzerland the temperature in working environments is limited to 28°C.

The following ventilation alternatives can be considered:

- Supply: supply with fresh air
- Extraction: extraction of polluted air (to be preferred for shotcrete applications).

---

[3]The unit cm^3/m^3 is also denoted as ppm (*part per million*). MAC-limits are released by several authorities and target various activities. Thus, their values may scatter across national borders.

[4]See also Chapter 12

The ductings are extendable tubes made of synthetic or steel sheets (Fig. 2.1). Their cross section amounts to ca 1/60 to 1/30 of the tunnel cross section. The required amount for fresh air is:

per person: $2{,}0 \, \text{m}^3/\text{min}$,
per kW of diesel motors: $4{,}0 \, \text{m}^3/\text{min}$.

To estimate the fresh air requirements one should take into account the nominal power of all available diesel motors for excavation, loading and transport (without accounting for simultaneity). On large tunnel construction sites one can easily obtain a total power of more than 1,000 diesel kW, which corresponds to a required fresh air supply of 4,000 m³/min. The Swiss standard SIA 196 prescribes the following fresh air supply to dilute the emissions of diesel motors:

	Machines for excavation and loading	Machines for mucking and concreting
Well maintained diesel motors without post treatment of exhaust gases	6 m³ per kW and min	3 m³ per kW and min
Diesel motors with particle filters and regular control of emmission limit of 10 mg/m³	4 m³ per kW and min	2 m³ per kW and min

The British standard BS6164 recommends a fresh air supply of $9 \, \text{m}^3/(\text{min} \cdot \text{m}^2)$ tunnel cross section area plus $1.9 \, \text{m}^3/\text{min}$ per kW diesel power.[5]

Leaking from the ductings should be taken into account when estimating the fresh air supply. Old ductings can loose up to 2/3 of the initially introduced air due to leaking.

In case of drill & blast heading (see Sect. 4.3) toxic blasting fumes are created. Therefore work should not be resumed after blasting for 15 to 20 minutes. The required fresh air supply is approx. $2 \, \text{m}^3/(\text{min} \cdot \text{kg explosive})$, and the air velocity averaged over the tunnel cross section should amount at least 0.3 m/s. Usually the fresh air is introduced near the face, so the most polluted air is found near the portal.

Note that the ventilation can reduce the moisture of the shotcrete and of the freshly excavated soil. A consequent loss of the 'apparent' (or capillary) cohesion of soil may cause collapse of the face if it is not supported.

---

[5] W.T.W. Cory, Fans for vehicular tunnels, *Tunnels & Tunneling International*, September 1998, 62-65

2.3.2 Design of construction ventilation

Let us consider a ducting with diameter d and length l (Fig. 2.2). Because of the small pressure variation we neglect the compressibility of the air, i.e. we consider the density ϱ as constant.[8] The ventilator should establish the pressure p_1 to overcome the losses due to aerodynamic drag in the ducting. From the known relation

$$\frac{dp}{dx} = \lambda \frac{1}{d} \frac{\varrho}{2} v^2$$

Fig. 2.1. Ventilation ducting[6]

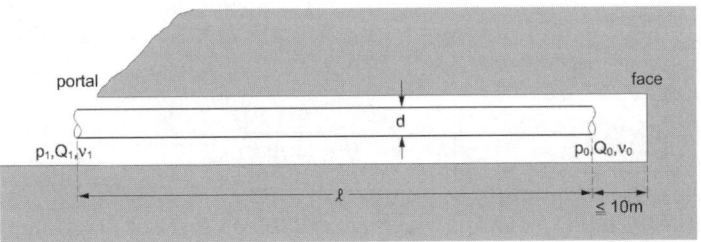

Fig. 2.2. Symbols for the design of ventilation[7]

---

[6] http://www.brescianigroup.com

[7] The end of the ducting should be as near to the face as possible – but not too near in case of drill & blast excavation.

[8] The air density prevailing at the height of the construction site above sea level and at the temperature of the site should be used. ϱ ranges between 0.98 and 1.36 kg/m^3.

follows

$$\Delta p = p_1 - p_2 = \lambda \frac{l}{d}\frac{\varrho}{2}v^2 = \frac{8\lambda l \varrho}{\pi^2}\frac{Q^2}{d^5} \qquad (2.1)$$

with

$$Q = \pi \frac{d^2}{4} v \quad .$$

Typical values for λ range between 0.015 and 0.024, depending on the quality of the ducting. Thus, the ventilator overpressure Δp depends quadratically on the air supply Q. The actual problem is more complex as leaking from the ducting has to be taken into account, hence $Q_1 \neq Q_0$. With the specific leakage surface f (= leaking surface per unit surface of the ducting) and with the outlet velocity v_L the equation of continuity reads

$$\frac{\pi}{4}d^2 \frac{dv}{dx} = f\pi d\, v_L \quad .$$

The outlet velocity v_L is, according to BERNOULLI, proportional to $\sqrt{p - p_0}$, where $p - p_0$ is the overpressure inside the ducting. Taking into account the loss coefficient ξ one obtains[9]

$$v_L = \sqrt{\frac{2(p - p_0)}{\varrho(1 + \xi)}} \quad .$$

If the ducting has a low pressure (in case of extraction ventilation), then v_L is oriented inwards. The resulting system of differential equations

$$\frac{dp}{dx} = \lambda \frac{l}{d}\frac{\varrho}{2}v^2$$

$$\frac{dv}{dx} = \frac{4f}{d}\sqrt{\frac{2(p - p_0)}{\varrho(1 + \xi)}}$$

has a rather complex solution which is available in form of diagrams.[10] When the quantities d, l, λ, Q_0 (= required fresh air supply), p_0 (= pressure at the end of the ducting, say $p_0 = 0$) are given, the values Q_1/Q_0 and $p_1/(\frac{\varrho}{2}v_1^2)$ can be taken from the diagrams. The ventilator pressure results from p_1 plus additional pressure losses due to entrance into the ventilator and possible strictures and curves of the ducting. These losses are expressed as $\xi_i \frac{\varrho}{2}v^2$. The power consumed by the ventilator reads

---

[9] Depending on the quality of the ducting, the value of $f/\sqrt{1+\xi}$ is between $5 \cdot 10^{-6}$ and $20 \cdot 10^{-6}$

[10] see SIA 196

$$N = \frac{1}{\eta_v \eta_M} Q_1 p_{vent} \quad ,$$

where η_v is the degree of efficiency of the ventilator and η_M is the degree of efficiency of the motor.

The aerodynamics of the ducting determine the relation between the supply Q_1 and the related pressure p_1. This relation is parabola-like and is called the 'characteristic of the ducting'. The aerodynamics of the ventilator determine another relation between Q_1 and p_1, the so-called characteristic of the ventilator. The intersection of the two characteristics determines the operational values of Q_1 and p_1 (Fig. 2.3).

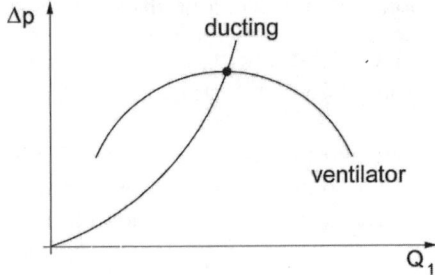

Fig. 2.3. Intersection of the two characteristics

With the advance of the face, the length of the ducting has to be adapted and, thus, also the operation characteristics change.

Removal of dust: Excavation and shotcreting (especially with dry mix) produces a lot of dust which has to be removed. Dry dust separation is based on filtering the air. Dust should be removed as close as possible to its source, while its propagation should be reduced with partition walls.

2.3.3 Ventilation of road tunnels

The ventilation is intended to guard against pollution, guarantee visibility and, in case of fire, secure the escape routes, ensure the entrance of rescue teams and reduce damage.[11] The required fresh air supply should be calculated in accordance with the anticipated traffic flow to guarantee that the following concentrations are not exceeded:[12]

---

[11] G. Teichmann: Fire Protection in Tunnels and Subsurface Transport Facilities, *Tunnel* 5/1998, 41-46
[12] *PIARC (Permanent International Association of Road Congresses)-handbook*

CO concentration	< 100 ppm
NO_x concentration	< 25 ppm
Opacity: extinction coefficient	$< 7 \cdot 10^{-3}$ m^{-1}
Air velocity (averaged over the cross section)	< 10 m/s

In rail tunnels (in particular metros), cooling, i.e. the removal of warm air (e.g. due to locomotives), is another task of ventilation.

Road traffic produces the maximum pollution at speeds of 10 to 15 km/h. The increase in the number of vehicles equipped with catalysers resulted in a considerable reduction of fresh air requirements of road tunnels which, depending on the ratio of trucks and on the slope, amounts to between 30 and 50%.

Four types of ventilation techniques can be distinguished:

Natural longitudinal ventilation: This ventilation is accomplished by the pressure difference between the portals and also by the piston action of the vehicles.

Fans: These have a spacing of ca 10 tunnel diameters and produce a longitudinal ventilation. The blowing direction is reversible. Large ventilators achieve a higher thrust related to installation costs.

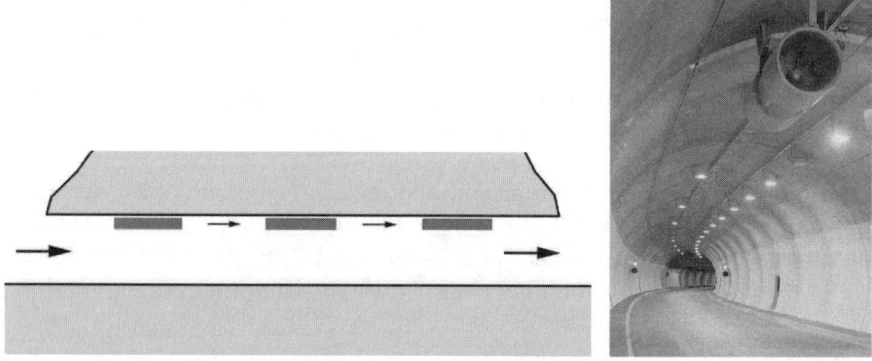

Fig. 2.4. Ventilation fans

Semi-transverse ventilation: Fresh air is supplied from special pathways perpendicular to the tunnel's longitudinal axis, whereas the polluted air escapes from the portals (supply system), see Fig. 2.5 left. This is appropriate for tunnels with 2 to 4 km length and medium traffic load. Alternatively, the used air is extracted through specials ducts (extraction system, Fig. 2.5 right), while fresh air enters through the portals. Thus, the worst air quality is found in the middle of the tunnel. The air velocity reaches its maximum near the portals.

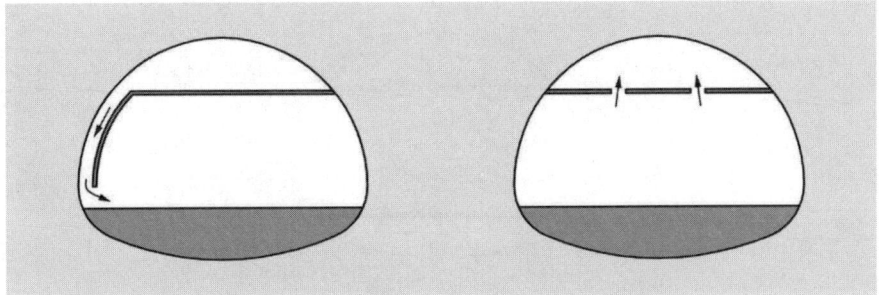

Fig. 2.5. Examples of semi-transverse ventilation (supply and extract)

Fig. 2.6. Supply and extract channels above the carriageway. The separating vertical wall has not yet been mounted.[14]

Transverse ventilation: Fresh air is introduced and polluted air (which moves upwards) is extracted perpendicular to the tunnel axis (Fig. 2.7).

Air ducts with cross sections up to $30\,m^2$ are used. Not only aerodynamic considerations are taken into account but also the necessity of access for maintenance. Long tunnels may require ventilation shafts (Fig. 2.8) or ventilation adits that are driven parallel to the tunnel.

According to the German standards for the equipment and service of road tunnels (RABT), the ventilation systems shown in Table 2.1 should be used with regard to construction type and tunnel length.

If the outlet of polluted air has an adverse environmental impact, cleaning (by means of electrostatic filters etc.) should be considered.

---

[14]Historische Alpendurchstiche in der Schweiz, Gesellschaft für Ingenieurbaukunst, Band 2, 1996, ISBN 3-7266-0029-9

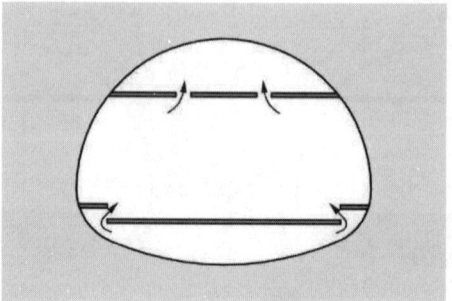

Fig. 2.7. Examples for transverse ventilation

Tunnel length in km		ventilation
bidirectional 1 tube	one-directional 2 tubes	
< 0,4	< 0,7	natural ventilation with CO-warning
< 2	< 4	longitudinal ventilation - fans
< 4	< 6	- fans and ventilation shaft
< 0,5	< 2	semi-transverse ventilation - reversible
< 1	< 2	- semi-transverse ventilation
< 2	< 6	transverse ventilation

Table 2.1. Ventilation systems according to RABT

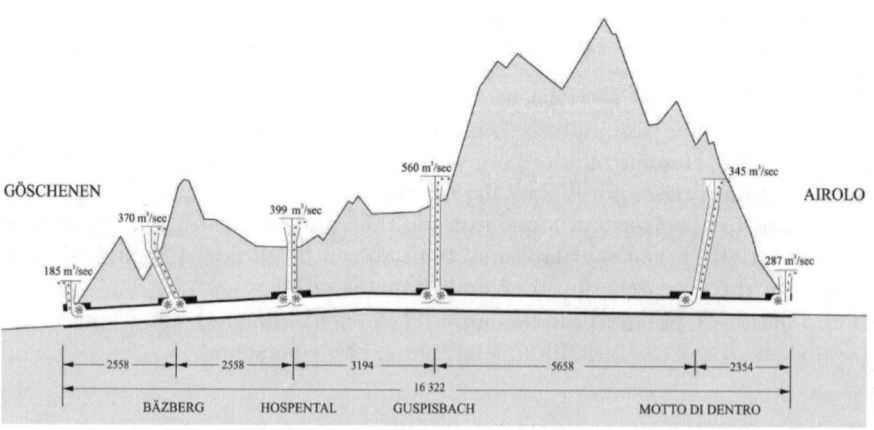

Fig. 2.8. Schematic longitudinal section of ventilation. Gotthard road tunnel[15]

2.3.4 Control of ventilation

The following quantities are monitored:

Air velocity: The average air velocity in the tunnel is measured with ultrasonic transducers which register the travel times t_v und t_r of sonic pulses from A to B and from B to A, respectively (Fig. 2.9). For simplicity we consider only the direction of the tunnel axis. Obviously, for $x \gg d$:

$$t_v = \frac{x}{c+v}, \quad t_r = \frac{x}{c-v},$$

where c is the propagation velocity of sound. Hence,

$$v = \frac{x}{2}\left(\frac{1}{t_v} - \frac{1}{t_r}\right).$$

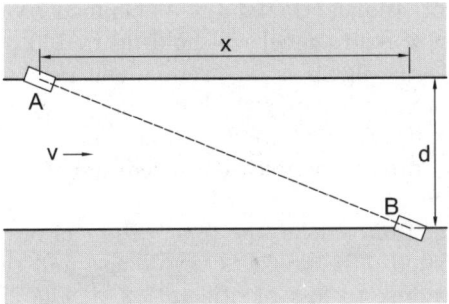

Fig. 2.9. Measurement of the average air velocity v

Temperature: The air temperature in the tunnel can be derived from the measured propagation velocity of sound.

Air opacity: Usually this quantity is measured via the intensity loss of a light beam in the tunnel air, opaqued by dust, diesel grime and aerosols from combustion motors. The measuring distance amounts at minimum 15 m. As measuring devices are very sensitive to installation and maintenance, it is advisable to measure opacity by determining the intensity of scattered light.[16] The required photometer is placed at some appropriate position, which can be up to 500 m off the extraction position.

CO-concentration: The specific absorption of an infrared beam is measured in order to determine the CO-concentration. If this is kept within prescribed limits, NO_x levels are also admissible.

---

[15]Tunnelling Switzerland, K. Kovári & F. Descoeudres (eds.), Swiss Tunnelling Society, 2001, ISBN 3-9803390-6-8

[16]G. Halbach und H. Rhyn: Sichtweitemessung in Straßentunneln. *Felsbau* 5/1995, 296-300

2.4 Fire protection

Because of the confined space, fires in tunnels can be disastrous.[17,18] In 1995 a fire in the metro of Baku caused 289 casualties. Other disastrous fires in metros occurred 1903 in Paris (84 casualties) and 1987 in London (Kings cross station, the fire cost 31 lifes). Between 1978 and 1999 97 casualties resulted from accidents in tunnels.[19] In 1999 12 people died in the Tauern-tunnel and the Montblanc-tunnel disaster claimed 41 lives. The following reasons of the Montblanc disaster have been reported:

- Obsolete ventilation
- Inefficient warning systems
- Insufficient communication between the French and the Italian sides
- When the fire broke out only one fireman was on duty.

The USA have not seen many fire disasters, probably thanks to the rigorous fire protection regulations and frequent application of double tubes. In 1982 a fire in the Caldecott tunnel (Oakland, CA) claimed 7 victims.[20]

In peak traffic times a road tunnel will hold up to 100 persons per km and lane. In rail tunnels the figure is on average 800 persons per train. These people are exposed to fire hazard, which can be due to motor conflagration or conflagration of truck freight, fire due to accident, overheated axes, arson etc. During heading, fire can be released by leakage of methane from rock or oil from machines.[21]

People are at danger from smoke, the toxic products of combustion, lack of oxygen, heat and panic. Fire smoke is very dense and can reduce visibility below one meter leading to loss of orientation.[22] The extraction of smoke must occur as close as possible to the fire source. Fume hoods (dumpers) should be provided with a sufficiently small spacing. It is important to open only the hoods nearest to the fire source. This can be achieved either with

---

[17] This is, however, not the case for all tunnel fires. Before the disaster of 1999, there have been in the Montblanc road tunnel 17 fires of trucks, in Gotthard road tunnel there have been from 1992 to 1998 42 fires of vehicles.

[18] A list of fire disasters in tunnels can be found in: U. Schneider et al., Versuche zum Brandverhalten von Tunnelinnenschalenbeton mit Faserzusatz, *Bautechnik* **78** (2001), Heft 11, 795-804. See also: Fire Protection in Tunnels, *Tunnel*, 2/2002, 58-63; K. Kordina, Planning Underground Transport Facilities to Cope with Fire Incidents. *Tunnel* 5/2004, 9-20

[19] J. Day, Road tunnel design and fire life safety, *Tunnels & Tunnelling International*, Oct. 1999, 29-31

[20] The US fire protection regulations are compiled in the *National Fire Protection Association (NFPA)* 502.

[21] Therefore, oil with a of low combustibility is now used.

[22] Illuminated hand-rails indicate the route to the nearest emergency exit, see e.g. http://www.nils.nl

2.4 Fire protection

remote control or with melting wires.[23] The localised extraction of smoke can be supported if the air is forced to blow towards the smoke source. For two-lane road tunnels with a longitudinal inclination $i < 3\%$ (for $i > 3\%$ special investigations are needed), the Austrian recommendation RVS 9.261 requires to extract at least 120 m^3/s over a length of 150 m through large dumpers.

For the design of the remedy measures, it is important to assume a realistic 'design fire' including temperature and smoke production. Based on recent disasters and field tests, fire scenarios have been modified. Now one has to assume smoke production of 240 m^3/s, fire power up to 100 MW, durations between 30 minutes and several hours and increases in temperature above 1,000°C within 5 minutes. The rise of temperature with time is specified in several standards:[24]

- RABT curve in Germany ($T_{max} = 1200°$ C within 5 minutes)
- Eurocode 1-2-2 (Hydrocarbon curve, $T_{max} = 1100°$ C)
- Rijkswaterstaat-curve ($T_{max} = 1350°$ C within 60 minutes; such high temperatures can appear when trucks carrying gasoline are involved)

To estimate the speed of events, the following facts should be taken into account:

- usual gasoil tanks resist fire for about 3 minutes
- thermal convection winds attain speeds of up to 3 m/s. Thus, smoke can spread out over a length of 180 m before being detected. The propagation speed of smoke equals the speed of escape of individuals.

The transition of smouldering to open fire (so-called 'flash-over') with the accompanied sharp rise of temperature can set on within 7 to 10 minutes.

A layering of heated gases over the cool air helps the efficiency of ventilation and facilitates escape. This layering can be perturbed by fast movements of air, e.g. due to moving cars, sprinklers and fire induced convection. Therefore, longitudinal air velocity should be reduced to 1 m/s in the case of a fire. The air velocity in normal tunnel operation has a strong influence on the convection and layering of smoke in case of fire and should, therefore, not exceed 2.5 m/s (although in naturally ventilated mountain tunnels values up to 14 m/s have been observed).[25] Consequently, the fire ventilation has to achieve the two following objectives: massive smoke extraction in a limited section around a fire and control of the longitudinal air velocity. When designing ventilators, it has to be taken into account that their power is reduced by up to 50% in heated air.

---

[23] C. Steinert, Dimensioning semicross ventilation system for cases of emergency, Tunnel 1, 1999, Tunnel 2, 1999, 36–52

[24] Kordina, K., Meyer-Ottens, R., Beton Brandschutz Handbuch, 2. Auflage, Verlag Bau + Technik, Düsseldorf 1999

[25] K. Pucher, P. Sturm, "Fire Response Management" bei einem Brand im Tunnel. Österreichische Ingenieur- und Architektenzeitschrift 146 Heft 4/2001, 134-138

One way to constrain the propagation of smoke could be the so-called tunnel stopper[26]: An inflatable bellows, made of reinforced rubber or plastic, plugs the entire tunnel cross section. This idea[27] is recommended for long train tunnels only used for freight transports like the Betuwe line in the Netherlands. By stopping the flow of fresh air, the fire will extinguish for lack of oxygen. The train driver, equipped with an oxygen mask and fire protective clothing can escape through the emergency sluice.

Tunnel users can largely contribute to safety with cautious driving e.g. by keeping a sufficient distance between cars. Also the safety training for tunnel operator personnel is important.[28]

The safety and rescue plan should be discussed in advance among the designer, contractor, owner and rescue services. Such an integrated procedure is advantageous even from a financial point of view because it helps to avoid expensive modifications. Given the remote locations of most tunnels, rescue plans should not rely primarily on the arrival of the rescue services but rather on enabling the endangered people to rescue themselves. This so-called self-rescue phase lasts only a few minutes but is crucial. The escape speed ranges from 2.5 to 5 m/s but is reduced to 0.5 to 1 m/s for older or infirm people. The best means of rescue is a second tube with a sufficient number of ventilated cross walks. Alternatively, a rescue adit parallel to the main tunnel can be provided.

Fire-protected niches are problematic in view of ventilation and also for psychological reasons. Experience derived from the fire disasters in the Mont-Blanc and Tauern tunnels is that the semi-transverse and transverse ventilations were unable to extract the smoke sufficiently fast. Safety equipment should be redundant and should work according to the fail-safe principle, i.e. upon failure of a safety relevant component the system enters a safe state. An outfall of electric power supply should release a substitute supply within 10 seconds.

Fire combat measures should exist actively and passively. Active measures aim at the extinction of the fire by means of fire detectors, fire extinguishers, sprinklers, emergency ventilation and telecommunication devices. Passive measures aim at minimising the damage, e.g. by means of fire resistant concrete, synthetic materials which do not produce toxic gases when burning, safe electric cables placed below the carriageway, transverse drain pipes that collect leaking fuel, use of materials of low porosity (that do not fill with fuel) and also clear signals indicating the escape routes.

---

[26] T. Thomas, Protecting tunnels against fire, *Tunnels and Tunnelling International*, Jan. 2000, 44-46

[27] G.L. Tan, Fire fighting in tunnels, *Tunnelling and Underground Space Technology*, 17 (2002), 179-180

[28] The company DMT operates a training centre for fire fighting in Dortmund.

2.4 Fire protection

Fire protection measures should by verified with tests. Cold-smoke tests are comparatively easy and inexpensive, but only real fire tests can provide a genuine verification of the safety level achieved.

2.4.1 Fire-resistant concrete

Fire causes damage of the concrete lining due to spalling. The depth of this spalling increases with the duration of the fire and can reach more than 30 cm. Submerged tunnels in weak rock are particularly vulnerable. E.g., the concrete lining segments of the Eurotunnel have been damaged by spalling in the 1996 fire up to 2/3 of their thickness.[29] The spalling is due to the fast rise of temperature combined with the humidity of the concrete and the structure of its pores. From 100°C onwards the water entrapped in the pores of the concrete transforms to vapour, whose increased pressure spalls the concrete. Concretes of high strength (C55/67) are particularly vulnerable. At higher temperatures also chemical transformations of the aggregates will occur.

Beyond 300°C the stiffness and strength of the reinforcement are reduced. Steel fibre reinforcement increases the thermal conductivity and thus accelerates the heating process. It should therefore be accompanied with ordinary rebars.[30]

Polypropylene fibres disintegrate beyond 380°C. The resulting pores provide escape routes for the vapour, relieve the pressure and thus reduce spalling. Short fibres (e.g. $l=6$ mm) are preferable to longer ones.

Besides the creation of vapour pressure and deterioration of reinforcement, heating affects concrete in the following ways:[31]

- Cement stone and aggregates exhibit a different thermal expansion, which leads to thermal stresses. Cement stone shrinks at elevated temperatures.
- Above 573°C the volume of the aggregates increases due to re-mineralisation in quartz components (e.g. granite and gneiss)
- Above approx. 800°C limestone ($CaCO_3$) transforms to calcium oxide (CaO), which disintegrates after cooling, and carbon dioxide (CO_2)
- Above 1200°C stones melt and entrapped gases escape.

To increase the fire resistivity of concrete, it is recommended to avoid minerals that disintegrate at elevated temperatures. Coarse aggregates (>16 mm) are particularly vulnerable in this respect and should be avoided.

---

[29] The subsequent support has been undertaken with steel arches, rockbolts and steel fibre reinforced shotcrete.

[30] ETV-Tunnel, Teil 2, Absatz 9.3.2. Dortmund, Verkehrsblatt-Verlag, 1995

[31] K. Paliga, A. Schaab, Vermeidung zerstörender Betonabplatzungen bei Tunnelbränden. *Bauingenieur* 77, Juli/August 2002

Fire resistant concretes such as "System Hochtief"[32], or "Lightcem"-concrete[33] have appropriately sized aggregates. The fire protection concrete "System Hochtief" contains 3 kg polypropylene fibres per m^3 concrete. This concrete has resisted a standard fire and suffered spalling of up to 1 cm depth at only 20% of the exposed surface. A concrete cover of 6 cm protected the reinforcement from temperatures higher than 300°C.

As an alternative to fire-resistant concrete, protective panels or spray-ons can be used. However, they do not offer protection during construction, hinder the visual inspection of the lining and require a slightly larger excavation cross section since their thickness ranges up to 10 cm. For instance, in the Westerschelde tunnel a heat-resistant cladding with a thickness of 45 mm was sprayed onto the lining with robots. Covers are intended to protect the lining against fire for 2 hours.[34]

The following versions of protective panels exist:[35]

- Boards made of glass fibre reinforced concrete whose aggregate is glass foam granulate.
- Perforated metallic plates. With a total thickness of approx. 35 mm they represent the thinnest panels.
- Silicate fire protection boards consisting of special concretes made of high-temperature resistant materials.

2.4.2 Fire detectors and extinguishers

Fire detectors are provided to enable fast fire fighting rescue and to prevent further vehicles from entering the tunnel.[36] Fire alarms should also be engaged manually by pressing a button in the emergency call niches and at the portals. In tunnels of over 1,500 m length, automatic fire detectors should be installed. The following detecting systems can be considered:

Heat sensors: These sensors should be activated along the entire tunnel length and not only at isolated points. Otherwise, fire detection can be delayed. Resistivity measurements are vulnerable to the harsh conditions within tunnels. The pneumatic principle is more effective: Higher temperatures increase the pressure of air entrapped in copper pipes. This pressure

---

[32] J. Dahl, E. Richter, Fire Protection: New Development to avoid Concrete Splintering, *Tunnel*, 6/2001, 10-22

[33] The "Lightcem" concrete can withstand a temperature of 1350°C up to two hours and doesn't split up at contact with extinguishing water. It contains microsilica as binder and aggregates of inflated clay.

[34] J. Heijboer, J. van der Hoonaard, F.W.J. van de Linde, The Westerschelde tunnel, Balkema, 2004

[35] A. Schlüter, Passive Fire Protection for Tunnels: Guidelines, Parameters, Reality and suitable Measures. *Tunnel* 7/2004, 22-33

[36] J.P. Emch and S. Brügger: Performance Requirements for Fire Detection Systems in Road Tunnels, *Tunnel* 4/1995, 36-43

increase is registered by pressure transducers. Fire detectors based on glass fibre and laser beams (which are increasingly scattered at higher temperatures) can localise accurately the source of heat and (with appropriate software) can even detect heat propagation (e.g. 'Fibro-Laser').

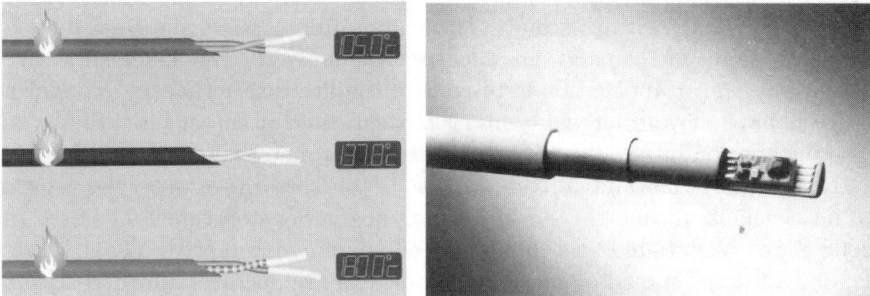

Fig. 2.10. Cables for fire detection[37]

Gas sensors: These register the reduction of the oxygen content and the presence of combustible gases and toxic combustion products.
Smoke sensors: In road tunnels these can easily cause false alarms and consequently release a full power ventilation, which is very expensive because of its high power consumption.
Flame sensors: These only register open flames.
Video control: The automatic detection of traffic jams, smoke and fire by means of video control and pattern recognition is a promising innovation and can respond much earlier than heat sensors.

Fire detectors should be highly resistant to break-down. They must be provided with an emergency power supply and the maximum length that may drop out should be limited to 50 - 500 m (depending on traffic density). The equipment must be robust and capable of withstanding e.g. cleaning brushes (Fig. 2.14) or loose goods falling from vehicles.

Firefighting: For tunnels with a length between 600 and 1,000 m a pressure pipe of at least 100 mm diameter is sufficient. This pipe does not need to be permanently filled with water but should be fed at the portals. At each portal water tanks, containing at least 80 m^3 water, must be provided.

In tunnels with more than 1,000 m length, a pressure pipe must be installed beneath the lateral strips of the carriageway and protected against freezing. The pressure at the taps should be between 6 and 12 bar. A supply of 1,200 l/min for at least one hour must be guaranteed. Water for firefighting is to be available at the emergency call niches with a hydrant and a 120 m long waterhose.

---

[37] left: Brochure Thermostick Elettrotecnica s.r.L.; right: Listec GmbH

In addition, the tunnel users should have access to two fire extinguishers kept in each emergency call niche.

When utilising sprinklers, it should be kept in mind that droplets fall through the fumes, if they are too large, and therefore do not bind smoke particles. A mist of fine water droplets (0.01 mm diameter) can be produced with high pressure (100 bar) and appropriate nozzles.[38] Thus the temperature can be lowered by 700 °C within seconds. The evaporation of water increases its volume by a factor of 1640 and thus chockes the supply of oxygen. The objective of the water mist application is to control the fire rather than to extinguish it. Tests have shown that water mist is efficient even at air velocities of 5 m/s. The installation of nozzles can be undertaken in metro stations, in on-board systems in the trains and in road tunnels.[39] E.g., spraying of water has proved to be beneficial during the shield heading of the Socatop tunnel in Paris. In 2002 a fire was initiated by a rubber tire of a hauling locomotive. Water spraying impeded the spread of smoke and cooled the temperature in the back-up section down to 80°C. This enabled the workers to escape into the excavation chamber, which happened to be accessible at that time.[40]

2.4.3 Example: Refurbishment of the Montblanc tunnel

The refurbishment of the 11.6 m long Mont-Blanc road tunnel after the fire in March 1999 demonstrates various aspects of modern fire protection.[41] The following arrangements have been made:

- Lay-by's in both directions every 600 m allow heavy goods vehicles to stop. Also every 600 m turn-around points are provided.
- Refuges (shelters) are situated on one side of the tunnel every 300 m. They are ventilated via fresh air ducts and put under light overpressure. They are connected with the tunnel by means of airlocks. In case of a fire in front of the refuge, the temperature inside should not exceed 35 °C after 4 h. The rescuers can get to the refuges and evacuate the victims from outside the tunnel via escape galleries situated in underground fresh air galleries (Fig. 2.11). In this case the fresh air ventilation is reduced to a minimum.
- Emergency recesses have been placed alternately at intervals of ca 100 m. They are equipped with emergency telephones, fire extinguishers and glass

---

[38] http://www.fogtec.com

[39] R.A. Dirksmeier, Fire Protection Solution with Water Mist, *Tunnel* 5/2002, 22-26

[40] M. Herrenknecht, U. Rehm, Bewältigung schwieriger geologischer Bereiche beim Einsatz von Tunnelvortriebsmaschinen, *VGE, ISBN 3-7739-5982-6*

[41] F. Vuilleumier, A. Weatherill, B. Crausaz, Safety aspects of railway and road tunnel: example of the Lötschberg railway tunnel and Mont-Blanc road tunnel. *Tunnelling and Underground Space Technology*, 17 (2002) 153 - 158; M. Bettelini, Mont-Blanc fire safety, *Tunnels & Tunnelling International* June 2002, 26-28

doors. Firefighting recesses are located every 150 m with a hydrant every 300 m.
- Fans provide up to 82.5 m³/s fresh air ventilation.
- The fire ventilation has an extraction capacity of 150 m³/s.
- Closed circuit TV monitors allow all refuges and bays to be monitored. 150 cameras (one every 150 m on each wall side) allow complete surveillance of the tunnel.

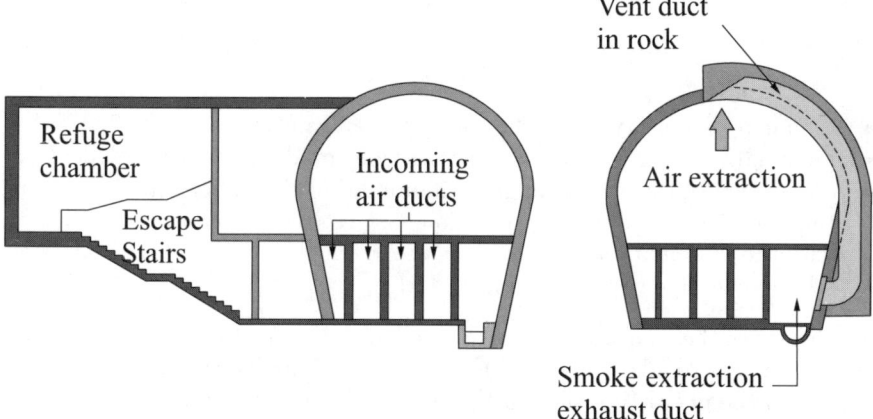

Fig. 2.11. Refuges in the refurbished Montblanc tunnel

2.5 Illumination of road tunnels

Road tunnels must be sufficiently illuminated. The illumination density is gradually reduced from the portal towards the interior of the tunnel. Figure 2.12 shows the three main sections of a tunnel which are distinguished with respect to illumination.[42]

The first two sections are the entrance and transition sections. The length S_H of the entrance section is that length needed to stop a vehicle. It results from the design velocity V (km/h) according to the equation

$$S_H \, (\mathrm{m}) = \frac{1}{3}V + \frac{(V/3.6)^2}{2(b \pm g \cdot s/100)} \quad ,$$

where $b=6.5\,\mathrm{m/s^2}$ (average braking retardation), s=longitudinal slope in %, g=gravity acceleration in m/s².

In the transition section the illumination density is gradually reduced. Its length S_{tr} depends on the following quantities:

---

[42] RVS 9.27, October 1991

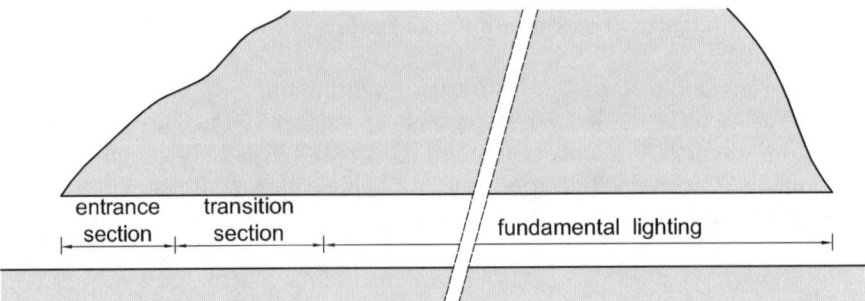

Fig. 2.12. Sections of tunnel illumination

- illumination density L_e at the end of the entrance section
- illumination density L_i in the inner section (fundamental lighting density)
- adaptation of the eye to changing illumination conditions and, thus, from the allowable maximum speed. Equ. 2.2 relates the adaptation time t_r with the illumination densities L_e and L_i. The length S_{tr} of the transition section can be determined from t_r and the velocity V.

The length of the transition section S_{tr} is determined by the time t_r needed by the human eye to adapt from the luminescence L_e at the end of the entrance section to the one within the inner section (fundamental lighting L_i):

$$L_i/L_e = [1.9 + t_r(\text{sec})]^{-1.423} \qquad (2.2)$$
$$S_{tr} = t_r \cdot V$$

The illumination densities L_e and L_i depend on the exterior illumination density L_{20} (cd/m^2) which is either averaged over the sector indicated in Fig. 2.13 or taken to 4,000 cd/m^2. L_e is obtained from L_{20} according to the following table:

V (km/h)	L_e
≤ 60	$L_{20}/30$
$60 < V \leq 80$	$L_{20}/25$
$80 < V \leq 100$	$L_{20}/20$

The fundamental lighting applies to the remaining tunnel section. Its density L_i should be 3 cd/m^2, where a reduction of 30% takes dirt and ageing into account.[43] The illumination density should be uniform within the tunnel, i.e. it should be $L_{\min}/L_{\max} \geq 0.55$. Flickering in the frequency range from 2.5 to 15 Hz should be avoided.

To facilitate the adaption of the eye, the portals should be as dark as possible. Therefore, galleries and illumination reduction constructions should be used,

---

[43] Specialists for human physiology recommend considerably higher L_i-values, e.g. 30 cd/m^2. However, the related costs are in most cases not affordable.

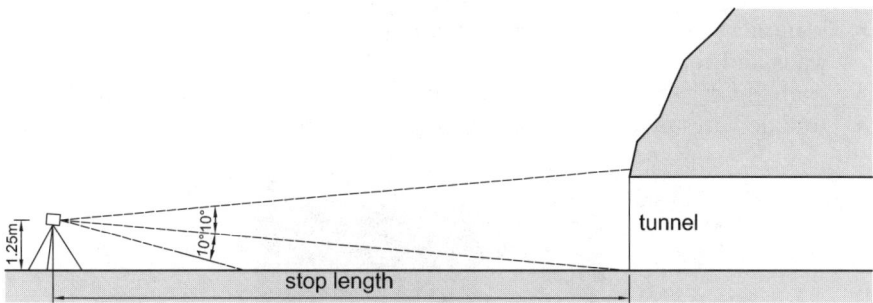

Fig. 2.13. Sector for the measurement of the exterior illumination density

Fig. 2.14. Truck for tunnel washing[45]

if affordable. To achieve maximum illumination and to enable easy cleaning, the tunnel inner walls should be covered with a bright and reflecting, but not dazzling coating, which should be cleaned at regular intervals.[44]

A good solution is to mount enamel panels on the tunnel wall (Fig. 2.15). The objectives of interior finish are:

---

[44] Before the fire disaster, the Mont-Blanc tunnel was washed once in a fortnight (in winter) and monthly (in summer). The costs for each washing amounted to € 17,000.

[45] Assaloni Commerciale s.r.L.

- identification of alignment by means of the difference in luminance between walls and road surface,
- enabling drivers to assess the distance from the walls,
- to hide wires and pipes that divert attention of the drivers.

Fig. 2.15. Enamel panels ('Smaltodesign')

To improve the driver's concentration, zones of ca 20 m length and $10\,\text{cd/m}^2$ illumination density should be provided at every second lay-by. During black outs of electric power the emergency call niches and some intermediate points as well as the cross-overs should be illuminated.

Regulations for illumination of road tunnels are continuously improving and take into account recent advances in physiology and illumination technique.

2.6 Drainage

The interaction of a tunnel with the groundwater via its drainage is a very important issue treated in Chapter 8. In this section we merely consider the installations for drainage. The following waters should be collected and diverted:

- Groundwater (for ecological reasons the withdrawal of groundwater should be as low as possible)
- Day water (precipitation or melting ice entering from the portals)
- Service water (e.g. from washing).

Contrary to the mixed system, where all types of water are put together, groundwater on the one hand, precipitation water (entering from the portals) and service water (used for maintenance or firefighting) on the other hand are withdrawn in separate pipes in the so-called separate system ('dual drainage system'). Provision for non-propagation of fire should be taken for the case of effusion of inflammable fluids.

Longitudinal drains are installed in the sides of the tunnel. Their diameter depends on the slope (at least 0.5%) and on the pipe material and should be at least 15 cm. Cleaning and flushing shafts should be provided with a spacing of 50-65 m. Water is removed from the carriageway by means of longitudinal drains with diameter $\geq$ 20 cm, slope $\geq$ 0.5% and cleaning shafts with spacing of, e.g., 110 m. A possible aggressivity of the water as well as the danger of freezing should be taken into account.

2.7 Examples for the equipment of modern road tunnels

Kaisermühlen tunnel:[46] The Kaisermühlen tunnel in Vienna is considered to be one of the safest among the 200 Austrian road tunnels. It belongs to the motorway A22 (Donauufer Autobahn) and has a length of 2,150 m. With a daily traffic load of 100,000 vehicles it is one of the most densely used tunnels in Austria. More than 2,000 lamps provide illumination; 104 fans are used for ventilation; 7.5 km sensor cable are installed for automatic fire monitoring; 2 ventilation buildings with ventilators assure a complete exchange of the tunnel air within 10 minutes. 22 emergency call niches with spacing of 200 m and 9 emergency call columns in front of the portals enable contact with the tunnel guard. Fifty robot video monitors control the tunnel. Radio amplifiers and antenna cables at the tunnel roof enable communication for the service teams. A special pipeline fed from the river Donau (Danube) is provided for extinguishing water. Numerous escape doors in the separating wall and several escape stairhouses make it possible to escape from the tunnel as fast as possible.

The tunnel control room operates day and night and the tunnel guard is assisted by 13 automatic fire programmes and more than 150 traffic programmes. A concise system of traffic guidance and information helps the users to react to unexpected situations. Three spare power stations and accumulators as well as a mobile and a stationary generator guard against total black outs.

Engelberg base tunnel:[47] Two tubes with 3 drive and one park lane each (maximum excavated cross section 265 m^2, maximum thickness of concrete inner lining 3.50 m) accommodate a daily traffic load of 120,000 vehicles. The tunnel has the following equipment:
Power supply by two loops of 20 kV
6.4 MW installed transformator power

---

[46] *Austria Innovativ*, 4/1999

[47] H. Gottstein: Engelberg Base Tunnel - Technical Operating Equipment, *Tunnel* 7/99, 18-26

2 Installations in tunnels

2 UPS[48]-plants with 125 kVA each
More than 800 km cable of various cross sections
1,200 sodium high pressure lamps 50 to 400 W, partially as dual shiners
6 videos measuring illumination density
2 illumination computers
4 axial fans with 650 kW each for fresh air supply
3 axial fans with 370 kW each for air extraction east tube
4 axial fans with 240 kW each for air extraction west tube
29 CO-sensors
4 NO-sensors
16 monitoring stations for haziness of air
4 monitoring stations for air flow
8 fire-detector cable with 625 m length
39 buttons for manual fire alarm release in the emergency call niches
30 buttons for manual fire alarm release in the control rooms
2 fire control rooms
2 fire-brigade operating fields
22 colour video monitors (11 in each tube)
4 dirigeable colour video monitors with zoom at the portals
2 video cross rails
6 colour monitors
39 emergency call niches with emergency call devices ANE 90
4 transposers for radio
3 transposers for rescue teams 4 m-band
4 transposers for rescue teams 2 m-band
2 antennas per tube, to emit and receive, separated in 5 sectors
2 external antennas for radio, BOS[49] and mobile phone
4 amplifier stations in the cross overs with 8 radio amplifiers each
mobile phone transposer and antennas for 5 mobile phone providers (C-net, D1, D2, E plus and E2-net)

---

[48] uninterruptible power supply
[49] *Behörden und Organisationen mit Sicherheitsaufgaben*, i.e. police, fire brigade, customs, relief organisations, etc.

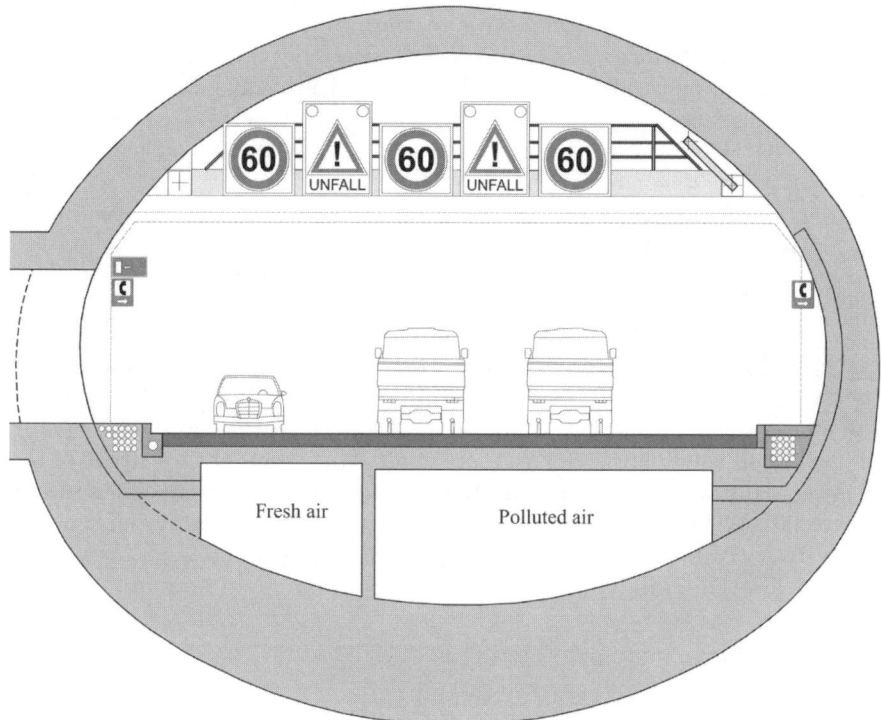

Fig. 2.16. Safety installations in the Engelberg base tunnel

2.8 Rating of safety in road tunnels

In 2001 the German automobile club ADAC checked the safety of 16 important European tunnels. The Farchant tunnel at Garmisch-Partenkirchen and the Kaisermühlen tunnel in Vienna came off best, whereas the recently finished Laerdal tunnel in Norway came third from bottom and the Monrepos tunnel in Spain came last. The following items have been rated (each with specification of the corresponding weighing factor):

Tunnel system (8.5 %): side lanes, break-down lay-by's, width
Actual state (6 %): illumination, labelling of the carriageway
Traffic marks (5 %): traffic jams, speed limits, limitations of transportation of dangerous goods
Traffic control (7.5 %): safety control room, nearby police guardhouse, video monitoring, automatic recognition of hazardous goods, traffic census and control
Communication (8 %): traffic radio, emergency telephones (in foreign languages), mobile telephones, loudspeakers
Escape, rescue ways (13 %): spacing of emergency exits

Case of fire (24 %): fire protective coating, safely installed electric power cables, fire detectors, fire extinguishers, supply of pressurized water, sufficient hydrants, gully slots for leaking combustible fluids, time for arrival of fire brigade

Fire ventilation (17 %): length of the sections into which smoke can propagate, reversible fans, extraction of smoke

Emergency management (11 %): alarm and activity plans, regular control of the safety equipment.

Any rating is more or less arbitrary. With respect to rating of tunnel safety, e.g., one has to determine which factors should be rated and which should be the weight of each factor. There is no way to avoid arbitrariness in fixing the factors and their weights. This, however, does not exclude objectivity in the application to individual tunnels, as long as the same criteria are equally applied to all candidates. Thus, a more or less true image of the safeties of the investigated tunnels can be expected. This largely contributes to higher safety. Since 2001, ADAC tunnel ratings have been annually carried out and several badly rated tunnels are in the meanwhile refurbished and have obtained a very good level of safety.

3

Investigation and description of the ground

3.1 Geotechnical investigations

The underground is a vast unknown that can hide many unpleasant surprises (e.g. weak zones, water inrushes etc.). Therefore, a detailed site investigation is necessary not only for technical purposes, but also for the contractual regulations of all involved parties.

Geotechnical investigations aim at collecting all ground properties that are relevant to the heading.[1] Usually, site investigation, design and construction are executed by different specialists, companies and authorities as well as at different times. It may thus happen that necessary data are not available when needed, i.e. when no more money or time are available for further investigations. The expenses for site investigation can make up to 3% of the construction costs.[2] Usually, the site investigation is released by the owner who should provide the bidders with data that are as complete as possible. Otherwise, the owner will have to face the possibility of increased claims. According to experience from the USA[3], over 55% of claims relate to unforeseen ground conditions and they decrease with increasing exploration, as shown in Table 3.1.

---

[1] see e.g. ETB (Empfehlungen des Arbeitskreises 'Tunnelbau'), RVS 9.241 und RVS 9.242, DS 853 (Eisenbahntunnel Planen, Bauen und Instandhalten), DIN 4020 (Geotechnische Untersuchung für bautechnische Zwecke), SIA 199 (Erfassen des Gebirges im Untertagebau), AFTES guidelines for caracterisation of rock masses useful for the design and the construction of unterground structures (www.aftes.asso.fr)

[2] According to Norwegian experience, the relative costs of geotechnical investigations (related to the construction costs) are 1% for simple tunnels, 3-4% for storage caverns, 5-6% for special projects.

[3] R.A. Robinson, M.A. Kucker, J.P. Gildner, Levels of geotechnical input for design-build contracts, In: Rapid Excavation and Tunneling Conference, 2001 Proceedings, 829-839, Society for Mining, Metallurgy, and Exploration, Inc., Littleton, Colorado

3 Investigation and description of the ground

Meter of exploration boring per meter of tunnel alignment	Claims relative to the bid price
0.5	30-40%
1	< 20%
1.5	< 10%

Table 3.1. Influence of exploration on claims

Information on the underground is attained step by step. Each step reveals the proper subsequent investigations, (GLOSSOP[4] said: *"If you do not know what you should be looking for in a site investigation, you are not likely to find much of value."*). In general, three steps or phases are applied:

Phase 1: preliminary investigation
Phase 2: main ground investigation
Phase 3: further investigation undertaken as part of the project itself

All investigations must be well documented. The geotechnical investigations should be done in close collaboration with Engineering Geology, a discipline that deals with the site investigation for civil engineering projects.[5]

3.1.1 Preliminary investigation

The preliminary investigation consist of

- preliminary appreciation of the site, i.e. an examination of existing information (also called 'desk-study') and comprises
    - topographical, geological, hydrological and other maps (e.g. aerial photos to find faults)
    - evidence and experience from nearby construction sites
    - walk over survey to detect jointing, weakness zones etc. and engineering geological mapping along tunnel alignment

    The desk study aims at determining
    - whether the envisaged tunnel is feasible
    - to what extent geology affects the excavation and support
    - further investigations

    The desk study is the basis for the provisional design and investigation of variants (alternative alignments).
- preliminary ground investigation, i.e. site work to confirm feasibility, and to enable plans for the next phase of the investigation to be formulated.

---

[4]R. Glossop, 1968 Rankine lecture
[5]P.N.W. Verhoef, Wear of Rock Cutting Tools, Balkema, 1997

3.1.2 Main site investigation

The main site investigation comprises information that must be acquired ad hoc and serves the design, tendering, assessment of environmental impact and execution, as well as damage analysis.
In the frame of site investigation, a geotechnical report has to be worked out (see Section 3.7).

3.1.3 Investigation during and after construction

It is never possible beforehand to obtain complete information on the strata which will be encountered by a tunnel, to anticipate their behaviour, and for this reason provision must be made for observation and any other necessary investigation during construction. Thus, this investigation aims at continuously updating the acquired information and to check the validity of the prognoses. It comprises mappings of the tunnel face (Fig. 3.1) and wall, measurements of deformations, settlements, stresses, vibrations and groundwater.

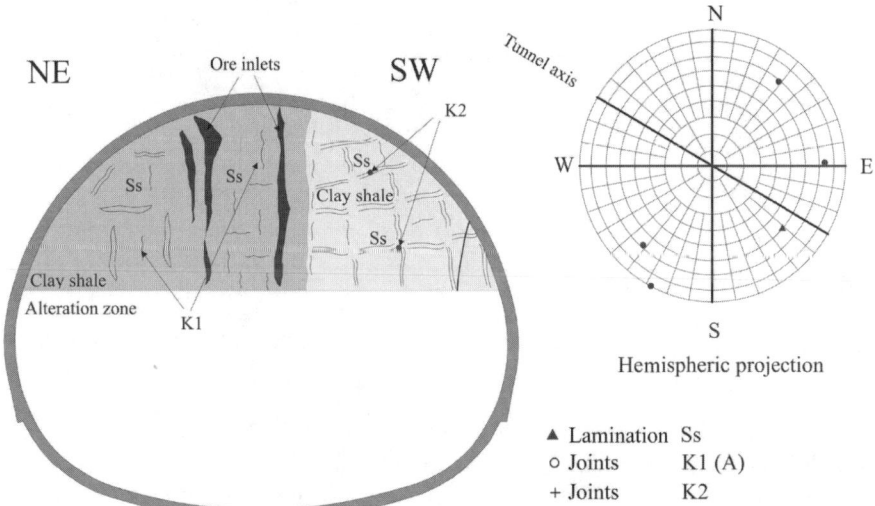

Fig. 3.1. Geological record of the face at Aegidienberg tunnel

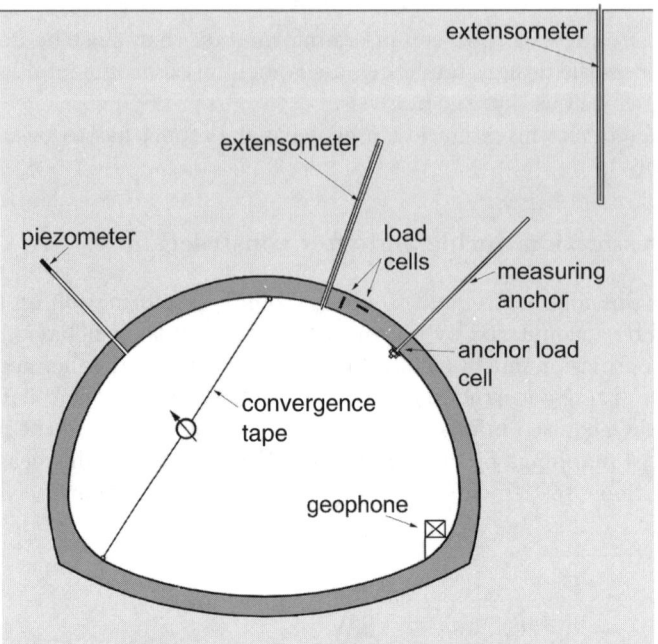

Fig. 3.2. Measurements for tunnelling[6]

3.2 Site investigation

Apart from the general methods used in geotechnical engineering, the following ones are applied in tunnelling:

- Exploration adits and test drifts
- Field tests to determine the air permeability
- Tests for the applicability of slurry and earth pressure support (EPB)
- Tests for abrasivity of drilling equipment

To determine the geological structure of the rock, core drillings are applied (diameter at least 100 mm). These have to be placed outside the planned tunnel at alternating sides. The boreholes should subsequently be re-filled or equipped as groundwater observation wells. They should reach ca one tunnel diameter below the tunnel invert. Near to the portals ca 3 exploration boreholes should be provided. Along shallow tunnels, such boreholes are drilled

---

[6]Convergence tapes are rather obsolete. Nowadays, convergence is mostly measured via remote sensing.

every 100 to 300 m in plan view. For deep tunnels, boreholes could prove as too difficult or even not feasible if they have to be started from inaccessible mountains. It should be added that in the early time of tunnelling the depth of exploration drilling was limited to a few metres and the structure of the underground had to be extrapolated from surface mapping.

In some cases cores with large diameters (0.6 - 1.0 m) can be extracted and tested in the laboratory (Fig. 3.3 and 3.4). The obtained experimental results are of high quality, since small scale inhomogeneities are automatically averaged. The extraction of large cores is laborious and presupposes direct access to the site.

Fig. 3.3. Drilling of large core sample ($\varnothing$ 60 cm)

---

[6]Institute of Soil and Rock Mechanics of the University of Karlsruhe

62   3 Investigation and description of the ground

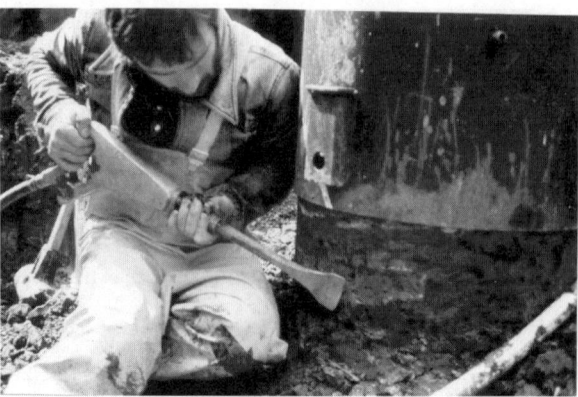

Fig. 3.4. Extraction of large sample by cutting

The CIRIA report 79 recommends:[7]

- Site investigation specialists are not necessarily tunnelling specialists, therefore the terms of reference for the ground investigation contracts should be explicit as to the data required.
- Selection of site investigation contractors should be primarily on merit, not on price — recognising the need for experience and the technical ability required.

3.2.1 Exploration drilling

The structure of the underground is judged on the basis of cores (Fig. 3.6), i.e. cylindrical samples that are extracted with special drilling equipment. The recovery of cores, as undisturbed as possible, requires special care that should be contrasted to the drilling of boreholes, e.g. for blasting or grouting. In the latter cases the excavated rock is broken into small pieces and chips.

Core samplers are driven into the rock by rotation and downwards thrust. They mill an annular gap, and the rock core is protruded into the sampler's interior. After a while the cores are pulled upwards. Doing this, the core is snapped and picked up by the aid of an annular wedge (Fig. 3.5).

The drive of the sampler has to be accompanied by flushing, which proves, however, to be detrimental for the core, in case of soft rock or clay: The core is continuously exposed to the flushing liquid, which washes away the loose constituents, and the core is thus eroded. Therefore, single tube core samples are inappropriate for the extraction of cores.

With double tube core samplers, only the external tube is rotating. The downlift of the flushing liquid occurs within the annular gap between external and internal tubes. Thus, the core gets in contact with the flushing liquid only at

---

[7]CIRIA Report 79, Tunnelling – improved contract practices, May 1978

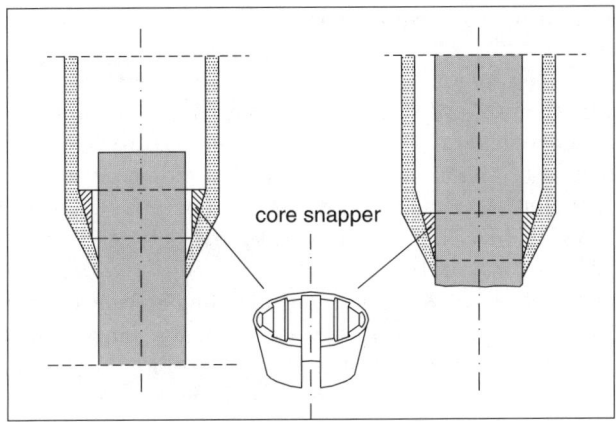

Fig. 3.5. Core bit and snapper

its lower end and is considerably less eroded. Even though, flushing should be interrupted as soon as the sampler enters an erodible layer of, say, silt or clay. When a core has obtained a length of 1.5 to 3 m it is hauled to the surface. For this purpose, the entire drilling tube has to be lifted up and dismounted piece by piece. This is time consuming, especially for deep boreholes. E.g., for a 100 m deep borehole and tubes of 3 m length a skilled team needs 30 to 40 minutes for removal and back-installation of the drilling rig. As an alternative, the drilling rig remains within the ground and the core is hauled by means of a wire (wire-line drilling). This method works up to depths of 400 m, and the usual core diameter is 132 mm.

Usually, core diameters of at least 100 mm are required for laboratory testing. In general, the quality (but also the price) of a core increases with its diameter. Note that pneumatically flushed drillings may still provide some information on the structure of the ground, even if they do not extract cores, being faster and cheaper than the core drilling. This information can be derived from the records of drilling progress, thrust, torque and flushing pressure. E.g., it is conceivable to drill 10 un-cored boreholes and one cored borehole for calibration. Another option to reduce exploration costs for, say, a 300 m deep tunnel is to drill the upper part of the borehole (say, 200 to 250 m) with a fast downhole hammer without coring and extract cores only from the lower 50 to 100 m part.

The progress of drilling and quality of cores also depends on the core bit. Its choice depends on many factors, but it can be generally stated the softer the ground, the sharper the teeth should be.

3.3 Geophysical exploration

The transmission and reflexion of seismic, electric, magnetic and gravity fields is influenced by the properties of the investigated media. Thus, measuring the transmission and reflexion of various waves enables to infer on these properties. Analysing the signals of several senders and receivers may even yield the spatial distribution of field properties (tomography). For example, with seismic tomography blocks and cavities, situated between two boreholes with a spacing of 5 to 10 m, can be detected.

One of the most unpleasant situations in tunnelling is to encounter unexpected zones or obstacles such as cavities, water bearing joints, inclusions, faults filled with soft soil etc. Therefore, it is tried to explore the ground behind the face during a tunnel excavation not only with exploratory drillings but also with geophysical methods. The latter are expected to image the ground on-line, i.e. without interruption of the heading. There is research running with the aim to emit e.g. waves and to analyse the obtained reflections on-line. Senders and receivers are mounted in the cutterhead of a TBM.[8] The results are promising but not yet satisfactory.

The mathematical procedures to attain the spatial distribution of field quantities from the measured data belong to the so-called inverse methods. These methods are 'ill-posed' in a mathematical sense with the consequence that the obtained results are not unique. Their interpretation presupposes, therefore, the availability of some additional information.

3.4 Joints

The spatial distribution of joints (discontinuities) is measured with a geological compass and a tape measure, nowadays also by analysing digital images. Joints are sometimes encountered in clusters. The following features are important:

- Form and dimensions of the individual blocks
- Density of joints. This density is designated by the average number of joints λ_i of a cluster i per meter. $a_i = 1/\lambda_i$ is the average spacing between the joints. λ is the sum over all clusters and can be determined by measuring all joints found on 1 meter of a borehole.

  An indirect way to determine λ is to measure the speed of propagation of elastic waves, because this speed is reduced by joints. If c_l is the propagation speed within intact rock (measured at rock samples in the laboratory) and c_f is the propagation speed within rock mass (measured in the field), then the ratio $k := c_f/c_l$ correlates with λ: [9]

---

[8]B. Lehmann, C. Falk, T. Dickmann: Neue Entwicklungen zur Baugrunderkundung für die 4. Röhre Elbtunnel, Baugrundtagung 1998 in Stuttgart, p. 189-200

[9]J.A. Franklin & M.B. Dusseault: Rock Engineering, McGraw-Hill Publishing Company, 1989

$$\lambda \approx \frac{5}{k^2} - 4$$

The so-called velocity index is defined as $100\,k^2$ and scatters from 0 to 20 for heavy weathered rock, and from 80 to 100 for intact rock.
- Core recovery. This is the length of all intact cores found in 1 m of sampling. Alternatively, one counts only those cores having a length of at least 10 cm. This number, introduced in 1963 by DEERE, is called RQD (Rock Quality Designation). RQD correlates with λ:

$$RQD \approx 115 - 3.3\lambda \quad .$$

For calcareous rock it was found:

$$RQD \approx 100 e^{-0.1\lambda}(0.1\lambda + 1)$$

Fig. 3.6. Rock cores

- The spatial orientation of joints is indicated by strike and dip. Strike is given by the angle α that denotes the bearing of a contour line. α is measured clockwise and obtains values between $0°$ and $180°$. Alternatively, one uses the angle α_F that indicates the bearing of a contour gradient (line of steepest descent). α_F varies between $0°$ and $360°$. The dip β is the angular deviation between the line of deepest descent and the horizontal. β varies between $0°$ and $90°$.

To plot the spatial position of planes and lines one can use the lower-hemispherical projection, which is the area-preserving projection of a lower halfsphere onto a horizontal plane (Fig. 3.8). Planar surfaces are denoted by their normal passing through the centre of the lower-hemisphere. This normal penetrates the lower-hemisphere at a point, the so-called pole, whose projection is plotted.

For analytical and numerical procedures the unit normal vector **n** of the joint can be expressed in a x-y-z-system of coordinates (Fig. 3.9):

$$\boldsymbol{n} = \begin{pmatrix} \sin\beta \sin\alpha_F \\ \sin\beta \cos\alpha_F \\ \cos\beta \end{pmatrix}$$

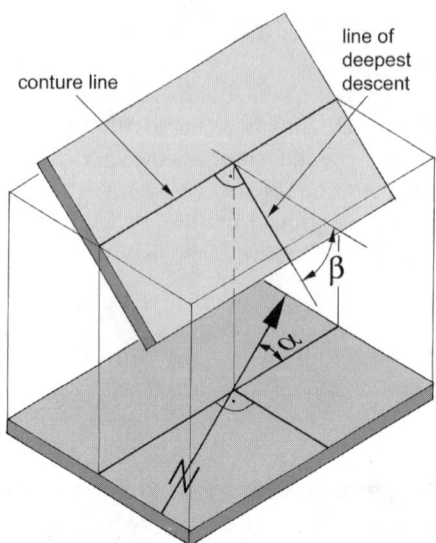

Fig. 3.7. Definition of strike and dip

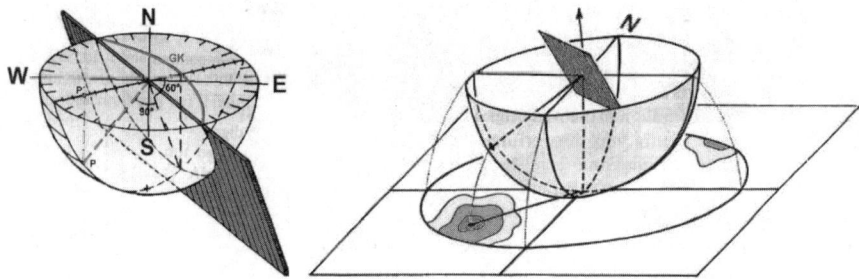

Fig. 3.8. 3D plot of the lower hemisphere

The joints of a cluster are not strictly parallel. Therefore, their poles do not coincide but form rather a cloud of points in the lower-hemispherical projection.

The spatial orientation of joints can be rated with respect to tunnel excavation according to BIENIAWSKI and WICKHAM[10] (Table 3.2).

- Degree of interruption: This property describes whether a joint is complete or interrupted by so-called bridges,
- Roughness
- Joint spacing
- Properties of the joint fill.

---

[10] E. Hoek, P.K. Kaiser, W.F. Bawden, Support of Underground Excavations in Hard Rock, Balkema, 1995.

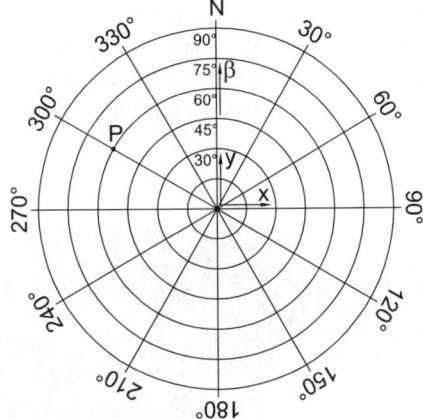

Fig. 3.9. Lower-hemispherical projection. Point P is the pole of a joint with $\alpha_F = 120°$ and $\beta = 60°$

Strike (α_F)	Dip (β)	Rating
parallel to tunnel axis,	$45° \ldots 90°$	very favourable
drive with dip	$20° \ldots 45°$	favourable
parallel,	$45° \ldots 90°$	fair
drive against dip	$20° \ldots 45°$	unfavourable
perpendicular to	$45° \ldots 90°$	very favourable
tunnel axis	$20° \ldots 45°$	fair
for all α_F	$0° \leq \beta \leq 20°$ (so-called coffin cover)	fair

Table 3.2. Rating of joint orientation

The following types of joints are distinguished:

Faults (shearbands): These are formed when two blocks slide relatively to each other (Fig. 3.10). The thickness ranges from microscopically small to several kilometres (cf. St. Andreas fault in California). Their fill can be disturbed, weathered or loosened. It is distinguished between the following types of fill:
- Kakirite: isotropic soil (i.e. loose material with low or no cohesion)
- Kataklasite: isotropic rock
- Mylonite[11]: rock (shale)

The fill can be more permeable than the adjacent rock. It also happens, however, that the fault is filled with quartz or other impermeable minerals thereby forming an aquitard. From a mechanical point of view it is observed that joints filled, e.g. with talc, have friction angles as low as 2

---

[11]Earlier, 'Mylonite' designated a disturbed and loosened rock and had, thus, the same meaning as has Kakirite nowadays.

- 4°. Sometimes the surfaces of the fault are polished due to gliding processes. Such a joint is called slickenside and is characterised by very low friction. Geologically formed slickensides, mainly encountered in overconsolidated clay, can be re-activated by engineering constructions and may cause damage.

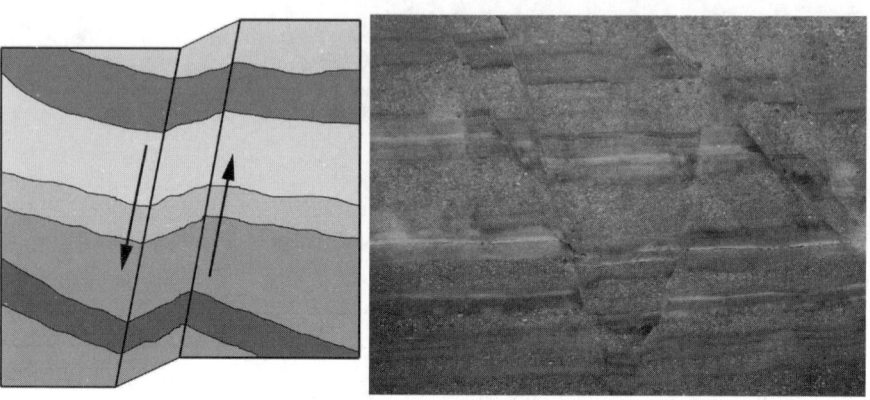

Fig. 3.10. Fault   **Fig. 3.11.** Faults in stratified rock

Beddings, separations of sediment strata: These do not necessarily constitute surfaces of reduced shear strength.

Foliation ('schistosity') These are formed by metamorphic re-crystallisation.

Fig. 3.12. Stratified sedimentary rock

A synonym of foliation is 'schistosity'. The generic term of bedding and foliation is 'lamination'. It should be added that this terminology is by no means unique. Foliation should not be confused with 'core discing', which occurs at high in situ stress in brittle rock and results in cores decomposed into thin chips (see Section 13.5.3).

3.5 Weathering

Weathering denotes the degradation of rock under the impact of water, wind, temperature, chemical decomposition etc. In arid and cold regions, weathering is mainly due to physical actions, whereas in warm regions it is mainly due to chemical reasons. It can reach considerable depths (e.g. up to 1,500 m in Russia). Attention should be paid to the fact that the weathering front does not propagate in a uniform way. It preferably follows joints and other weak zones. Note also that the weathering front can be very irregular. There are cases reported (e.g. in Porto/Portugal), where weathered rock was found beneath the unweathered one.

The effect of weathering can be very strong. A granite can, for example, show the behaviour of a soft soil, while retaining its original appearance.

3.6 Rock rating and classification

We still lack a satisfactory mathematical model capable of describing the behaviour of jointed and weathered rock mass. One tries, therefore, to proceed pragmatically and develop a rating of rock mass based on some easily observable properties. This rating addresses specific applications like excavation or support. The rating can either be expressed verbally ("This soil is very permeable.") or with numbers ("The soil's permeability is 10^{-5} m/s."). Several rock classification schemes have been proposed[12], whose aims are:

- To create notions that support the communication of engineers and geologists and that describe the rock mass
- To determine homogeneous regions with more or less constant properties
- To provide numbers useful for the design of tunnels
- To formulate recommendations for excavation and support
- To create the possibility to transfer experiences from one site to another.

When using rock classification schemes, one should avoid:

- To use rock classification as a 'recipe' and to neglect analytical and observational methods

---

[12] Z.T. Bieniawski: Classification of Rock Masses for Engineering: The RMR System and Future Trends. In: Comprehensive Rock Engineering, Vol. **3**, 553-573, Pergamon Press 1993

- Making use of one classification scheme only
- Insufficient input data
- Neglecting the conservative character and limitations of statements based on rock classification.

The oldest classification scheme was developed by TERZAGHI. Later classifications were proposed according to the stand-up time of the unsupported face by LAUFFER (1958), the *Rock Quality Designation* of DEERE, the RSR-system *(Rock Structure Rating)* of WICKHAM, the RMR-system of BIENIAWSKI, the Q-System of BARTON, the GSI-index *(Geological Strength Index)* and others. Statements and recommendations based on rock classification are, virtually, theories. As any other theory, they aim at extracting experience and transfer it to other situations, thereby making predictions. In contrast to the rational theories of mechanics, however, these are rather arbitrary and should therefore be used with caution. Their main disadvantage is that statements based on rock classification can hardly be comprehended or checked.

The classification according to the Austrian code ÖNORM B 2203-1 refers to heading. The heading classes are characterised by two numbers, the first of which indicates the advance step and the second one results from the support rated according to a given scheme.

3.6.1 RMR-System

The RMR (Rock Mass Rating) classification system[13] by BIENIAWSKI rates the rock mass according to 6 criteria. A certain number of points is assigned to each of these criteria:[14]

Unconfined strength q_u in MPa	<1	1-5	5-25	25-50	50-100	100-250	>250
Strength index I_s from point load test in MPa	-	-	-	1-2	2-4	4-10	>10
RMR-points	0	1	2	4	7	12	15

RQD	<25%	25-50%	50-75%	75-90%	90-100%
RMR-points	3	8	13	17	20

Spacing of joints	<6 cm	6-20 cm	20-60 cm	0.6-2 m	>2 m
RMR-points	5	8	10	15	20

---

[13] There are versions of 1976, 1989 and 2002.

[14] The assessment of rock mass strength based on Rock Mass Rating is presented in Section 13.11.

3.6 Rock rating and classification

State of joints	RMR-points
Continuous opening	0
Slickensides or continuous separation 1-5 mm	10
Separation <1 mm, slightly rough, highly weathered walls	25
Very rough, unweathered walls, no separation, discontinuous joints	30

Spacial orientation of joints, rating (see Table 3.2, p. 67)	unfavourable	fair	favourable	very favourable
RMR-points	−10	−5	−2	0

Groundwater ingress in l/min per 10 m tunnel	0	0-10	10-25	25-125	>125
RMR-points	15	10	7	4	0

Table 3.3. Rating points for the RMR-system

The sum of the points is the total rating of the rock mass and is designated as RMR-value. RMR can be consulted in several empirical estimations.

3.6.2 Q-system

BARTON's Q-system takes was introduced in 1974 in Norway and takes the following 6 criteria into account:

1. RQD (core recovery)
2. Number of joint clusters J_n
3. Joint roughness J_r
4. Weathering of joints J_a (joint alteration)
5. Joint water reduction factor J_w
6. Stress reduction factor SRF

The rating Q is then obtained as

$$Q = \left(\frac{RQD}{J_n}\right) \cdot \left(\frac{J_r}{J_a}\right) \cdot \left(\frac{J_w}{SRF}\right) \quad .$$

In this equation RQD/J_n rates the block size, J_r/J_a rates the interblock shear strenth. Typical values of these parameters are shown in Table 3.4.
Several relations between Q and RMR have been proposed such as $RMR \approx 9 \ln Q + (26 - 42)$.[16]

---

[16] Z.T. Bieniawski, Rock Mechanics Design in Mining and Tunnelling, Balkema, 1984, 126 ff

Quantity	Range
Q	0.001 ('exceptionally poor') - 1,000 ('exceptionally good quality rock')
RQD	0 - 100 %. Values lower than 10 % should not be used
J_n	0.5 (massive rock with no or few joints) - 20 (crushed, earth like rock)
J_r	0.5 (slickensided planar joints) - 4 (discontinuous joints)
J_a	0.75 (unaltered joint walls) - 20 (thick zones of swelling clay)
J_w	1.0 (dry excavation) - 0.05 (exceptionally high inflow)
SRF	1.0 (medium rock pressure) - 20 (heavy rock pressure)[15]

Table 3.4. Ranges of values of parameters for Q-System

3.7 Reports

It should be clearly distinguished between the raw data from laboratory and field test results and their interpretation. Referring to so-called design-build contracts, the owner should not present interpretations of construction behaviour that might define or limit construction means and methods. However, he should present sufficient data from investigations to avoid that the bids are too diverse and difficult to compare.[17]

According to SIA 199, the following documents have to be worked out:

1. **Description of the ground**
   - Geology: Geological and tectonical layout, formation of the relevant soils and rocks. Classification into homogeneous sections. In this respect, the most important parameters for soils are: structure and density, for rocks: structure, joints, weathering.
   - Hydrological conditions: Groundwater flow and its relation to surface water, circulation within pores, joints, karst, permeability. Influence of groundwater on tunnel (e.g. aggressivity against concrete) and influence of the tunnel on groundwater and sources on the ground surface.
   - Geotechnical conditions: This item contains all geotechnical properties relevant for excavation and support.
   - Possible additional statements: They refer to possible gases such as CH_4, CO_2, H_2S, radioactivity (radon), rock temperature, earthquakes, active joints, rockslides, dangerous substances such as quartz, asbest, contaminations.[18]

2. **Rating the ground** This item contains predictions on excavation and usage of a tunnel, based on the above stated ground properties. One of the

---

[17] R.A. Robinson, M.A. Kucker, J.P. Gildner, Levels of geotechnical input for design-build contracts, In: Rapid Excavation and Tunneling Conference, 2001 Proceedings, 829-839, Society for Mining, Metallurgy, and Exploration, Inc., Littleton, Colorado

[18] Contaminated soil was, e.g., encountered during the construction of the Baltimore metro, *Tunnels & Tunnelling*, September 1995.

main aims is to recognise hazards (e.g. cave ins, daylight collapses, rock bursts, gas explosions, surface settlements) and to plan countermeasures. The hazards and countermeasures have to be compiled in the so-called safety plan. Hazards have to be classified into the following degrees:

Degree of hazard	1	2	3
Occurrence	improbable	possible	probable

Judgement of the ground comprises statements on excavation and support, i.e.
- method of excavation, length of advance step, overexcavation, drilling, support of tunnel wall and face.
- draining and stabilising measures such us grouting, freezing, support with compressed air.
- reusage or disposability of the muck.

3. **Geological, hydrological und geotechnical reports**
   The above items should be presented in reports that have the following structure:
   - contract and relevant questions
   - provenience of data, used documents
   - executed experiments with documentation of method
   - description of the ground
   - conclusions and summary on the geological units
   - judgement of the ground
   - conclusions, recommendations, open questions, missing data, supplementary investigations
   - annexes with plans (drawings)
   - annexes with data, profiles of sounding and drilling, results of laboratory and field measurements.

   The conditions encountered during excavation have to be assembled in the final report.

4. **Presentation of data**
   The provenience of all data should be clearly documented. Hence, it should be distinguished between measured data and data taken from bibliography, experience, estimation and assumptions (so-called soft data). When referring to measured data, one should add the measuring method, the number of measurements, scatter and uncertainty of the values.

It should also be distinguished between the measured data (factual information) and the therefrom extracted conclusions (interpretation). Thus, it has to be distinguished between the Geotechnical Data Record that contains the exploration data, and the Geotechnical Baseline Record that summarises and interprets the data. This distinction has often contractual implications. However, factual information and some degree of interpretation are, virtually, inseparable.

The behaviour of rock is described with reference to some distinct properties. These can be assessed verbally by judgement (e.g. 'weak rock'), but in the sense of a rational approach, properties should be quantifiable. Attention should be paid to the definition of properties. To be more specific, some properties are rigorously defined by means of constitutive equations. E.g., YOUNG's modulus is defined by means of HOOKE's law $\sigma = E\varepsilon$. Referring to E, it should be asked whether the underlying constitutive equation is realistic for a particular material (e.g., reference to E makes no sense for sand or water). Some other properties, such as 'abrasivity' cannot be described in the frame of equations. In such cases one has to resort to some whatsoever arbitrarily prescribed procedures. Apart from rock abrasivity, there are also other properties which cannot be specified properly. This is the case, e.g., with rock toughness, which is understood as the resistance to cutting. All attempts to further specify the toughness failed so far.

The specification of measurable quantities occurs on the basis of:

- Direct laboratory or field measurements
- Empirical correlations with other quantities
- Consideration of scale effects
- Judgement from similar situations
- Back analysis.

4
Heading

The heading of a tunnel comprises the following actions: excavation, support of the cavity and removal of the excavated earth (mucking). It is distinguished between conventional (also called incremental or cyclic) heading and continuous heading. This chapter introduces the several methods applied. A rigorous classification is difficult, as these methods are often combined. It has become usual to distinguish between conventional (or incremental) heading on the one hand and continuous (or TBM) heading on the other hand. This is, however, not reasonable: also TBM-heading consists of several steps and is, thus, incremental (cf. Sections 4.2 and 4.4.4).
The main characteristic of conventional heading is that it proceeds in small advance steps, whose length ranges between 0.5 and 1.0 m in soft ground. This length is an important design parameter, as the freshly excavated space has to remain stable for a while (at least 90 minutes) until the support has been installed. The length of advance steps influences also the settlement of the ground surface: reducing the advance length decreases considerably the surface settlement.

4.1 Full face and partial face excavation

Large cavities are less stable than small ones (cf. Section 16.3). Therefore, in many cases the tunnel cross section is not excavated at once, but in parts. For cross sections $> 30 - 50$ m^2 in weak rock the face must be supported (with a heap of soil, shotcrete, nails) or excavated in parts. In the early era of tunnelling many different schemes of partial face excavation were developed. The terminology is neither unique nor systematic. To give examples, the core-heading *(Kernbauweise)* and the Old Austrian Tunnelling Method could be mentioned. Such partial excavation methods were developed when support was made with timbering and masonry. In contrast, contemporary support is accomplished with steel and sprayed concrete.
The various types of partial face excavations are:

76   4 Heading

Core heading: This is also known as the German heading method (although it was first used in France). It consists of excavating and supporting first the side and top parts of the cross section and subsequently the central part (core). The ring closure at the invert comes at the end. The first gallery also serves for exploration. The crown arch is founded on the side galleries thereby keeping the related settlements small.

Old Austrian Tunnelling Method: This method is schematically represented in Figure 4.1. Its characteristic feature is the crown slot. The simultaneous work in several excavation faces allows a fast advance. E.g., the old Arlberg (rail) tunnel was built with the Old Austrian Tunnelling Method. Its construction time was comparable to the one of the new Arlberg (road) tunnel, which has roughly the same length and was built ca 100 years later.

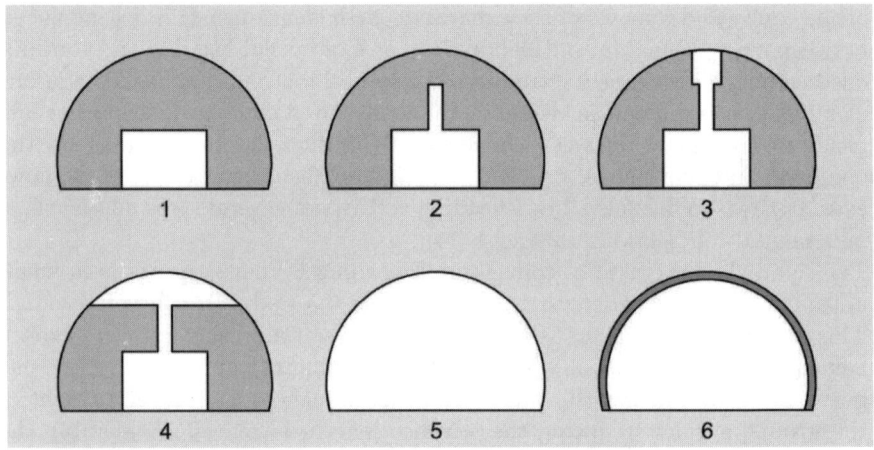

Fig. 4.1. Excavation sequences of the Old Austrian Method

Nowadays the most widespread methods of partial face excavation are (i) top heading and (ii) sidewall drift. Variants are shown in Fig. 4.2

Top heading: The crown is excavated before the bench (Fig. 4.4, 4.5). The temporary support of the crown with shotcrete can be conceived as a sort of arch bridge (Fig. 4.6). This explains why the abutments are prone to settlements, which induce settlements of the ground surface. Countermeasures are to enlarge the abutments (so-called elephant feet), to strengthen them with micropiles or the construction of a temporary invert. The latter must be constructed soon after the heading of the crown, i.e. not more

---

[2]Wu, W. and Rooney, P.O., The role of numerical analysis in tunnel design. In: Tunneling Mechanics (edited by D. Kolymbas), 87-168, Logos, Berlin, 2002

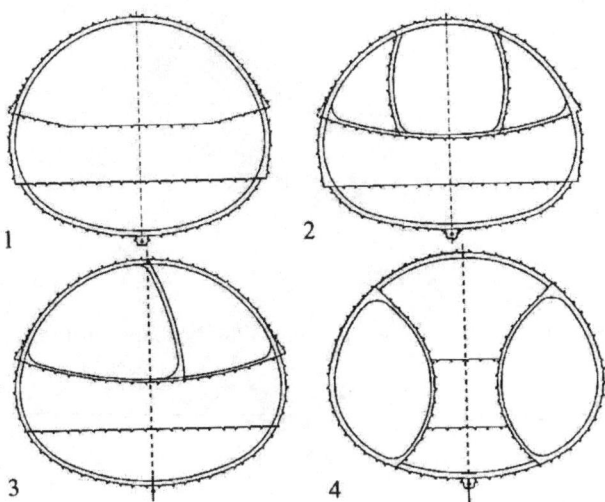

Fig. 4.2. Several types of partial face excavation[2]. The depicted parts of the cross section are driven and supported consecutively.

than 2 to 5 advance steps beyond it.[3] A soon construction of a temporary invert of the crown section or, better, the soon excavation and support of the bench and invert helps avoiding large settlements of the abutments of the crown arch. This means that the length $a = a_1 + a_2$ (Fig. 4.5) should be kept as small as possible. On the other hand, a_1 should be sufficiently large to enable efficient excavation and support works in the crown.

If the crown and the bench are excavated simultaneously, then the ramp must be 'continuously' moved forward (i.e. every now and then). Alternatively, the ramp is not placed at the centre (as shown in Fig. 4.5) but on the side of the bench. Then, the other side of the bench can be excavated over a longer distance. If the excavation of a ramp may cause instability, then the ramp must be heaped up after excavation and support of the bench. In 1985 a collapse occurred during the heading of the Kaiserau tunnel in Germany.[2] It was caused by a slit which was made to construct the ramp and which rendered the temporary invert of the calotte ineffective (Fig. 4.3).

Sidewall drift: The side galleries are excavated and supported first. They serve as abutment for the support of the crown, which is subsequently excavated (Fig. 4.7 and 4.10). This type of heading is approx. 50% more expensive and slower than top heading. Therefore it is preferred

---

[3]Richtlinien und Vorschriften für den Straßenbau (RVS 9.32)

[4]Tunnelling Switzerland, K. Kovári & F. Descoeudres (eds.), Swiss Tunnelling Society, 2001, ISBN 3-9803390-6-8

4 Heading

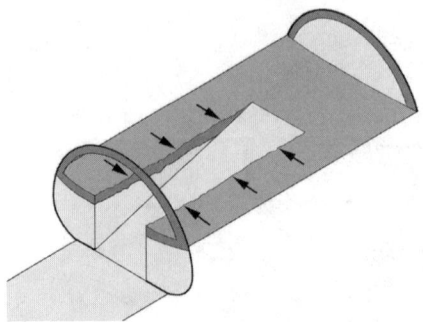

Fig. 4.3. The slit in the temporary lining caused collapse in the Kaiserau tunnel

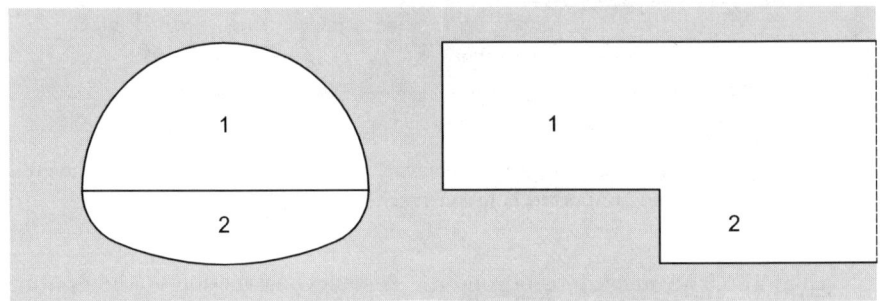

Fig. 4.4. Top heading, cross and longitudinal sections. 1: calotte, 2: bench

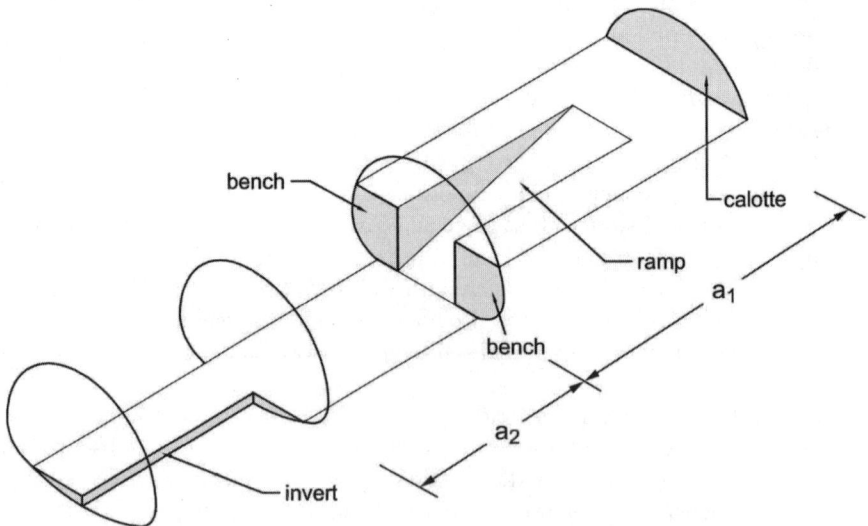

Fig. 4.5. Schematic representation of top heading

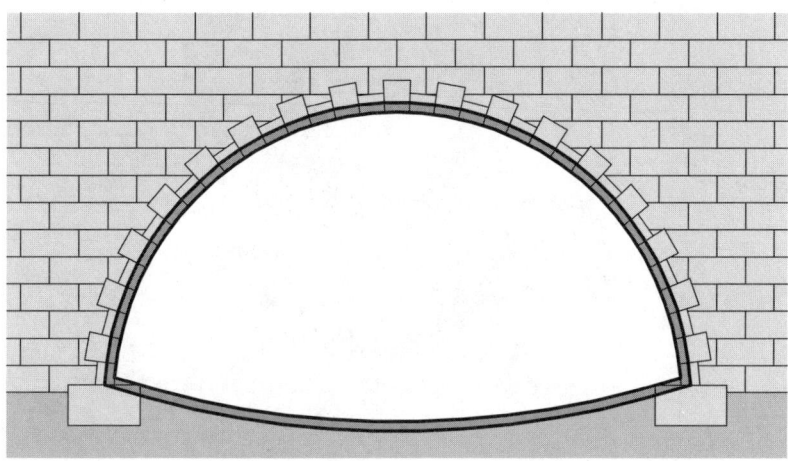

Fig. 4.6. Concept to explain the support of crown. The excavation support acts in a similar way as an arch of a masonry bridge. The weight of the overburden is concentrated in the tow abutments.

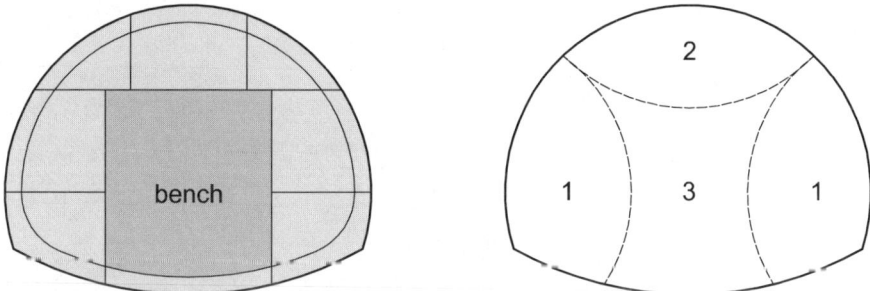

Fig. 4.7. left: core heading, right: sidewall drift

in soils/rocks of low strength. Note that a change from top heading to sidewall drift is difficult to accomplish.

In all types of partial excavation attention should be paid to the connections of the lining segments constructed at different steps. The final lining (including its reinforcement) should be continuous without any weak points (interruptions or slits). Considering e.g. the connection of the elephant foot with the temporary invert, the following method can be applied. The reinforcement (wire mesh or rebars) to provide the connection with the bench lining (to be placed later) protrudes from the elephant foot and is folded within a protective sand bed below the temporary invert (Fig. 4.9). The subsequent demolition of the temporary invert and the further excavation should be executed with caution, so that this wire mesh is not damaged.

---

[5] *Tunnel*, **9**, 2/2000, p. 19

Fig. 4.8. Sidewall drift[4]

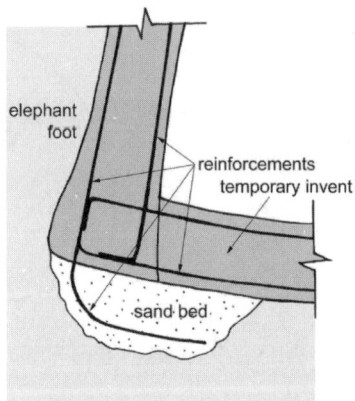

Fig. 4.9. Construction detail of the reinforcement at the connection of the elephant foot with the temporary invert

4.2 Excavation

Excavation is the process of detaching the rock using the following methods and tools:

Hammer: Pneumatic and hydraulic hammers can be applied in weak rocks and achieve performances comparable with drill & blast (see later in this section). In addition, they avoid the vibrations caused by drill & blast. E.g., the hydraulic hammer HM 2500 of Krupp Berco has the following excavation performance:

---
[6]Krupp Berco

Fig. 4.10. Niedernhausen tunnel, sidewall drift[5]

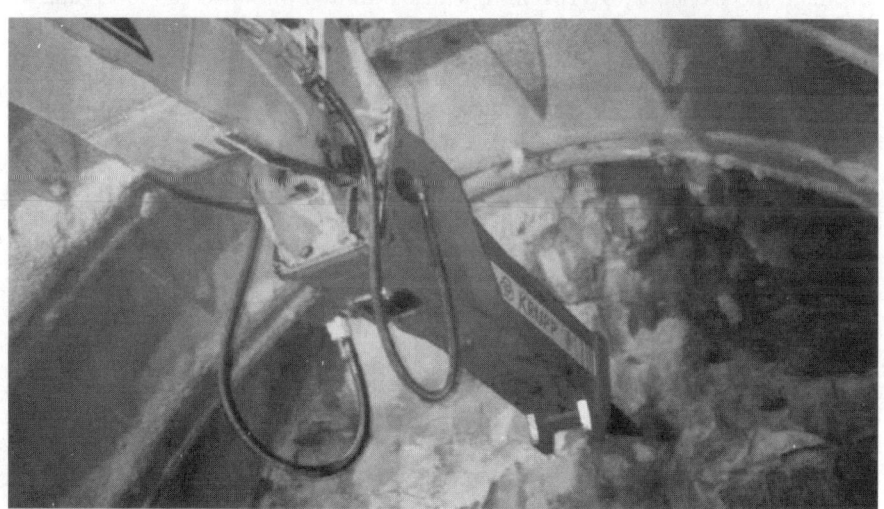

Fig. 4.11. Hydraulic hammer[6]

Unconfined strength of the rock (MPa)	Excavation capacity (m^3/h)
40-50	40
70-80	26
80-100	20

These performances can be improved by loosening blasts. Hydraulic hammers are carried by wheel or track vehicles. Dust is treated (not completely successfully) with spraying water from nozzles or water-hoses.

Excavators: Boomed backhoe buckets excavate weak rock, while thin rippers and hydraulic chisels are applied whenever hard rock inclusions are encountered. In order to exactly follow the prescribed tunnel profile these tools must be sufficiently free to rotate. Excavators can exhibit a high performance if the strength of the rock is moderate, i.e. if the rock is either soft or jointed ($RMR<30$). Ripping is applied for a RMR between 30 and 60. Another criterion for the applicability of ripping is a propagation velocity of P-waves between 1 and 2 km/s.

Roadheaders (boom cutters): These tools are used for moderate rock strengths and for laminated or joined rock. The cutter is mounted on an extension arm (boom) of the excavator and millcuts the rock into small pieces. Thus, overprofiling can be limited and also the loosening of the surrounding rock is widely avoided. One has to provide for measures against dust (suction or water spraying). The required power of the motors increases with rock strength.

Fig. 4.12. Excavator[7] (left); Roadheader[8] (right)

Tunnel boring machines (TBM): TBMs are applicable to rock of medium to high strength ($50 < q_u < 300$ MN/m^2) if its abrasivity is not too high. TBMs excavate circular cross sections with a rotating cutterhead equipped with disc cutters. To press the cutterhead against the rock, the TBM is

---

[7] Liebherr 932
[8] Voest Alpine AM 100

propped at the tunnel wall by means of extendable grippers. Therefore, the rock must have a sufficient strength.

The support can be installed soon after the excavation. The classification of TBM as 'continuous' heading instead of 'incremental' (also called 'cyclic') is misleading, because a TBM advances in strokes. Regular stops are needed, mainly for the maintenance of the excavation tools.

The design of a TBM comprises determination of thrust, torque, size and spacing of discs. The advance rate (m/h) is given by the product penetration rate (m/h)×TBM utilisation (%).

Since the TBM fills up the excavated space more or less completely, the systematic support can only be installed beyond it (i.e. beyond a working space of ca 10 to 15 m length). In weak rock, however, support measures such as rock bolts (Fig. 4.13 left) and wire meshes (Fig. 4.13 right) have to be applied adjacent to the cutterhead.

The minimum curvature radius is 40 - 80 m, with back-up equipment the minimum radius is 150 - 450 m.

TBMs are often protected against cave-ins by cylindrical steel shields. Therefore, more details on TBMs are given in Section 4.4, whereas the rock excavation by means of rotating cutterheads is described in Section 4.6.2.

Fig. 4.13. left: Anchoring behind the cutterhead; right: Installation of wire mesh behind the cutterhead[10]

---

[10]Herrenknecht AG

Drill & blast: Drill & blast was first applied in 1627 by the Tyrolean KASPAR WEINDL in a silver mine in Banská Štiavnica (former Schemmnitz, Slovakia). It is suitable for hard rock (e.g. granite, gneis, basalt, quartz) as well as for soft rock (e.g. marl, loam, clay, chalk). Thus, it is applicable for rocks with varying properties. Moreover, drill & blast is advantageous for:

- relatively short tunnels, where a TBM does not pay
- very hard rock
- non-circular cross sections.

To keep drill & blast economical, the involved steps (drilling, charging, tamping, igniting, ventilating, support) must be coordinated in such a way that downtimes are avoided. The most time-consuming operations are drilling and charging.

More details on drill & blast excavation are presented in the next section.

Fig. 4.14. Gripper TBM[12]; 1 Shield, 2 arch segments, 3 annular errector, 4 drilling equipment for rock bolting, 5 protection canopy, 6 protection girder, 7 grippers

4.3 Drill & blast

The drill & blast method consists of several subsequent steps (drilling, charging, tamping, ignition, extraction of fumes by ventilation, mucking, support), which will be described subsequently.

---

[12]Brochure Herrenknecht TBM

4.3.1 Drilling of blastholes

Rotary and percussion drilling is applied to drive blastholes within a diameter range from 17 to 127 mm (mostly being ca 40 mm) with drilling rates of up to 3 m/min into the rock. The prescribed positions, orientations and lengths of blastholes must be kept precisely, therefore the drilling equipment is mounted on tire carriages, called jumbos, with 2 - 6 booms (Fig. 4.15). The length of the blastholes corresponds to the advance step (usually 1-3 m). To achieve good blast results, the advance step should not exceed the minimum curvature radius of the tunnel cross section (for parallel cut — see section 4.3.5 — the advance step can be longer).

Fig. 4.15. Jumbo (also called 'boomer')[13]

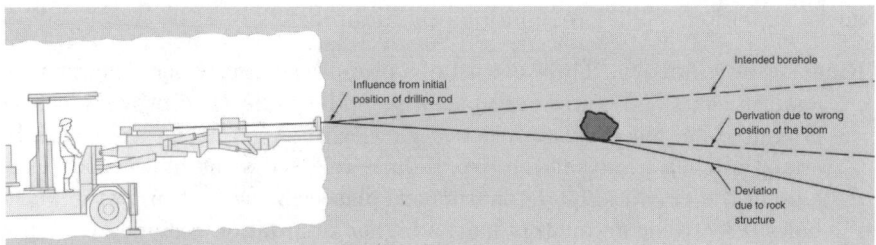

Fig. 4.16. Advance drilling. Intended and real borehole positions.

---

[13] Hydralift

4.3.2 Charging

Lengthy charges are applied in tunnelling. Charging depends on the type of explosive. Cartridges are pushed with the help of rods, powders (such as "ANFO", a mixture of ammonium nitrate and fuel oil) and emulsions are cast or pumped in into the boreholes.

4.3.3 Tamping

Explosion is the 'instantaneous' transformation of the solid (or fluid) explosive into a mixture of gases called fume (see Appendix A). To achieve pounding, the fume must be contained, i.e. its expansion must be hindered. This is why blastholes are tamped, i.e. plugged. Since the impact of the fume is supersonic, the strength of the plug is immaterial. Thus, a sufficient tamping is obtained with sand or water cartridges. In long blastholes even the inertia of the air column provides a sufficient tamping. Tamping increases pounding of the rock and reduces the amount of toxic blast fumes by improving the chemical transformation of the charge. Tamping, in particular with water, is also a countermeasure against dust production.

4.3.4 Ignition

Detonation is oxidation, where oxygen is present in compound form within the explosive. The reaction front propagates within the explosive with a detonation velocity which amounts up to 8 km/s and depends on the chemical composition, size, containment and age of the explosive. The detonation front leaves behind the fume, which is a highly compressed gas mixture. 1 kg explosive produces gas volume of nearly 1 m^3 under atmospheric pressure. The highly compressed fume exerts a large pressure upon its containment. The energy content of an explosive is not overly high, but the rate at which this energy is released corresponds to a tremendous power. Modern explosives are inert against hits, friction and heat. They can only be ignited with a (smaller) initial explosion. Therefore, ignition occurs through

Electric detonators: These consist of a primary charge, which is susceptible against heat, and a less susceptible secondary one. The primary charge is ignited by means of an electric glow wire. A retarding agent can be added in such way that the explosion is released some miliseconds after closing the circuit. The detonators are placed in the bottom of the blastholes. Electronic detonators have a higher retardation accuracy (which is important for smooth blasting) and can be ignited with a coded signal.

Detonating cords: These cords (⌀ 5 to 14 mm) have a core made of explosive and are ignited with an electric detonator. The detonation propagates along the cord with a velocity of approx. 6.8 km/s. Modern variants ("Nonel", "Shockstar") are synthetic flexible tubes, whose inner walls are coated with 10-100 g/m explosive (Nitropenta). Detonating cords allow bunched ignitions with only one electric detonator.

The power of detonators reduces with time, but appropriate storage ensures a long life.

Fig. 4.17. Detonators

4.3.5 Distribution of charges and consecution of ignition

The explosion aims at (i) breaking the rock into pieces which are manageable for haulage, (ii) avoiding overbreak or insufficient excavation profile (so-called smooth blasting) and (iii) not disturbing the surrounding rock. To this end, several schemes (drilling and ignition patterns) have been empirically developed for the distribution of charges and the consecution of ignition. It is distinguished between production and contour drillholes. The most efficient excavation is obtained if the fume pushes the rock against a free surface. This can be achieved e.g. with a V-cut ('edge' or 'fan' cut)[14]: The blastholes in the central part of the face are conically arranged and ignited first (Fig. 4.18). The surrounding blastholes are ignited consecutively with a delay of some milliseconds. Thus the rock is progressively pared, from the cut to the contour. Parallel blastholes ('parallel cut') are easier to drill precisely and enable longer advance steps but require more explosive than conically arranged ones (V-cut). Several unloaded drillholes are provided in the parallel cut, creating thus a cavity against which the detonation pushes the rock. Thus the efficiency is

---

[14] J. Johansen, Modern Trends in Tunneling and Blast Design, Balkema, 2000

88   4 Heading

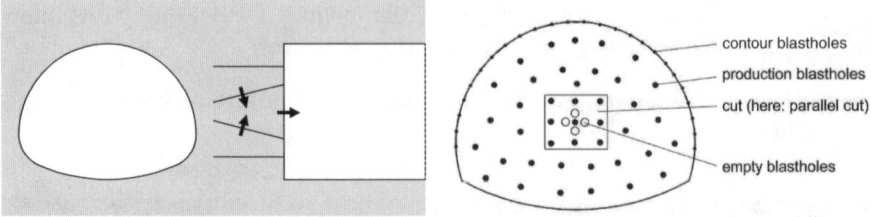

Fig. 4.18. Wedge cut (left), distribution of blast holes (right)

increased. For smooth blasting, the contour holes have a small spacing (e.g. 40-50 mm) and are charged with detonating cords (Fig. 4.19). Smooth blasting helps to minimize the costly after-treatment (post-profiling).

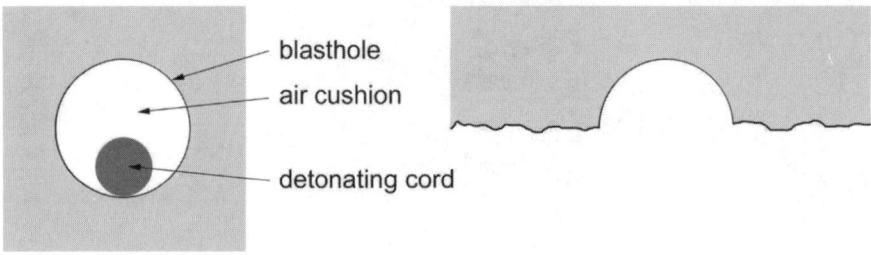

Fig. 4.19. Charging of contour blastholes (left), rock surface after the detonation (right)

In case of anisotropic, jointed or stratified rock, the distribution of blastholes and the ignition pattern have to be adapted to the structure of the rock.

The various work steps for drill & blast are repeated cyclically. Adjusting and optimization of the blasting scheme with regard to the individual rock properties are therefore recommended. This can be accomplished by the aid of computers.[15]

4.3.6 Explosives

There are several types of explosives. Explosive gelatines like Ammongelite or Gelamon are plastic, have a high strength and develop large gas volume (approx. 800 l/kg). They are water resistant and allow safe handling. The detonation speed is 6.5 km/s.

---

[15] "Computerised drill & blast tunnelling", *Tunnels & Tunnelling*, July 1995, 29-30, and G. Girmscheid "Tunnelbau im Sprengvortrieb – Rationalisierung durch Teilrobotisierung und Innovation", *Bautechnik* 75 (1998), 11-27

In explosive emulsions, oil serves as the combustible agent. Oxygen is delivered from salt (nitrate) solutions available within droplets of 10^{-6} m diameter. These droplets are covered with a 10^{-9} m thick oil film. They develop a gas volume of 1,000 l/kg and a detonation speed of 5.7 km/s. They are water resistant, safe in handling, and can be pumped into the blastholes or placed within cartridges. The non-explosive components are separately delivered to the construction site, where they are mixed together. Compared with explosive gelatines, the strength of explosive emulsions is reduced and, therefore, ca 10% more blastholes are needed. On the other hand, their fumes are less toxic (having, though, a pronounced smell of ammonia), so a shorter ventilation is needed and the spoil is less contaminated.

Explosive powders are also safe but not water resistant. With a gas volume of 1,000 l/kg and a low detonation speed of approx. 3 km/s they push rather than shatter the rock.

A recent development are the so-called non-explosive explosives like BRISTAR or CALMMITE. These are expansive cements. Mixed with water and cast into boreholes they expand, thus developing pressures up to 600 bar and tearing up the surrounding rock.

Fig. 4.20. Explosive catridges

4.3.7 Explosive consumption

Explosives are an important cost issue. The required amount of explosives increases with decreasing tunnel cross section and increasing rock strength. It ranges between 0.3 and 4.5 kg/m^3.

4.3.8 Safety provisions

Only licensed individuals are allowed to handle explosives. Receipt, consumption and return of explosives must be registered in a tractable way. Transportation must conform the regulations for transportation of dangerous materials.[16] The electric circuit for ignition should be checked. The blasted rock can be spread over large distances. Therefore a safety distance of 200 to 300 m off the face should be maintained. The removal of non exploded charges is particularly difficult. Sometimes the tamping can be extracted and a new detonator be placed. Another method is to overlay a new charge and let it explode together with the old one.

Electromagnetic fields (e.g. radio-frequency radiation from mobile telephones) may cause inadvertent initiation of electro-explosive devices. Therefore, a safe distance must be kept from radiation sources.[17]

4.3.9 Ventilation

Contemporary explosives have a positive oxygen balance but still do contain toxic gases such as CO_2, CO and nitrogen oxids. Also dust is dangerous, in particular quartz dust. Therefore, following an explosion, work should only be resumed after a ventilation time of at least 15 minutes with an air velocity (averaged over the largest cross section) of at least 0.3 m/s. During ventilation, the personnel should stay either in the open air or within a fume protection container. This is also advisable because parts of the shotcrete lining can be detached (due to the explosion shock) and fall down.

4.3.10 Backup

A new development is the hanging backup system as applied in the Loetschberg base tunnel (Fig. 4.21). Rockbolts are used to mount two overhead rails on which the backup runs. The benefits are faster production, greater safety, free access to the face, easier construction of invert and cross-passages.[18]

4.3.11 Shocks and Vibrations

Vibrations due to drill & blast (as well as due to TBM)[19] propagate in the underground as elastic waves and can affect constructions and human well-being. The disturbance is controlled (for frequencies > 10 Hz)[20] by the maximum vibration velocity, which can be measured by means of so-called geophones. Some typical values of $v_{\max}$ are shown in Table 4.1.

---

[16] www.sprenginfo.com

[17] P. Röh, Beeinflussung elektrischer Zündanlagen durch mobile und stationäre Funkeinrichtungen. *Nobelhefte* 2002, 5-14

[18] M. Knights, P. Hoyland, Policies and projects - Swiss style, *Tunnels & Tunnelling International*, April 2002, 28-31

[19] R.F. Flanagan: Ground vibration from TBMs and shields, *Tunnels & Tunnelling*, Vol. 25, No. **10**,1993, 30-33. J. Verspohl: Vibrations on buildings caused by tunnelling, *Tunnels & Tunnelling*, 1995, BAUMA Special Issue, 81-85

[20] Values of the peak velocity for domestic structures exposed to transient vibration can be seen in *Tunnels & Tunnelling International*, June 2002, 31-33

Fig. 4.21. Hanging backup system in drill & blast heading in the Loetschberg base tunnel (perspex model).

v_{max} in mm/s			
Source	distance		
	5 m	10 m	25 m
Hydraulic-hammer	1.7	0.6	0.1
truck	3.0	1.1	0.3
large bulldozer	3.9	1.4	0.3
TBM (hard rock)	5.5	2.2	0.6
vibrated pile	30	12	2.8

Table 4.1. Typical values of v_{max}

Rating criteria for disturbance and hazard due to vibrations can be found in standards such as DIN 4150 and ÖNORM S 9020. Transient vibrations (e.g. due to drill & blast) can be noticed from $v_{max} = 0.5$ mm/s and cause complaints from $v_{max} = 5$ mm/s. Humans are particularly susceptible and less tolerant of continued vibrations, especially during the night. Vibration hazard is usually over-rated and this may cause pre-existing fissures in buildings to be attributed to drill & blast. Therefore, residents in the neighborhood should be informed about possible hazards. In addition, a perpetuation of evidence should be carried out. Vibrations due to drill & blast can be reduced by splitting the explosion into several smaller explosions with a consecution of some miliseconds. E.g., Nonel detonators enable 25 explosions within 6 seconds.[21] TBM vibrations are usually not noticeable at distances greater than 45 m.

---

[21] *Tunnels & Tunnelling*, July 1995, p. 8

4.4 Shield heading

Shields are used for heading in weak rocks and soil.[22] The shield is a steel tube with a (usually) circular cross section (Fig. 4.22 and 4.23). Its front is equiped with cutters. The shield is pushed forward into the ground by means of jacks. A so-called blade shield is equipped with blades that can be protruded separately (Fig. 4.24).

Fig. 4.22. Modern shield heading[23]

Jacks: The jacks have a stroke between 0.8 and 1.5 m, they operate with pressures up to 400 bar and apply forces up to 3 MN. Their abutment is the already installed lining. The shield can be steered by applying different pressures to the jacks distributed along its circumference. On this, shields

---

[22]see also B. Maidl, M. Herrenknecht , L. Anheuser, Maschineller Tunnelbau im Schildvortrieb, Ernst & Sohn Verlag, Berlin 1995; M. Kretschmer und E. Fliegner, Unterwassertunnel, Ernst & Sohn Verlag, Berlin 1987

[23]Tunnelling Switzerland, K. Kovári & F. Descoeudres (eds.), Swiss Tunnelling Society, 2001, ISBN 3-9803390-6-8

should be not too lengthy ($L < 0.8\,D$) and the cavity diameter should be slightly larger than the outer diameter of the shield, i.e. a shield-to-rock-clearance of ca 25 mm should be provided for.

Fig. 4.23. Historic shield tunnelling beneath the St. Clair River

Fig. 4.24. Shield with blades and roadheader[24]

After each advance stroke the hydraulic pistons of the jacks are retracted in such a way that an additional slice of lining can be added under the protection of the shield tail (Fig. 4.25). During the subsequent advance

---

[24] Herrenknecht

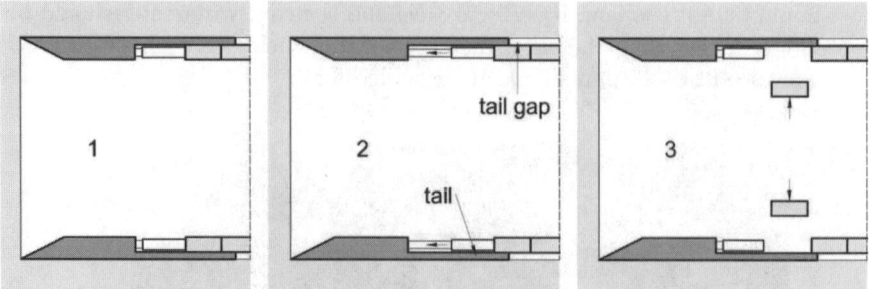

Fig. 4.25. Sequences of shield heading

stroke the tail gap is filled with grout. Coordination of the individual steps is important. Typical advance rates are 0.5 to 2 m/h.

An overly large thrust can damage the lining segments. The thrust is applied to overcome mainly the wall friction, which can be estimated by $\mu \sigma U L$, where L and U are the length and the circumference of the shield, respectively. The normal stress σ is taken simply as the effective vertical stress at the depth of the tunnel axis and the friction coefficient μ is usually assumed between 0.7 and 0.9. If the face is supported by breasting flaps (Fig. 4.26 right), a force of 8 to 10 MN has to be added.

Fig. 4.26. Platform-shield[25]; shield with face breasting flaps[26]

Excavation: The excavation is either full-face, with rotary cutterheads (Fig. 4.27), or selective, with boom headers or roadcutters (roadheaders) that excavate the face in a series of sweeps (see also Section 4.2). The cutterhead can

---

[25] Les Vignes-Tunnel, Herrenknecht
[26] Wayss & Freytag

be driven either electrically or hydraulically. Electric drive has a higher efficiency but is more difficult to control. Hydraulic drive is more flexible. The necessary thrust and torque are determined empirically. Note, however, that in general the cutterhead is not expected to support the face. At standstill the openings should be closed with panels. In sticky soils the wheel should be as open as possible. Adhesion of sticky soil (e.g. plastic clay) is a considerable handicap and can be combated with flushing.

The cutterhead is equipped with the following tools to excavate the ground:

- disc cutters, drag bits (for rock and embedded blocks)
- chisels, scrapers (for sand and gravel)
- scrapers (for cohesive soil)

Fig. 4.27. Cutterheads with scrapers and discs

Some maintenace works to the cutterhead can only be undertaken from ahead. To this end, the adjacent soil can be stabilised by grouting or freezing. Then the shield is retracted so that a space is created from which the inspection can proceed.

Cutterheads also have facilities for advance drilling that serves either for soil exploration or for grouting. The Herrenknecht cutterhead utilized in the 4th Elbe tube and later in Moscow (Lefortovo), see Fig. 4.27 (right), was equipped by Amberg with a device to detect blocks and other obstacles with sonic pulses.

Face support: An unstable face can cause surface settlements and should, therefore, be supported e.g. with panels (Fig. 4.26 right) or with the cutterhead (if $5 < c_u < 30$ kN/m^2). Alternatively, the soil is allowed to run out onto platforms to form supporting heaps (platform-shield, Fig. 4.26 left). Very soft soil can be laterally displaced. Unsupported faces are

---

[27] *Felsbau*, 3/2001, p .26
[28] Herrenknecht

Fig. 4.28. Double shield[27]

Fig. 4.29. Double shield TBM[28], 1: telescopic shield, 2: gripper, 3: jacks

very problematic or impossible, especially below water level. See also Section 4.4.1 and Chapter 17.

If the cutterhead has to support the face, it must cover a part of the face which should be as large as possible. The muck enters through slots. Alternatively, panels can be elastically mounted on the spokes of the cutterhead.

Lining segments: These are made of precast reinforced concrete or of cast iron and have secant lengths up to 2.2 m and widths between 0.6 and

2.0 m (Fig. 4.32). Several segments form a ring. Each ring is closed with a so-called key segment (Fig. 4.30).

Wider segments speed up the rate of lining but are disadvantageous for curved tunnels. The segments are moved with vacuum erectors and temporarily fixed with straight or curved bolts (Fig. 4.31). Alternatively plugs and studs can be used.

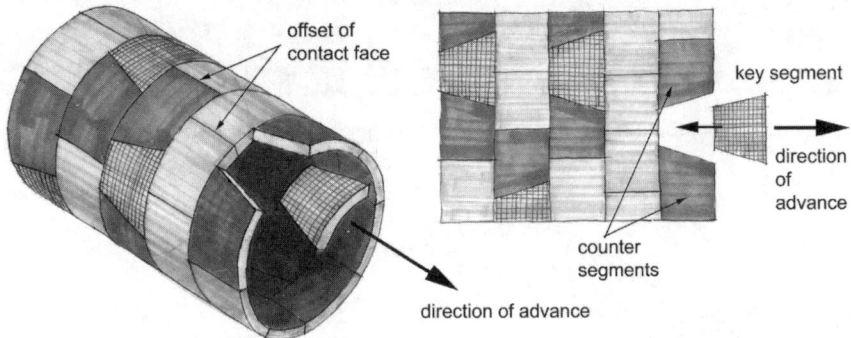

Fig. 4.30. Array of segments. Left: perspectively, rigth: unfurled

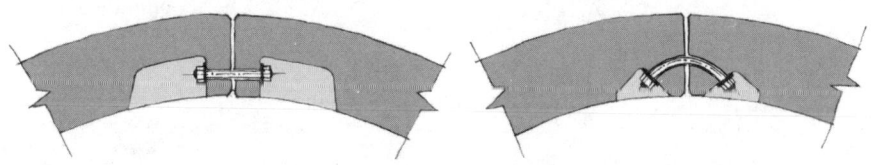

Fig. 4.31. Segment assemblies with straight or curved bolts

Usually, segments are made of concrete C35/45. A higher concrete strength is not necessary and renders the segments more brittle (edges can flake). The previously used hollow or ribbed segments are no longer in use. The segments must also be designed to carry the loads during transport and installation. The use of steel fibre reinforced concrete (e.g. with 30 kg fibres per m^3 concrete) simplifies the manufacturing process and reduces the hazard of edge flaking. Steel fibres are often used together with bar reinforcement. With the introduction of spheroidal graphite iron (SGI), cast

---

[29] Metro Madrid

[30] J. Heijboer, J. van den Hoonaard and F.W.J. van de Linde, The Westerschelde Tunnel, Approaching Limits, Balkema, 2004

98   4 Heading

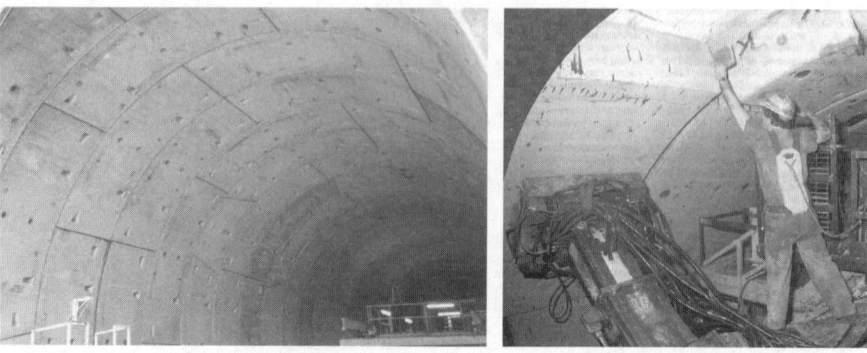

Fig. 4.32. Segmental tunnel lining[29]; Placing of small discs of wood in the joint

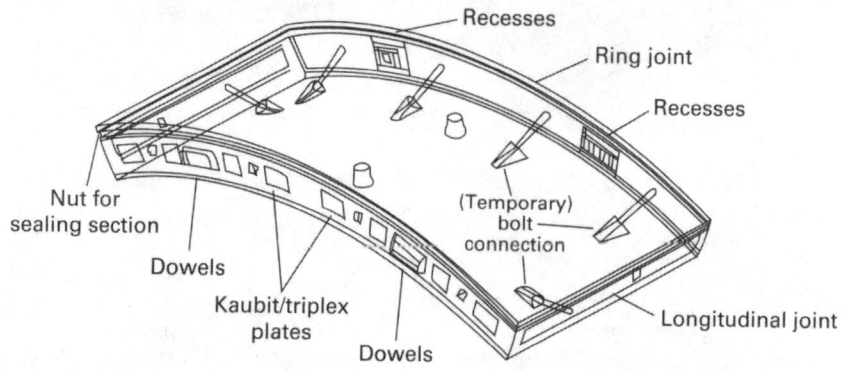

Fig. 4.33. Lining segment[30]

iron segments are now increasingly used.[31] Their tensile strength allows to carry bending moments and they are up to 30 - 40% lighter than reinforced concrete segments. In addition they occupy a smaller part of the tunnel cross section and their joints are sufficiently waterproof. Corrosion is small (0.4 mm per year).

To assure a good force transmittance, the segments must be produced with high accuracy. The tolerance in longitudinal direction should be no more than 0.3 to 1 mm and 2 mm for the thickness.

Planar joints with roughness less than 1 mm have been successfully applied, whereas groove and tongue (tenon and mortise) can be easily damaged. As the lining segments are compressed against each other, they do not need to be connected. However, they have to be bolted together for installation. If the transverse forces are high, the individual rings can slip

---

[31] *Tunnel & Tunnelling*, January 1998
[32] Unterirdisches Bauen in Deutschland, published by STUVA, 1995

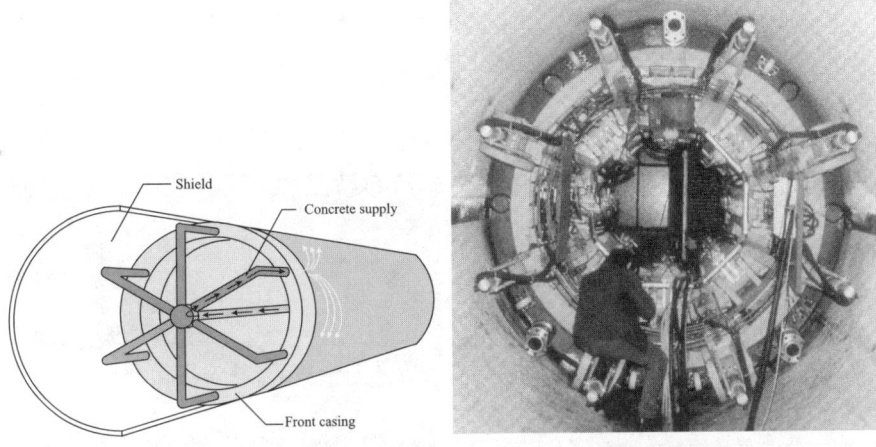

Fig. 4.34. Supply of extruded concrete[32]

relative to each other, thereby reducing these forces. To preserve this mobility, stuffings, e.g. thin plates of wood, are placed in the joints (Fig. 4.32). The lining is waterproofed by means of gaskets placed between the segments (Fig. 4.33). The gaskets are compressed and thus become watertight. There are also water-expansive gaskets.

A promising but not yet mature development is the lining with extruded concrete, which is a non-reinforced (or reinforced with steel fibres) concrete cast in the cover of the shield tail into a casing (Fig. 4.34).

In most cases the lining is much more expensive (approx. 80% of the tunnel construction costs) than the shield machine and should, therefore, be carefully designed.

Cover: If the tunnel is designed for the purpose to lead a surface lifeline underneath an obstacle, ramps must be provided for. To avoid lengthy ramps, there is a tendency towards small covers. In this case, however, the face support must be very accurate. It is recommended that a minimum cover of $0.8\,D$ (D=shield diameter) is maintained.

Mucking: With conventional excavators, mucking is done with trucks. With use of a cutterhead, mucking is done with the following methods (Fig. 4.35):
- transport with conveyor belt and subsequently with trucks or trains
- mix with slurry and pump
- mix with water or foam to form a mush and lift with screw conveyor or pump.

In tunnel heading only the soil that occupies the intended cavity has to be removed, whereas the surrounding soul should remain in place. However, it cannot be completely avoided that a part of the surrounding soil moves during excavation into the intended cavity and is removed too. This part of soil constitutes the so-called volume loss. In shield heading, the motion of soil into the cavity can occur via the face and can cause substantial settlements or even cave-ins, if it is excessive. Therefore it is tried to control the discharge of muck by weighing.

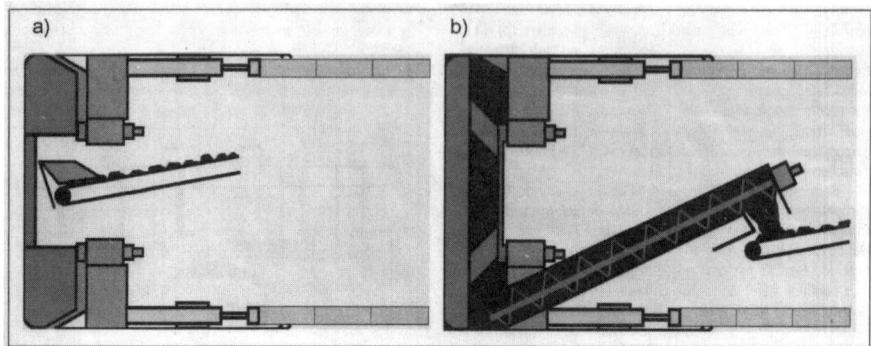

Fig. 4.35. Removal of the waste material with conveyor belt (a) or with screw conveyor (b), Herrenknecht

Tail void closure: The outer diameter of the shield is larger than the outer diameter of the lining in such a way that the moving shield leaves behind a tail void, whose thickness is up to 20 cm. Such a large void can cause the lining rings to shift and/or large surface settlements and must, therefore, be filled (closed) with mortar ('tail gap grouting' or 'back grouting'). The mortar should set as fast as possible but not too fast (otherwise it cannot be pumped in). To avoid settlements, the ring tail closure should be done as soon as possible after excavation and the grouting pressure should be equal to the primary normal stress. Of course, there is no exact way to fulfil this requirement, because the primary stress varies from point to point and is, at that, hardly known. In addition the grouting pressure field cannot be precisely controlled. The slot between lining and shield tail must be plugged, otherwise the grouted mortar can escape. The plugging is achieved with steel brushes, whose bristles are filled with grease (Fig. 4.36).

A new method of keeping the grouting pressure constant within the tail void is to provide a compliant sealing lip which yields only if a threshold pressure is reached, so that the void can be filled with a constant pressure.[33] It turns out that tail gap grouting cannot completely reverse settlement, even if the volume of the grout considerably exceeds the vol-

---

[33] S. Babendererde, Grouting the shield tail gap, *Tunnels & Tunnelling International*, Nov. 1999, 48-49

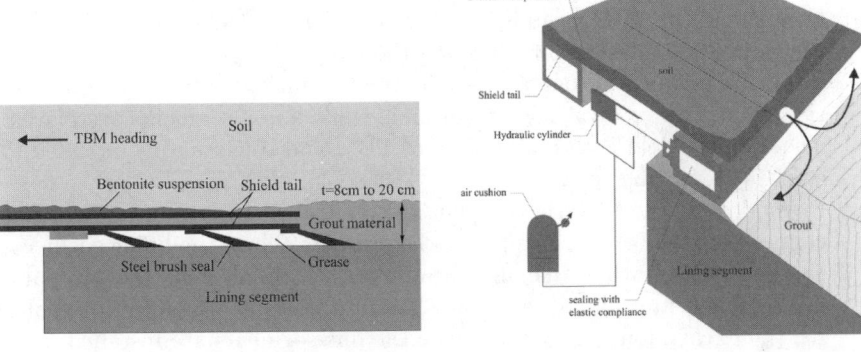

Fig. 4.36. Sealing of the tail gap

ume of the gap. This can be explained if one considers the mechanical behaviour of ground at loading-unloading cycles.[34]

Guidance: If the tunnel has a curvature radius less than 300 m, it is advisable to articulate the shield. The lining segments must be adjusted to the tunnel curvature otherwise they can be damaged. Tapered segments, as shown in Fig. 4.37, are universally applicable. To follow a prescribed curve, the jacks must be appropriately loaded and mounted in a slidable way, otherwise they can be excessively strained. In shields with a cutterhead the jacks are inclined to counteract the torque of the wheel. This torque is transmitted to the surrounding ground via wall friction of the lining. If this friction is not sufficient, then the shield will rotate.

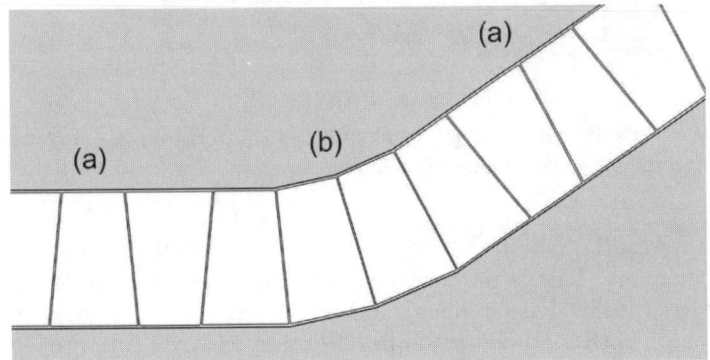

Fig. 4.37. Lining with conical segments, (a) straight drive, (b) curved drive

---

[34]M. Mähr, Settlements from Tail Gap Grouting due to Contractancy of Soil, *Felsbau* 22 (2004), No. 6, 42-48

Double shield: This shield consists of two parts (Fig. 4.28, 4.29). At the front shield the cutterhead is installed, at the rear shield ('gripper shield') the grippers for lateral bracing and the device for placement of the segments are installed. Front and rear parts are connected via a telescopic section. Thus, the excavation can continue while lining segments are placed. The double shield is a combination of TBM and shield and is intended to operate in varying rock.

In good rock, the cutterhead is buttressed against the rear part, which is connected to the adjacent rock via the grippers. Thus, the lining segments can be installed while the cutterhead works. In weak rock, the two parts of the shield are jointed, and the shield operates as a conventional one. I.e. the excavation has to stop while the line segments are installed.

Tunnel heading machines: Many shields are equipped with a cutterhead, which is the main feature of a TBM (Fig. 4.38).

Fig. 4.38. Shield with TBM (Herrenknecht), 1: cutterhead, 2: conveyor belt, 3: jacks, 4: erector for lining segments, 5: lining

This is why the notions 'shield' and 'TBM' are often confound, i.e. taken as synonyms. This is, however, wrong: shields are used in loose ground whereas (unshielded or 'open') TBMs are used in hard rock. A combination of both methods is applied in rocks with varying properties. As a generic term for shield and TBM the word 'tunnel heading machine' *(Tunnelvortriebsmaschine)* has been launched. A generally accepted classification is

4.4 Shield heading

TBM
- open (for rock tunnels)
- shielded (for weak or jointed rock/soil)
  - open face
  - closed face
    - slurry
    - EPB

A TBM needs to be protected with a shield if the rock is caving in.[35]

4.4.1 Shield heading in groundwater

If a shield operates below groundwater level, a sufficient safety against uplift must be assured for all situations to be encountered. In addition, the shield must be protected against inrush of water and soil. This can be achieved by supporting the face with a pressurised fluid (or air). The underlying mechanism is explained in Appendix B. The following variants exist:

Shield heading under compressed air: Compressed air provides support against soil and water. Either the front part of the shield or the entire tunnel can be pressurised. In the later case the large volume of compressed air improves safety against air losses, but all works are more difficult since they must be done under compressed air. Blow-outs can endanger the workers and create large craters in the ground surface. By means of a bulkhead the compressed air can be isolated in the front part of the shield. The muck has to pass an air lock.

The support of the soil is due to a seepage force, which presupposes a high pressure gradient. This can be achieved by coating the soil surface with sealing materials such as bentonite or shotcrete. The pressure is then reduced within these thin coats, thereby creating high gradients. The effect is the same as if the soil surface were covered with an impermeable membrane. Note that the coating can leak if it gets fissures due to shrinkage. Compressed air heading is applicable up to water depths of 30 m, because the pressure in the working space is limited to 3 bar for health reasons. A relatively impermeable cover of sufficient height is required to avoid blow-outs. The air supply Q, needed to replace the losses, for soils with a permeability between 10^{-5} and 10^{-3} m/s can be estimated with the empirical formula $Q\,(\mathrm{m^3/s}) \approx 4\text{-}8\,A\,(\mathrm{m^2})$, where A is the tunnel cross section.

---

[35]Recommendations for selecting and evaluating tunnel boring machines. Deutscher Ausschuß für unterirdisches Bauen, Österreichische Gesellschaft für Geomechanik, SIA-Fachgruppe für Untertagebauten, *Tunnel* 5/1997, 20-35. See also: Taschenbuch für den Tunnelbau 2001, Glückauf Verlag.

Slurry-shield: Some of the disadvantages of the support by compressed air are avoided if the support of the face is accomplished by a pressurized slurry, which in most cases is a bentonite suspension. There is no danger of blow-outs, and all works can be done under normal atmospheric pressure. The support of the ground is achieved by a seepage force which presupposes the formation of a mud cake (made of bentonite) on the soil surface.

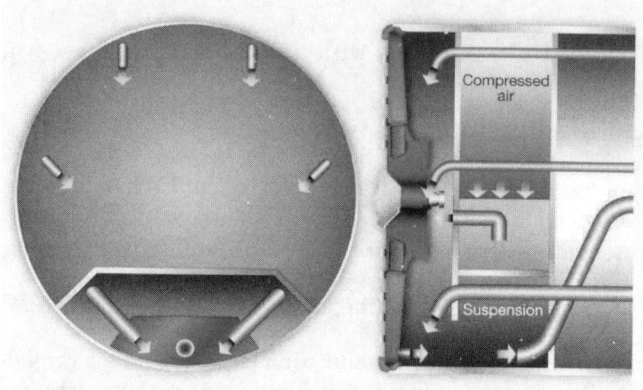

Fig. 4.39. Slurry shield[36]

The soil is excavated with a cutterhead. If the ground is very soft, excavation can even be accomplished with a water jet. The muck is mixed with the slurry at a ratio of 1:10 and pumped away to a separation plant (Fig. 4.40). Therefore, stones and blocks must be crashed first. The costs for separation rise at increased content of silt and clay.

The slurry pressure needed to support the ground must be estimated or calculated (see Chapter 17). To maintain the prescribed pressure and to replace any losses due to mucking, a reliable control of the pressure must be guaranteed. An air cushion is provided in a part of the pressurised space (Fig. 4.39). Due to the high compressibility of air, the pressure of this air cushion is much less susceptible to small volume changes. By tuning the air pressure and balancing the removed and added slurry, the pressure can be controlled with an accuracy between 0.05 and 0.1 bar.

For maintenance reasons and in order to remove blocks, the excavation chamber can be entered via an air lock. During maintenance, the slurry is replaced by compressed air. Note, however, that pressurized air is a risky method of support, since it can easily escape (blow-outs). For this reason daylight collapses occurred during the headings of the Grauholz and of the Westershelde tunnels. An alternative is to freeze the soil ahead of the

---

[36] *Tunnels & Tunnelling International*, March 2004, p. 15

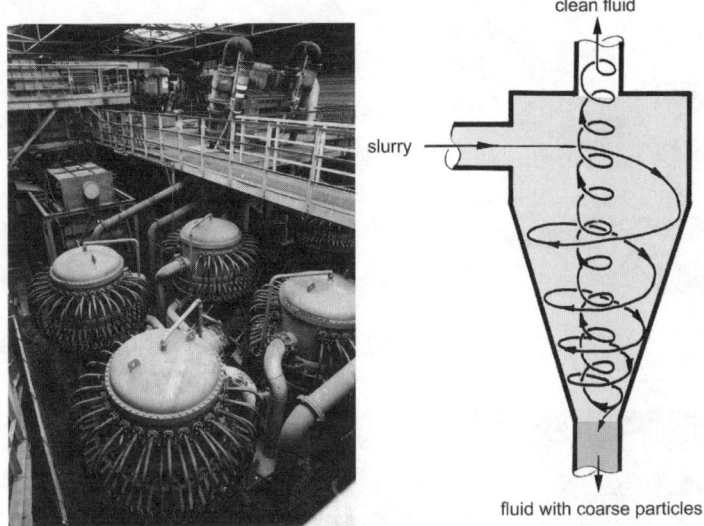

Fig. 4.40. Left: Separation plant, right: working principle of a hydrocyclon

face and carry out the maintenance works under the protection of frozen soil.

One of the largest slurry shield machines is the 'Trude' of Herrenknecht, which was utilised for the fourth tube of the Elbe tunnel (3.1 km long) and later in Moscow. The cutterhead is 400 t heavy, has a diameter of 14.2 m and operates with 2.5 rotations per minute. It can excavate soft sediments, gravel and rock.

With the 'Thixshield' technique (Fig. 4.43) the face is excavated with a movable roadheader which also sucks and extracts the muck. The roadheader has some advantages compared with a full face cutterhead:

- smaller torque
- not limited to circular cross sections
- obstacles can be easier removed

Earth-pressure-balance (EPB) shield: Instead of a slurry, the face is supported with a mud, formed of the excavated soil. The soil enters the excavation chamber through openings in the cutterhead. In most cases, water and some other additives (e.g. polymer foams) are added to render the excavated soil supple. Otherwise heat will be developed due to friction with the rotating cutterhead. The thick consistence of the mud (compared with a slurry) calls for a higher cutterhead torque (ca 2.5 times higher than for slurry shields). On the other hand, the torque is limited by

---

[37] *Tunnels & Tunnelling International*, October 2001, p. 26
[38] Herrenknecht
[39] Philipp Holzmann

106   4 Heading

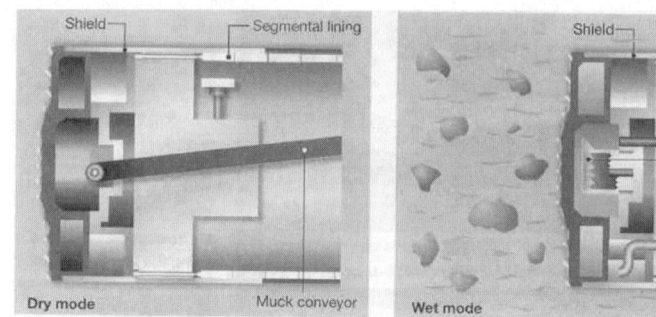

Fig. 4.41. Conversion of Mixshield mode from dry open (left) operations to slurry face support with a crusher (right)[37]

Fig. 4.42. Mixshield[38]; 1 cutterhead, 2 shield, 3 pressure bulkhead, 4 air cushion, 5 divider (dive wall), 6 hydraulic conveyor (feeder), 7 stone crusher, 8 slurry supply, 9 lining segments, 10 erector for segments, 11 jacks, 12 operator's cabin

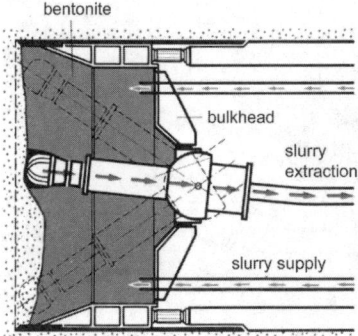 

Fig. 4.43. Thixshield: excavation chamber with cutter and bulkhead[39]

the available friction between lining and rock. Thus, the diameters of EPB shields are limited to ca 12 m. The control of the mud pressure, needed to support the face, is achieved by tuning the following quantities:
- rotation speed of the cutterhead (approx. 2-3 revolutions per minute)
- rotation speed of the screw conveyor which removes the muck from the front chamber (approx. 4-5 revolutions per minute). The mud should be sufficiently thick to plug the conveyor screw otherwise the pressure in the pressure chamber will drop
- advancing the shield by the jacks.

Due to the compressibility of the mud and the inhomogeneous non-hydrostatic pressure distribution, the pressure control is not precise, it fluctuates by ± 0.5 bar.

The mud contains 50 to 70% solids and can thus be mucked with trucks or conveyor belts. In general, its dumping capability can be easily assured, if the bentonite content is not too high.

The shield cannot start working until the pressure chamber is filled. This is much simpler with slurry shields than with EPB shields.

The appropriate conditioning of the muck with foam etc. makes the application of EPB shields possible in a large variety of grounds (including gravel). Attention should be paid to heterogeneous grounds, where the various soil layers need different support pressures.

The drive of a EPB-TBM in the underground of Porto (Portugal) proved to be particularly difficult. The encountered granite was in some places weathered and exhibited the behaviour of soft soil. So when the TBM encountered partly weathered and partly unweathered granite, the mud pressure could not be controlled. As a result, several collapses occured. The remedy was to let pressurized slurry act upon the mud. In this way, the stabilizing pressure on the face was much better controllable.

---

[40] Wirth Howden Tunnelling
[41] Herrenknecht

108   4 Heading

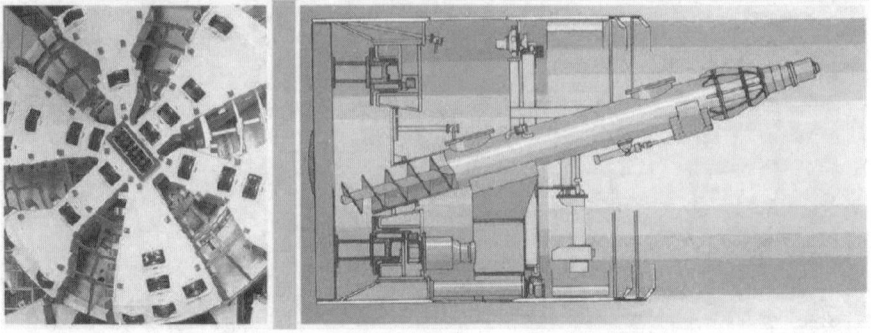

Fig. 4.44. EPB-shield[40]

Fig. 4.45. EPB-shield[41]; 1 face, 2 cutterhead, 3 pressure chamber, 4 bulkhead, 5 jacks, 6 conveying screw, 7 segment erector, 8 lining with segments

4.4.2 Tunnelling with box- or pipe-jacking

The tunnel heading proceeds with jacking of precast support elements (pipes or boxes/frames) while the ground is excavated or pushed away at the face (Fig. 4.46, 4.47). Box-jacking is usually applied to build subways under existing roads or rail tracks without interrupting the traffic.

Fig. 4.46. Pipe-jacking

Fig. 4.47. Pipe-jacking with hydraulic mucking[42]

---

[42]Herrenknecht Microtunneling

110   4 Heading

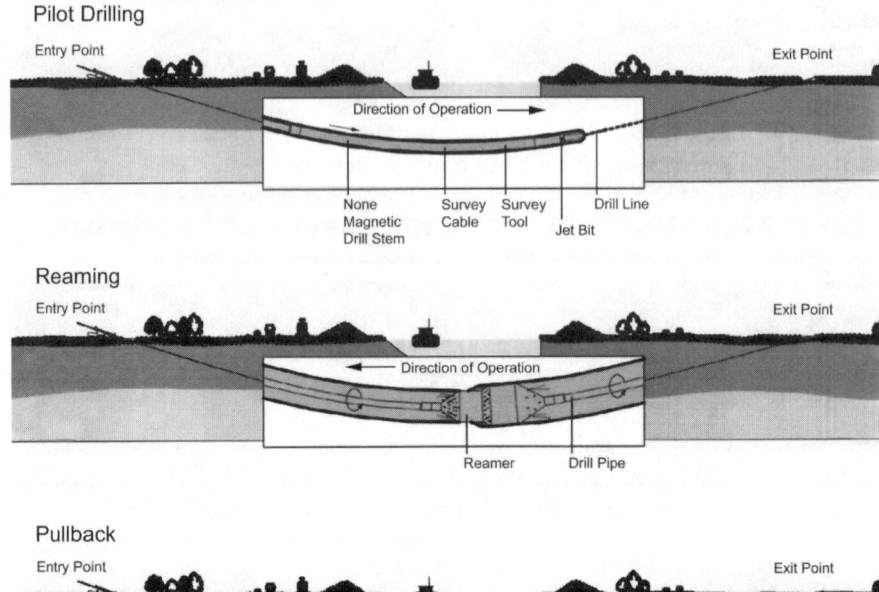

Fig. 4.48. Horizontal directional drilling with reaming[43]

The only difference between pipejacking and shield heading is the position of the jacks: for pipejacking they are situated in the start shaft and/or at intermediate jack stations, whereas for shield heading they are placed directly behind the shield.

The excavated diameter is slightly larger than the pipe diameter, so that the resulting gap can be grouted with bentonite to reduce friction.

4.4.3 Microtunnels

Microtunnels are used for the trenchless laying of cables and other lifelines. Since it is not allowed to work in spaces with ⌀ < 0.8 m, the heading is accomplished uncrewed by means of jacks and hammers that are installed at the excavation face. The soil is either displaced laterally, removed by screw conveyors or flushed. An important issue is the control of heading direction. A possible method is to use an asymmetric hammer which is permanently

---

[43]Herrenknecht HDD Rigs, February 2003

rotated to drive straight on, whereas the rotation is stopped whenever a curve is to be traced.

4.4.4 Speed of advance

The tunnel heading is composed of various steps, some of which are consecutive. Therefore it is important to adjust the individual steps relatively to each other.

Let the rate of excavation be v_a and let t_a be the daily operation time of a TBM. The rate of support is v_s and the time (per day) needed for support is t_s. Let the rates v_a and v_s be prescribed by the available machines and the ground. The question is now, how to determine t_a und t_s in such a way that a maximum advance rate is achieved? This means that the daily advance length z or the advance rate $V = z/24\,\mathrm{h}$ has to be maximized. We first determine z in dependence of t_s and take into account that $t_a + t_s = 24\,\mathrm{h} - t_w$. Herein, t_w is the daily downtime for maintenance. Thus we have $t_a = 24\,\mathrm{h} - t_w - t_s$, and the daily advance length is $z = \mathrm{Min}\{v_a t_a; v_s t_s\}$ or

$$z = \mathrm{Min}\{v_a(24\,\mathrm{h} - t_w - t_s); v_s t_s\} \qquad (4.1)$$

The relation 4.1 is shown in Fig. 4.49 (left). Obviously, for

$$t_s = t_{s0} := (24\,\mathrm{h} - t_w)\frac{v_a}{v_s + v_a}$$

the daily advance length z becomes a maximum $z_{\max}$:

$$z_{\max} = \frac{v_a v_s}{v_a + v_s}(24\,\mathrm{h} - t_w) \quad . \qquad (4.2)$$

Thus, the maximum advance rate $V_{\max}$ results to:

$$V_{\max} = \frac{z_{\max}}{24\,\mathrm{h}} = \frac{v_a v_s}{v_a + v_s} \cdot \frac{24\,\mathrm{h} - t_w}{24\,\mathrm{h}} \quad . \qquad (4.3)$$

From the plot of equation 4.3 in Fig. 4.49 it can be seen that the increase rate of $V_{\max}$ decreases with v_a.

In reality, the advance rate depends on many factors and cannot be given 'across the board'. Some reference points are:

top heading	5 m per day
heading with side galleries	3 m per day
TBM in rock[44]	8 m per day
shield heading in soil	20 m per day

---

[44] A record advance rate of 48 m/day has been achieved with a TBM at the Lötschberg tunnel construction site.

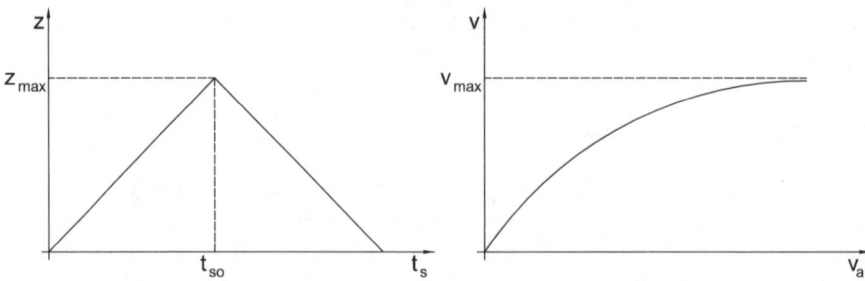

Fig. 4.49. Daily advance length z vs. the support time t_s; speed of advance v vs. excavation rate v_a

4.4.5 Drive-in and drive-out operations

Often, the shield moves between two shafts, the start (or launch) and the target shaft (Fig. 4.51). The back wall of the start shaft serves as abutment for the jacks. A lining length of at least 30 to 40 m is required to take on the jack thrust and transmit it to the surrounding ground by means of wall friction.

To drive the shield into the ground, starting from a shaft, a retaining wall has to be penetrated first.[45] This can be difficult if the retained soil has a very low strength or if the tunnel is driven below the groundwater level. Usually the wall is removed in strips of 0.5 to 1 m width and the face is temporarily supported with a soil heap. As a support measure, the adjacent soil can be grouted or frozen. If the heading has to proceed below groundwater level, the launch shaft can be put under compressed air. Alternatively, a membrane (Fig. 4.50) creates an intermediate partition between the ground and the shaft (caisson). The shield is then lowered in front of the membrane wall. The pressure difference is successively adjusted until the shield can be driven into the soil.

A drive-out operation is illustrated in Fig. 4.51.

4.4.6 Problems with shield heading

The fact that a shield protects against cave-ins of the tunnel wall does not mean that shield heading is free of difficulties. It should be noted that the costs of a shield or a TBM are low compared with the heading costs. Nevertheless, a jam can delay the heading progress for a long time and can, thus, cause immense consequential costs.

One should also consider the following items:

---

[45] The corresponding opening at the wall is called 'eye'.
[46] Herrenknecht Mixshields September 2002
[47] Tunnelling Switzerland, Swiss Tunnelling Society, Bertelsmann 2001

Fig. 4.50. Membrane between soil and caisson wall[46]

Fig. 4.51. TBM breakthrough[47]

Drillability: The advance speed in rock is limited by the fact that the rotational speed of the cutterhead cannot be increased beyond some limit. The bearings and the seals of the disc cutters allow a maximum velocity of 150 m/min. This imposes a limit to the rotational velocity of the wheel. An increasing number of disc cutters on the wheel increases also the downtime for maintenance. Thus, an efficient translation velocity is $\geq$ 3 to 4 mm/rotation. If a disc cutter is loaded more than allowed it vibrates and can be damaged. Its repair or replacement is difficult.

Stability of the face: Problems can appear when it encounters weak zones filled with soft soil. The cutterhead can jam due to accumulated or blocky muck. If the shield retracts to remove the problems at the face, then more soil or blocks can collapse into the resulting cavity. Advance grouting is tedious and can immobilize the cutterhead. Often, conventional heading is used until more favourable rock is encountered.

Penetrating squeezing rock: It is recommended that in squeezing rock (Sect. 14.10) the overbreak is increased from the usual 6-8 cm to 14-20 cm by means of a sufficiently large cutterhead. In addition, more powerfull jacks should be used that are capable of pushing the shield even if it is squeezed with a pressure of 2-5 MPa. Of course, the lining segments must be appropriately designed, and any standstill should be avoided.

4.5 Comparison of TBM with conventional heading

The choice between TBM and conventional excavation is an often faced dilemma. All over the world, TBM's are increasingly applied whereas conventional excavation techniques are rather used in difficult or varying ground, in short tunnels and in tunnels of varying cross sections.

Recently, the range of TBM applicability has been enlarged. Improved drilling techniques can now penetrate hard rock (q_u >400 MPa). Also jointed and soft rock can be penetrated with retracted cutterhead, appropriate shields and NATM support.

By means of its construction, a shield guards against the cave-in of the tunnel roof and, possibly, also of the face, whereas drill & blast can be very problematic whenever unforeseen weak zones are encountered.

The Los Rosales Tunnel in Bogotà was, for example, excavated in a hard sandstone which, however, was locally so weak that it rushed into the tunnel through the grouting drillholes (in amounts up to $6\,\text{m}^3$). The tunnel could only be completed with a TBM.

Referring to work safety: The advantage of a better support by the TBM can be partially overridden by the crowded working space. The uniform work in TBM heading is easier to learn than in drill & blast heading. Another argument in favour of the TBM technique is the avoidance of overbreak, which, on average, amounts to 10% for drill & blast and increases to 25% of the tunnel cross section for jointed rock or improper blasting (see also Section 4.7).

Comparing costs one should take into account that TBM heading is connected with high installation costs, so that it is profitable only for considerable tunnel lengths (Fig. 4.52).

Advantages of conventional heading with shotcrete support

- Heading of varying cross section (not only circular ones)
- Equipment can also be used for other purposes and can be easily replaced
- Low installation costs
- Adaption to geologic conditions is easy.

Disadvantages of conventional heading with shotcrete support

- Personnel are relatively unsecured close to the excavation face

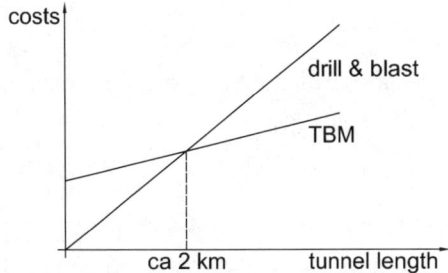

Fig. 4.52. Comparison of costs TBM – drill & blast

- Advance rate is limited to approx. 5-7 m/day
- Heading in difficult ground conditions (especially below groundwater level) is only possible if combined with expensive precautions
- In most cases an inner lining has to be added.

Advantages of shield heading with segmental lining

- Soft soil, also below groundwater level, can be excavated
- Sufficient safety, as the face is supported immediately
- Prescribed cross section is precisely excavated
- High advance rates, in particular after a learning phase of ca 1-2 months
- High quality lining due to pre-fabrication, a supplementary lining is not needed
- Low costs if the tunnel is sufficiently long.

Disadvantages of shield heading with segmental lining

- Limited to circular cross sections of constant diameter
- High installation costs
- Long learning phase of the crew
- Expensive drive-in operations
- Adjustment to varying ground conditions is difficult
- Machine damages cause total downtimes.

4.6 Rock excavation

The words 'drilling', 'boring' and 'cutting' are more or less synonymous in denoting rock excavation.

4.6.1 Drilling of boreholes

In tunneling, boreholes are drilled on the following purposes:

- exploration (site investigation)
- drill & blast
- grouting
- installation of bolts, spiles and other types of reinforcement.

Drilling comprises breakout, removal of the rock and cooling of the core bit.[48] Removal of the drill dust and chips as well as cooling is accomplished by flushing (Fig. 4.53). Exploration drillings are flushed with water, whereas percussion drillings are flushed with compressed air. Viscous fluids are used if the wall of the borehole has to be supported.

Drilling of blastholes requires high speed, low wear of the core bit and high precision (an accuracy of 0.1° is required for accurate blasts), whereas exploration drillings aim at a good core recovery and stable borehole walls. Drillings for oil and gas production aim at minimizing disturbance of the surrounding rock.

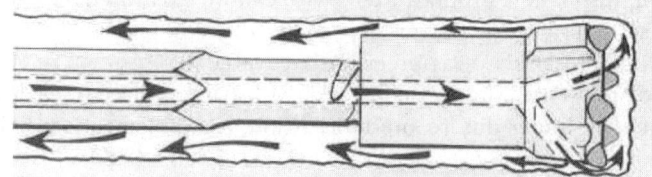

Fig. 4.53. Drilling with flushing[49]

Drilling is performed with percussive and rotary motion of the core bit. In the early days of mining, chisel and sledge hammer were used achieving rates of 1 m/h, whereas nowadays hydraulic percussive drilling with tungsten-carbid core bits can achieve rates up to 300 m/h (Fig. 4.54). Loose soils are drilled with augers; so-called continuous flight augers are increasingly used.

Percussive drilling is either driven pneumatically or hydraulically. Hydraulic drive is advantageous for several reasons: the boreholes are more precise, and they are driven twice as fast with only 1/3 of the pneumatically required power. Hydraulic compressors (electrically or diesel powered) are not as heavy and bulky as pneumatic ones. Moreover, the absence of compressed air discharges guarrantees visibility.

---

[48] See also: J.A. Franklin and M.B. Dusseault: Rock Engineering, McGraw Hill, 1989

[49] AtlasCopco Rock Tools

[50] AtlasCopco, Underground Rock Excavation

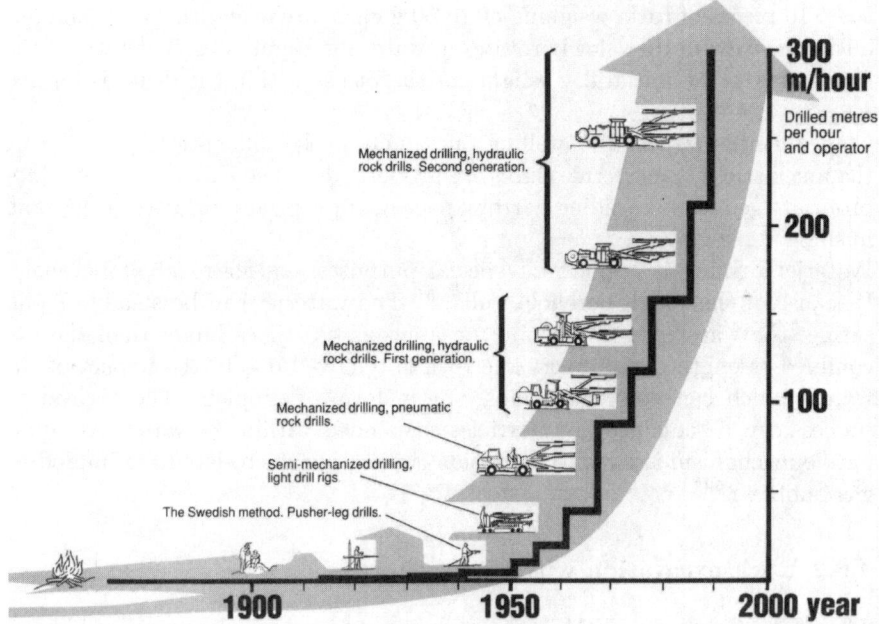

Fig. 4.54. Rock drilling technology[50]

The percussion is obtained by the impact of a piston. The impact momentum propagates along the drilling rod and eventually hits the rock. For long rods the energy is substantially reduced by wall friction. For this reason the hits have to be applied near the tip of the drill rod by means of a downhole hammer, if the drilling rod is longer than ca 120 m. For directional drilling downhole turbines are used.

To maximize the drilling speed, the following parameters have to be adjusted: thrust, rotation speed, rheological properties of the flushing fluid and the hydraulic drive. For soft rock a higher rotation speed (60-100 revolutions per minute) is advised than for hard rock (40-70 revolutions per minute). With increasing rotation speed the thrust should be reduced. For weak rock ($q_u <$ 200 MPa) the core bit should be equipped with teeth and for hard rock with buttons. Disc cutters cause chipping of the rock (see Section 4.6.2). Tungsten carbide and artificial diamonds increase the lifetime of drilling tools.

To choose the appropriate drilling tool, the following parameters have to be taken into account: rock properties, permeability, presence of dispersive or swelling clay minerals, rock temperature and stress state and rheological properties of the flushing fluid. The proper choice is more difficult for stratified rock with changing properties.

A difficulty arises if the excavated rock disintegrates to mud. Dispersivity (mainly due to clay minerals) means that water disintegrates the rock. The susceptibility of a rock to dispersivity is measured with the **slake durability**

test: 10 pieces of rock, weighing 40 to 60 g each, are oven-dried and then put into a sieve drum that slowly rotates in water for 10 minutes. The ratio of the remaining to the initial dry weights of the pieces within the drum is termed I_{d2}.

Clay minerals, especially swelling ones, may cause difficulties by rendering the muck sticky. Countermeasures are properly designed excavation tools (appropriate geometry avoiding narrow spaces), appropriate surface coatings and flushing.

Waterjet drilling is applied for special purposes, e.g. for rockbolt boreholes (because of the rough borehole walls).[51] The waterjet can be steady or pulsating, the water pressure should be approx. 100 times larger than the unconfined strength of the rock. The rock is removed due to the impact of the water, which can flow continuously or in form of droplets. The excavation process can be enhanced by particles suspended within the water. An alternative mechanism is cavitation erosion (impacting micro-jets from imploding gas bubbles).[52]

4.6.2 Rock excavation with disc cutters

The advance rate of a TBM is a very important item for planning and bidding tunnel projects. It should therefore be predicted as precisely as possible. So-called TBM performance prediction models have been developed, among them the ones of NTH (Norwegian Institute of Technology) and CSM (Colorado School of Mines).

The excavation of rock is a very complex process that depends on factors, which are hardly controllable. Thus, most of the TBM prediction models are based on empirical correlations of the several controlling parameters. The lack of rational analysis of the underlying processes limits their applicability to already existing methods and machines, i.e. extrapolations to new technical developments are questionable.

The statements in this section, being based on mechanical concepts rather than regression analysis of row data, are intended to give a rough orientation and are not meant as a tool for precise predictions.

Disc cutters (also called 'discs', Fig. 4.55) exert a high pressure and thus fragment the rock, see Fig. 4.56. The discs exert forces up to $F = 250$ kN that fragment the rock into flat chips (fragments). Their bearings (Fig. 4.55) are designed in such a way that friction is reduced when high thrust is exerted against the rock. Cutters with V-shaped cross sections are no more used, because wear progressively changes the contact area with the rock. They have been replaced with 'constant cross section' discs (Fig. 4.56).

---

[51] Waterjets are also applied in coal mining to avoid sparks.

[52] A.W. Momber, Wear of rocks by water flow, *Int. J. of Rock Mechanics and Mining Sciences*, **41** (2004), 51-68

[53] AtlasCopco, Underground Rock Excavation

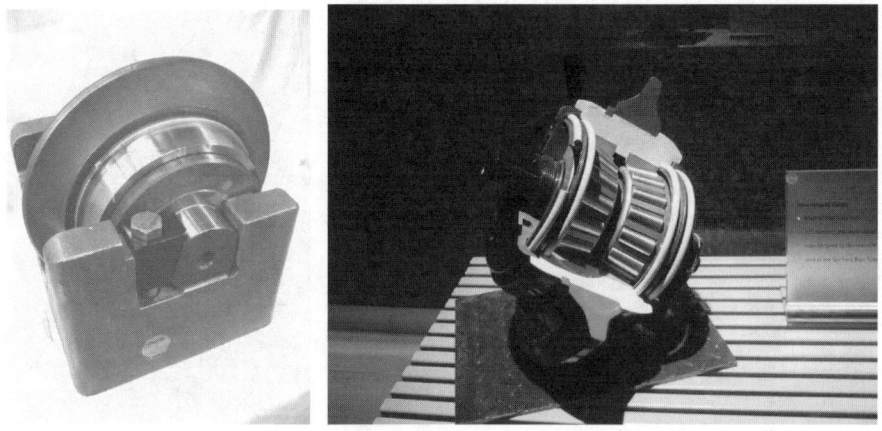

Fig. 4.55. Left: cutter, right: cutter bearing (Herrenknecht)

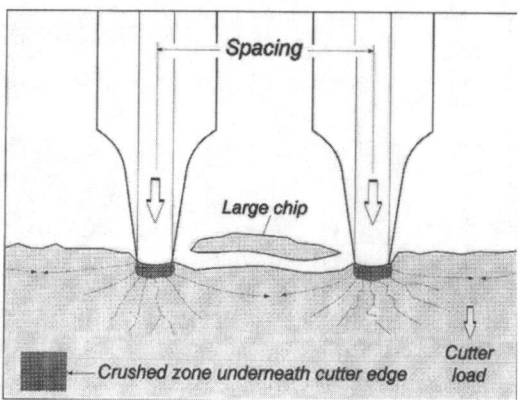

Fig. 4.56. Working principle of cutters[53]

As shown in Fig. 4.57, a cutter has a distance (radius) R from the TBM-axis. The TBM cutterhead rotates with N revolutions per time unit. The cutter has the radius r (typical cutter diameters vary between 38 and 48 cm) and rotates with n revolutions per time unit. Neglecting slip we have

$$n = \frac{2\pi R}{2\pi r} N = \frac{R}{r} N \quad .$$

Thus, the linear velocity of the cutter amounts to $v = 2\pi NR$. To control vibrations, v is limited to approx. 150 m/min. Another reason to limit the rotation speed (i.e. N) is to avoid overly large centrifugal forces of the rock chips. Therefore, for given R_{max} values, N is also limited. The forces acting upon a cutter and the instantaneous velocity distribution are shown in Fig. 4.58.

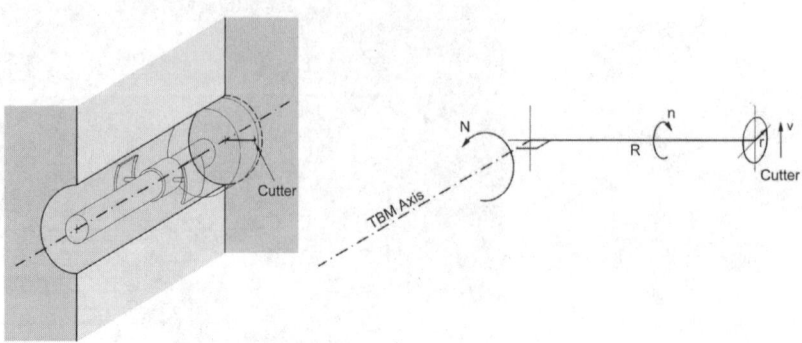

Fig. 4.57. Position of a cutter in a TBM

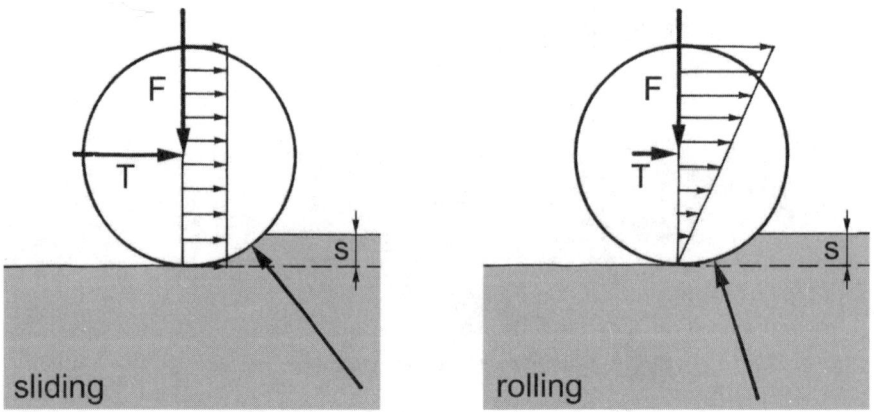

Fig. 4.58. Forces (T and F) and velocity distributions in the cutter. Left: sliding, right: rolling.

Clearly, the total torque of the cutterhead amounts to

$$M_t = \Sigma_i(T_i R_i) \tag{4.4}$$

where summation is run over all cutters. Note that also in case of rolling (Fig. 4.58, right) there is an abrasion of the wheel due to the relative slip between cutter and rock (Fig. 4.59).

---

[54] J.B. Cheatham,(1958). An analytical study of rock penetration by a single bit tooth. Proc. 8th Annual Drilling and Blasting Symp., Univ. of Minnesota, Minneapolis; Pariseau and Fairhurst (1967). The force penetration characteristic for wedge penetration into rock. *Int. J. Rock Mech. Min. Sci. & Geomech. Abstr.* 4: 165-180.

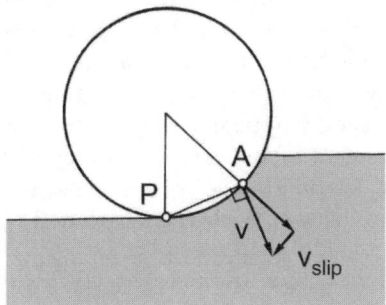

Fig. 4.59. P is the instantaneous pole of rotation. The instantaneous velocity v is normal to AP and has, therefore, a slip component v_{slip} that causes abrasion.

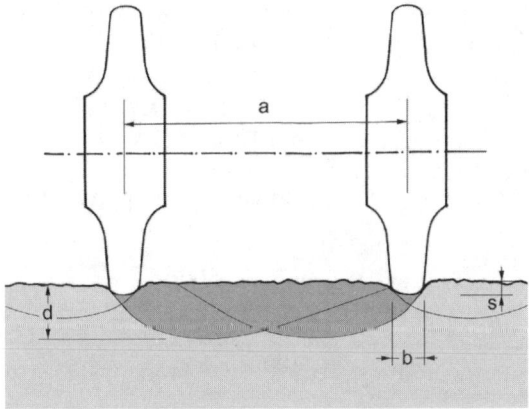

Fig. 4.60. Chipping of rock viewed as punching problem of plasticity[54]

The rock is chipped between two adjacent cutters, as shown in Fig. 4.60. A small part or rock directly adjacent to the cutter is not chipped but crushed into powder. In the elastic regime the relation between punching force F and indentation s is linear according to the equation of HERTZ:

$$s = \frac{F}{bE} \kappa$$

with

$$\kappa = \frac{1}{\pi}\left[\lambda(1 - \mu_{steel}^2) + 1 - \mu_{rock}^2\right] ,$$
$$\lambda = E_{rock}/E_{steel} ,$$
$$E = E_{rock} .$$

We assume that this relation holds for $F < F_l$. At the limit force F_l, plastic punching according to PRANDTL's theory is assumed to set on. Clearly, the application of this theory is limited to isotropic non-brittle materials. It is, however, common practice in rock mechanics to use plasticity theory and characterize the strength of rock by φ and c values. To this extent, the application of plasticity theory appears to be reasonable, at least for a rough, though rational, assessment of the considered fragmentation process.

For vanishing internal friction of the rock, i.e. $\varphi_{rock} = 0$, the punching load F_l can be estimated as follows. With the length a (Fig. 4.61) estimated as

$$a = 2\sqrt{r^2 - (r-s)^2} \approx 2\sqrt{2rs} \tag{4.5}$$

we obtain from PRANDTL's solution:

$$\frac{F}{ab} \approx 5c,$$

where c is the cohesion of the rock (for $\varphi = 0$ we have $c = q_u/2$, where q_u is the unconfined (uniaxial) compression strength).

$$\leadsto F \approx 5abc$$
$$\approx 10\sqrt{2rs}\, bc$$

With (4.5) we obtain the punching force, i.e. the required thrust per cutter, as

$$F \approx 200\kappa r b c^2/E$$
$$= 50\kappa r b q_u^2/E$$
$$= 100\, \kappa\, r\, b\, \epsilon$$

with ϵ being the fragmentation work per unit volume (Fig. 4.62): $\epsilon = \frac{1}{2}q_u \varepsilon_l$. ϵ can be obtained from a uniaxial compression test. Regarding the specific

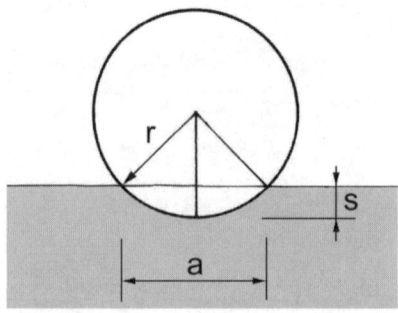

Fig. 4.61. To the estimation of the punching length a

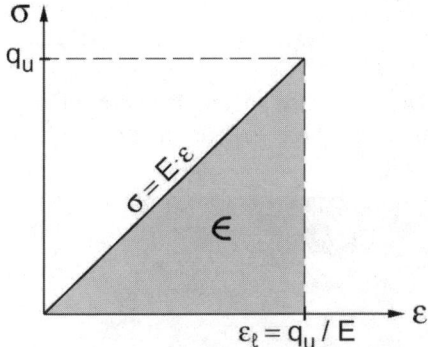

Fig. 4.62. Fragmentation work per unit volume $\epsilon = \int \sigma \, d\varepsilon = \frac{1}{2} q_u \, \varepsilon_l$

fragmentation work, i.e. the fragmentation work per unit volume, it is often argued that it must increase with increasing surface of debris. This is a reasonable but yet unproven assumption. Assuming the specific fragmentation work ϵ as a given quantity and neglecting the work for linear penetration in direction of the tunnel axis, we obtain the penetration work per revolution as

$$W = 2\pi M_t/(Ad)$$

with the applied torque M_t, the tunnel cross section A and the penetration depth d according to Fig. 4.60. With $A = \pi D^2/4$ and M_t according to Equ. 4.4 we obtain $M_t \propto D^3$. The net penetration rate is $N \cdot d$.

It should be taken into account that the position of a cutter is fixed on the cutterhead. Thus, s cannot be always adjusted to the limit punching force. Whenever the cutter touches muck (from previous punching), which has lower E and c values, the prevailing force F drops considerably. The bearing is designed in such a way that the rolling friction increases with decreasing thrust. As a result, abrasion dramatically increases. This leads to a reduction of indentation (since the position of the bearing is fixed) and, thus, to a further reduction of force. The cutter keeps sliding and abrasion proceeds very fast. It has been observed that a good muck removal reduces abrasion.

It should be added that hard rock boring is not yet completely understood. The equations derived above are merely an attempt to analyse the underlying processes and to give some qualitative insight into the interrelations between the involved quantities. A quantitative check against experiments is questionable.[55] Regarding e.g. the punching force as derived from the theory of plasticity, one should pay attention to the fact that the friction angle φ may sensitively affect the results. However, the experimental determination of φ

---

[55] R. E. Gertsch, Rock Toughness and Disc Cutting. PhD thesis, University of Missouri-Rolla, 2000.

124   4 Heading

is difficult. Moreover, φ is stress-dependent and thus not a material property. The prediction of TBM performance (i.e. penetration rate and abrasion depending on thrust, torque, rotation speed and TBM design) is therefore still empirical.[56]

An alternative application of disc cutters is undercutting (Fig. 4.63 and 4.64), where the disc cutter pares the rock from a pre-bored space.

Fig. 4.63. Undercutting at the Uetliberg tunnel[57]

4.6.3 Abrasion

Abrasion of the excavation tools is an important issue. Several methods have been proposed to rate the abrasivity of rock. Among them are:

CERCHAR (Laboratoire du Centre d' Etudes et Recherches des Charbonnages): The tip of a steel conus is loaded with 7 kg and scratched 6 times over a length of 1 cm along a fresh fracture surface of the investigated rock. The flattening of the tip due to the abrasion is then measured. CERCHAR indices, varying from 1 to 6, are then assigned to diameters of the truncated tip (varying from 0.1 to 0.6 mm).

---

[56]see e.g. University of Trondheim (NTH), 1994, "Hard Rock Tunnel Boring", Project report 1 - 94.

[57]S. Mauerhofer, M. Glättli, J. Bolliger, O. Schnelli: Uetliberg Tunnel: Stage reached by Work and Findings with the Enlargement Tunnel Boring Machine TBE, *Tunnel* 4/2004.

[58]L. Baumann, U. Zischinsky, Neue Löse- und Ausbautechniken zur maschinellen "Fertigung" von Tunneln in druckhaftem Fels. In: Innovationen im unterirdischen Bauen, STUVA Tagung 1993 (ISBN 3-87094-634-2), 64-69.

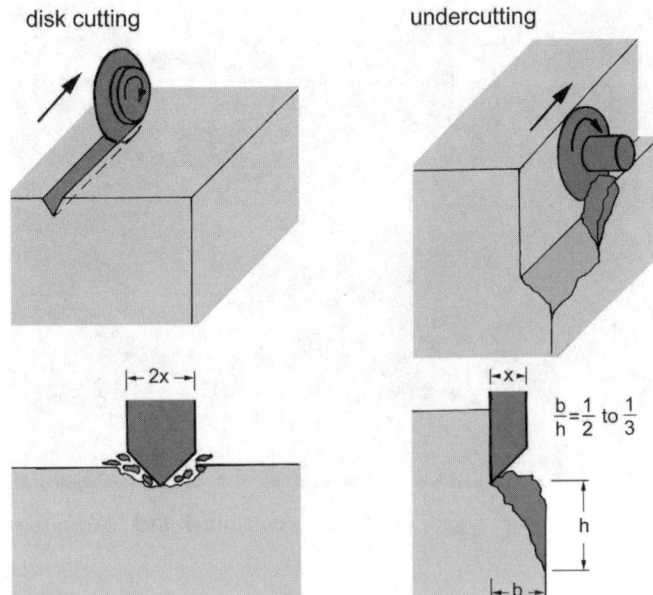

Fig. 4.64. Principles of rock cutting. left: usual technique, right: undercutting[58]

LCPC (Laboratoire Central des Ponts et Chaussées): First the rock is broken down (a sample of 500 g with grains sizing from 4 to 6.3 mm is whirled with a standardised steel propeller at 4,500 revolutions per minute). The index ABR is defined as the weight loss of the propeller (due to abrasion) per 1 t of rock.

Schimazek's coefficient of abrasivity $F = V d \sigma_z / 100$ with
  V: volumetric percentage of quartz
  d: mean size (in mm) of quartz grains
  σ_z: tensile strength (MN/m^2) of the rock.
  Examples of rating:
  $F < 0.05$: not very abrasive
  $F > 2$: extremely abrasive

4.6.4 Drilling: history review

In ancient times rock was bored with hammer and chisel. From the 16th to the 19th century 'stoking' was applied: The tunnel face was heated by fire and subsequently cooled with water. As a consequence, cracks appeared that made the rock easier to excavate. Rock blocks were removed by splitting: Wooden spikes were placed into drilled holes, added water caused swelling of the spikes and split the rock. The drill & split method is also applied nowadays, where the rock is split with the aid of steel splines. This method is applied where

Fig. 4.65. Disc cutters from Lötschberg base tunnel. Left: Worn disc cutters

vibrations due to drill & blast have to be avoided. The maximum excavation rate with this technique is 1 m per 24 h.

4.7 Profiling

For practical reasons the excavated profile does not coincide with the intended one. Underprofile (i.e. defficient excavation) can be detected by means of templates or geodetic devices. Subsequently the remaining rock has to be removed ('scaling'), e.g. with excavators or cautious blasting.
Overprofile (also called overbreak), i.e. surplus excavation, can be due to bad geological conditions, and is thus inevitable, and/or due to improper excavation. It causes additional costs for the contractor, as it has to be filled with shotcrete. Thus, the distinction between both types of overbreak is of economic importance. A certain amount of overbreak, represented by a strip of the width d (Fig. 4.66), should be allowed for, as required by the technical equipment. d should be specified by the contractor. A strip of the width D (Table 4.2) is to specify the 'geological' overbreak, i.e. the overbreak imposed by the geological conditions and is to be payed to the contractor.[59]
If not otherwise specified, SIA 198 recommends the following expressions for D, with A being the theoretical tunnel cross section:

---

[59] Swiss Code SIA 198

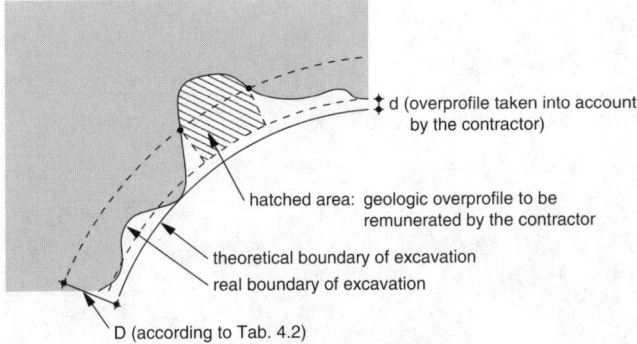

Fig. 4.66. Refundable overbreak (according to SIA 198) for tunnels excavated conventionally or with roadheader

excavation method	D
drill & blast	Max{$0.07\sqrt{A}$, 0.40 m}
roadheader	Max{$0.05\sqrt{A}$, 0.40 m}
soil, without shield	Max{$0.05\sqrt{A}$, 0.40 m}
shield	Max{$0.03\sqrt{A}$, 0.25 m}
TBM	Max{$0.03\sqrt{A}$, 0.20 m}

Table 4.2. Strip width D referring to Fig. 4.66 (SIA 198)

4.8 Mucking

The removal of the excavated rock/soil is known as mucking and consists of loading up, transport and unloading of the muck (also called 'spoil'). For transport (haulage) the following variants are available:

Trackless transport: Usual earthwork trucks are used. Of course, provision should be taken for the reduced production of harmful combustion products (cf. Section 2.3.1). A good carriageway is important to enable a speed of ca 50 km/h. The inclination is limited to approx. 7%, the necessary width of the tunnel is 7 to 8 m. The requirements for fresh air supply are very high: A dumper of 300 PS (220 kW) needs 2,000 m^3 fresh air per minute.

Railbound mucking (track transport) is also applicable in crowded spaces, i.e. for spans <6 m. Usual track gauges are between 600 and 1435 mm, the maximum inclination is 3 %, the minimum curvature radius is ca 7 m, the velocity is limited to ca 20 km/h. The locomotive is powered by diesel

---
[60] http://www.schoema-locos.de

Fig. 4.67. Railbound mucking[60]

or electric motors. In the latter case, the power is delivered by accumulators. The power consumption is lower than with trucks.

Continuous conveyors: Conveyors have a very large transport capacity (up to 200 t/h).[61] A separate transport for personnel is needed as well as a stone crusher and a dosing device. Apart from the high transport capacity, their main advantage is safe, clean and silent transport. But intensive maintenance is needed, because their outfall implies downtime of all other works. A minimum curvature radius of 500 m is required. Conveyors can be mounted on rail waggons.

Fig. 4.68. Mucking with continuous conveyor[62]

---

[61] S. Wallis: Continuous conveyors optimise TBM excavation in the Blue Mountains and Midmar, *Tunnel* 3/95, 10-20

[62] Rock-Machines Sweden AB

Tunnel muck is utilised, if possible, as aggregates for shotcrete and cast concrete and for fills. The utilisation as aggregates requires an appropriate mineralogical composition and, possibly, crushing and/or washing to remove the fines. In general, muck from drill & blast in igneous rock is more appropriate as aggregates than muck from TBM's.

If muck cannot be integrated immediately after heading, it has to be stored in a disposal dump. While the unloading of trucks is straightforward, rail tracks need special unloading facilities (an example is given in Fig. 4.69).

Fig. 4.69. Unloading from bottom outlets

5
Support

5.1 Basic idea of support

The lining of a tunnel is never loaded by the stress which initially prevailed in the ground. Luckily, the initial (or primary) stress is reduced by deformation of the ground that occurs during excavation but also after installation of the lining (here 'lining' is understood as the shell of shotcrete, which is placed as soon as possible after excavation). Here we shall consider the important phenomenon that deformation of the ground (soil or rock) implies a reduction of the primary stress. This is a manifestation of *arching*. Since the deformation of the ground is connected with the deformation of the lining, it follows that the load acting upon the lining depends on its own deformation. This is always the case with soil-structure interaction and constitutes an inherent difficulty for design as the load is not an independent variable. Thus, the question is not 'which is the pressure acting upon the lining', but rather 'which is the relation between pressure and deformation'.

The consideration of deformation in tunnelling is a merit of NATM[1] and is schematically shown in Fig. 5.1. The rock is symbolically represented by a beam. Excavation and installation of the lining is here represented by removal of the central column which is replaced by a lower column. In other words, the central column (which symbolises the lining) is displaced downwards and, therefore, receives a reduced load. Of course, the principle 'pressure is reduced by deformation' is to be applied cautiously. Exaggerated deformation can become counterproductive (Fig. 5.1 c) leading to a strong increase of pressure upon the bearing construction. To point this out was another merit of NATM: softening (and the related loosening) of geomaterials is an important issue. It should be emphasised, however, that this softening does not refer to the gentle stress reduction subsequent to the peak, as it is obtained in laboratory tests on dense soil samples. In contrast, the drastic strength reduction observed in poor rock due to loss of structural cohesion is meant.

---

[1] see Chapter 7

Civil engineers, by tradition, distinguish between deformation and failure (collapse) of a structure. It is, however, impossible to find a genuine difference between these two notions. Virtually, failure is nothing but an overly large deformation. At any rate, large deformations have to be avoided. How can this be achieved in underpinning/tunnelling? There are two ways: Either early and rigid support (which is not economic) or by keeping the size of the excavated cavities small. The latter option is pursued in tunnelling. There are two ways to do this:

- partial excavation instead of full face excavation
- small advance steps.

Of course, too small excavation steps would not be economic. So, the art of tunnelling consists in keeping the excavation steps as large as possible and exploiting the strength of the ground.

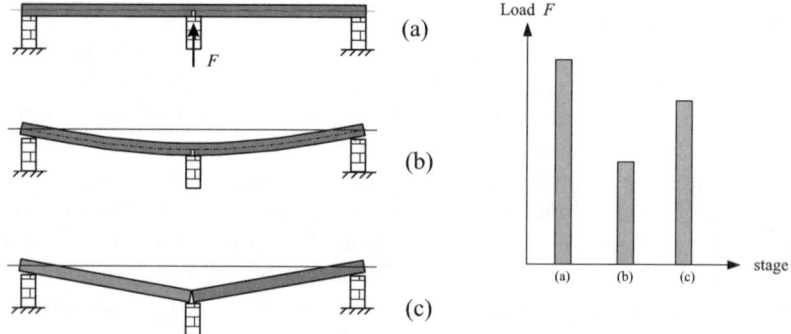

Fig. 5.1. Explanation of support principles. The increase of load at the transition b→c is due to the loss of strength (rupture) of the beam.

5.2 Shotcrete

What distinguishes shotcrete (or 'sprayed concrete') from cast concrete is not the strength of the final product but the process of its placement. In tunnelling, shotcrete is applied (i) to seal freshly uncovered surfaces (in thicknesses of 3 to 5 cm) and (ii) for the support of cavities.

The characteristics of sprayed concrete (shotcrete) are almost the same as those of usual concrete.[2] However, YOUNG's modulus is somewhat lower than with conventional concrete.

Up to the age of 28 days the stiffness and the strength of shotcrete develop approximately as they do with cast concrete. Afterwards, with sufficient humidity, the strength increases considerably due to post-hydration. Up to the

---

[2] P. Teichert: Sprayed concrete. Published 1991 by E. Laich SA, CH - 6670 Avegno

age of two years it increases by ca 50%. The strength of fast-setting sprayed concrete increases with time as follows:

Age	Strength (N/mm^2)
6 min	0.2 - 0.5
1 hour	0.5 - 1.0
24 hours	8 - 20
7 days	30 - 35

There are two methods to spray shotcrete:

Dry mix: dry cement and aggregates are pneumatically conveyed, water is added at the nozzle (Fig. 5.2).

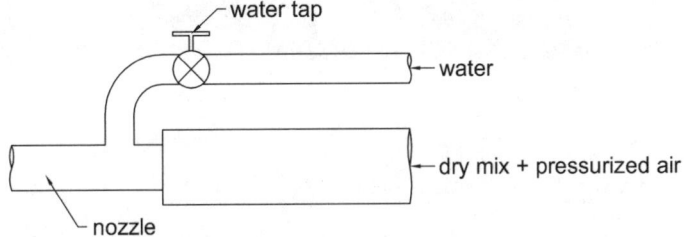

Fig. 5.2. Dry mix nozzle

Wet mix: ready mixed concrete is pumped to the nozzle, from where it is driven by compressed air. Due to the increased weight of the nozzle, a wet mix is better sprayed with robots (Fig. 5.3).

Advantages of dry mix:

- machines are smaller and cheaper
- lower costs for cleaning and maintenance
- stop and re-start of shotcreting is simpler
- longer conveying distances (up to 150 m)
- more precise dosage of additives
- better concrete (pumping of wet mix requires a higher water content)
- water content can be manually reduced, e.g. when spraying against a wet background.

Advantages of wet mix:

- reduced dust production
- reduced rebound
- reduced scatter of concrete properties
- higher capacity.

Fig. 5.3. Spraying concrete – wet mix mode

Excavated cross sections up to 50 m^2	$\geq$ 20 cm
Excavated cross sections between 50 and 100 m^2	$\geq$ 25 cm
Cross-overs, shafts	15 cm

Table 5.1. Usual thickness of shotcrete lining

Depending on the discharge, shotcrete is sprayed from a distance of 0.5 to 2 m, as perpendicular to the wall as possible, in layers of up to 4 cm (on vertical walls) and 2 to 3 cm (on the roof) thickness. Starting from lower parts, shotcreting moves to the roof. Care should be taken to shotcrete beyond reinforcement, i.e. to avoid 'shadows'. As the impact velocity is high (20 - 30 m/s), the rebound usually amounts to 15 - 30% for vertical walls and 25 - 40% for the roof and consists mainly of coarse grains. The aggregates of shotcrete are $\varnothing \leq 16$ mm, and their diameter should not exceed 1/3 of the layer thickness. The rebound can be reduced by increasing the proportion of fine grains, e.g. by adding cement or silica fume (i.e. SiO_2 powder). The high specific surface of the latter attracts water and thus reduces the consistency of shotcrete. Setting accelerators (such as sodium silicate) may also help, but the resulting concrete has a lower strength. Therefore the chosen accelerator should be tuned with respect to the cement. To achieve sufficient bonding, the target surface (rock or previous shotcrete layer) must be appropriately cleaned and moistened. This is achieved by spraying air and/or water.

Usual additives, aiming to reduce rebound and customise the setting, are alkaline and are therefore hazardous. Recently, non-alkaline additives have been developed. Dust production, rebound and presence of etching materials renders shotcreting an arduous job, which can be mechanised by the use of robots and remote controlled spaying arms. This can considerably speed up the heading. A promising idea to reduce rebound is the rollover shutterbelt shown in Fig. 5.4. Another idea is to replace pressurized air with centrifugal skidding in shotcreting.

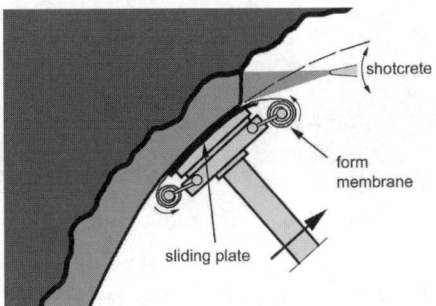

Fig. 5.4. Rollover shutterbelt

The surplus consumption of shotcrete is due to rebound (rebound material should not be re-used), overprofile and cleaning of devices. It amounts up to 200%.

Shotcrete sealing of the freshly excavated rock surface is also applied with TBM heading. In this case shotcrete is sprayed within a hood (Fig. 5.5)

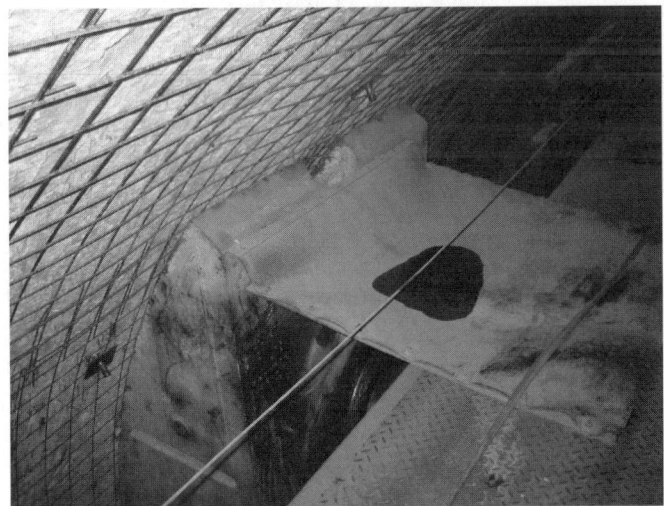

Fig. 5.5. Shotcrete hood in TBM heading of Loetschberg base tunnel

---

[3]Herrenknecht

Fig. 5.6. Shotcrete robots[3]

5.2.1 Steel fibre reinforced shotcrete (SFRS)

Adding steel (and, recently, also synthetic) fibres increases the tensile strength and ductility of the shotcrete.[4] Thus, traditional steel reinforcements become dispensable and spraying shadows are avoided. The length of the fibres should not exceed 2/3 of the minimum hose diameter. Usual sizes are 45-50 mm length, 0.8-1.0 mm diameter. The steel fibre content should be ≥ 30 kg/m^3, while the aggregates should not be coarser than 8 mm. The water/cement ratio should not exceed 0.5. If the shotcrete surface is to be covered with an impermeable membrane, then a layer of fibre-free shotcrete should be applied first, otherwise the membrane could be damaged.

Fig. 5.7. Steel fibres and steel fibre reinforced shotcrete[5]

---

[4]The creep behaviour of synthetic fibres still has to be checked.
[5]La Matassina

5.2.2 Quality assessment of shotcrete

The following controls help to assure a sufficient shotcrete quality:[6]

- Control of appropriate composition, packing, designation and storage of the ingredients.
- Control of the strength of fresh shotcrete. The extraction of core samples is not possible for compressive strengths < 10 N/mm^2 and shotcreting into moulds does not yield representative samples. Therefore, several indirect methods have been proposed. They are based on penetration of pins or on the pull-out of bolts or plates. Ultrasonic and hammer blow tests are inappropriate due to the rough surface of shotcrete.
- Stiffness, strength and permeability of hard shotcrete is tested on extracted cores of 10 cm diameter. For tunnelling, a shotcrete with an age of 28 days is usually required to have a compressive strength of 23 N/mm^2.
- The mechanical performance of steel fibre reinforced shotcrete is tested with bending of beams. The content of steel fibres is usually reduced as compared with the initial mixture. It can be measured by smashing a shotcrete sample and extracting the steel fibres with a magnet.
- The thickness of shotcrete is measured at random points (roughly one measurement every 100 m^2) either with stencils or via coring.

5.3 Steel meshes

Steel meshes (mesh size $\geq$ 100 mm, $\varnothing < 10$ mm, concrete cover ≥ 2 cm) are manually mounted and should, therefore, be not too heavy. A usual weight is 5 kg/m^2. Mesh installation is labour intensive and relatively hazardous, as the personnel are exposed to small rock falls. For drill & blast heading the mesh adjacent to the face (proximity < 1 m) can be damaged by the subsequent blast.

5.4 Rock reinforcement

The mechanical properties of rock (be it hard rock or soft rock and soil) in terms of stiffness and strength can be improved by the installation of various types of reinforcement.[7] Steel bars[8] can be fixed at their ends and pretensioned against the rock. In this way, the surrounding rock is compressed

---

[6] Österreichischer Betonverein, Richtlinie Spritzbeton 1998, http://homepages.netway.at/beton

[7] A good introduction to rock reinforcement is given in the book 'Rock Engineering' by J. A. Franklin and M. B. Dusseault, McGraw-Hill, 1989, which is, however, out of print.

[8] Anchors from synthetic material (e.g. fiberglass) or wood are used as temporary reinforcement of the face. They can be easily demolished during the subsequent excavation.

and, as a consequence, its stiffness and its strength increase. Such reinforcing bars are called anchors or bolts[9]. An alternative type of reinforcement consists of bars that are connected with the surrounding rock over their entire length, e.g. by grout. Such bars are not pre-tensioned and are called nails.[10] Rock with nails is a composite material, whose stiffness is increased as compared to the original rock. A third action of reinforcement is given when a steel bar (dowel) inhibits the relative slip of two adjacent rock blocks. In this case the bar is loaded by transverse forces and acts as a plug. The usage of names, stated here (anchor, bolt, dowel), is, however, not unique and they are often interchanged. The reinforcing actions of plugging (dowelling), pre-tensioning and nailing are illustrated in Fig. 5.8.

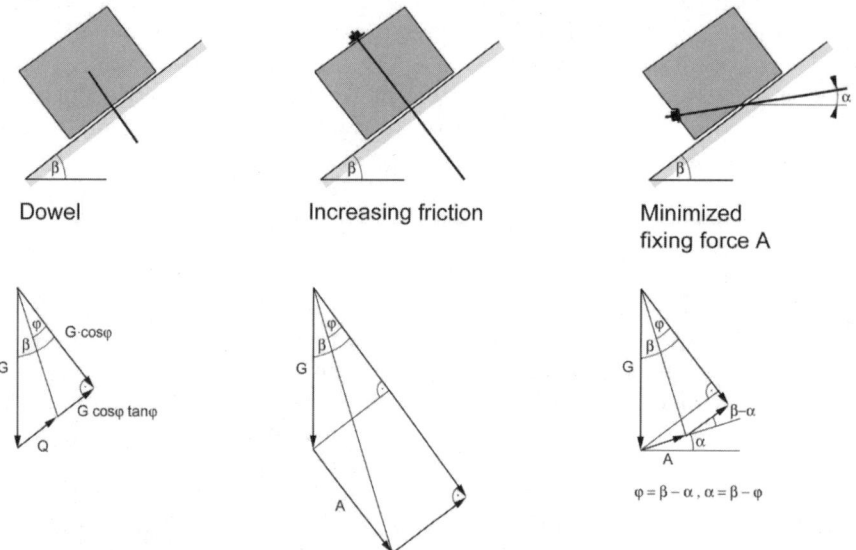

Fig. 5.8. Reinforcing actions of inlets.

Note that plugging and nailing occur simultaneously (Fig. 5.9). As a result, the reinforcing force A has not necessarily the direction of the reinforcement bar.

5.4.1 Connection with the adjacent rock

The connection, i.e the force transfer between reinforcement and surrounding rock, is either achieved mechanically or by means of grout (cement mortar

---

[9]The two words are synonymous. Some authors, however, use 'bolt' for forces $<$ 200 kN, typically achieved with bars with diameter $\leq$ 25 mm and length less than 2 - 3 m.

[10]In ground engineering, this type of reinforcement is called nailing.

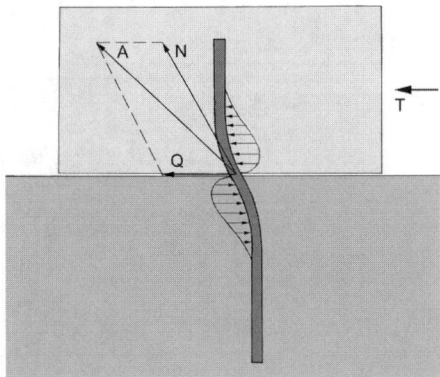

Fig. 5.9. Forces acting upon and within a dowel

or synthetic resin). Mechanical connections can be loaded immediately after installation. They comprise:

Wedges: Conical wedges are placed at the end of the borehole. They can be moved in longitudinal direction either by hammering ('slot and wedge anchors', Fig. 5.10) or by rotating a thread ('expanding shell anchors', Fig. 5.11) in such a way that they force their containment (which is either a slotted bar or a shell) to grip into the rock. The transmission of a concentrated force is only possible in sufficiently hard rock (compression strength > 100 MPa). Slot and wedge anchors can be loosened by shocks and vibrations (e.g. due to blasting).

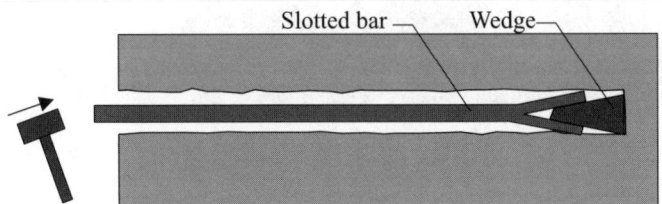

Fig. 5.10. Slot and wedge anchor

Tubular steel rockbolts: A contact over the entire bolt length is achieved by the expansion of a tubular steel (also called hollow anchor) against the borehole wall. The expansion is either elastic (Fig. 5.12) or is achieved by means of water pressure ('Swellex' rockbolt by Atlas Copco, Fig. 5.13). The shear force is transferred to the rock by friction.

Alternatively, the connection with the rock can be achieved by means of cement mortar or resin. The obtained support is not immediate, since setting and hardening needs some time.

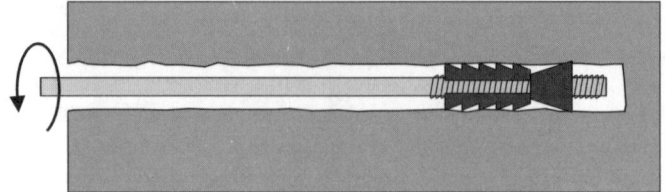

Fig. 5.11. Expanding shell anchor

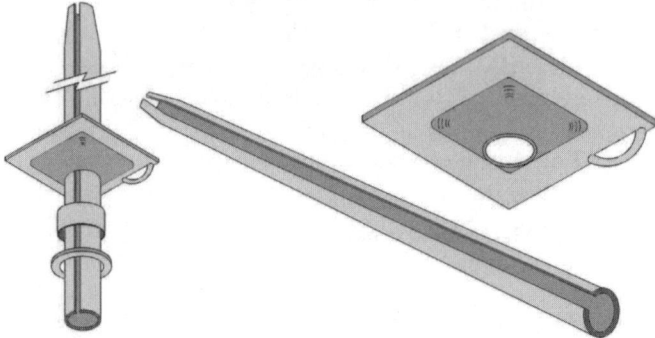

Fig. 5.12. Hollow Anchor Systems

Fig. 5.13. Swellex-Anchor

Grouted rockbolts[11]: The annular gap between rebar and drillhole wall is filled with cement or resin grout. Before grouting, the drillhole must be thoroughly flushed with water or air to ensure a clean rock surface. It should also be ensured that the rock does not contain wide open joints into which the grout may disappear. This can be avoided by using geotextile containments of the grout (Fig. 5.14). Cement grout (mortar) consists of well graded sand and cement in ratios between 50/50 and 60/40. To obtain a sufficient strength, the water-to-cement ratio should be $\leq 40\ \%$

---

[11] so-called SN-anchors. This name originates from their first application in Store Norfors in Sweden. An alternative etymology attributes SN to Soil & Nail.

5.4 Rock reinforcement

by weight. The mortar should set (i.e. obtain the required strength) within 6 hours. To increase plasticity, a bentonite fraction of up to 2 % of the cement weight can be added. Note that cement grout can be damaged by vibrations due to blasting. In some cases, grouting is not allowed until the heading has advanced by 40-50 m.

Synthetic resins harden very quickly (2-30 minutes) by polymerisation when mixed with a catalyst. The two components are either injected or introduced into the drillhole within cartridges which are subsequently burst by introducing the rebar.

Cement mortar can be introduced in several ways:

- Grouting into the annual gap between rebar and drillhole wall.
- Perforated tubes filled with mortar are placed into the borehole. The subsequent introduction of the anchor (by means of hammer blows) squeezes the mortar into the remaining free space ('perfobolts', Fig. 5.15). The perfobolts are now obsolete.
- 'Self-boring' or 'self-drilling' anchors (SDA)[12]: the rod is a steel tube of 42 - 130 mm diameter driven into the rock with rotary-percussion drilling equipment, flushing and a sacrificial drill bit. Standard delivery lengths vary between 1 and 6 m. Two rods can be connected with couplers. Grouting of mortar occurs through the tube with pressures up to 70 bar.

Fig. 5.14. Anchor rod within geotextile containment

At the head of the anchor the tendon (steel rod) is fixed against a bearing plate or faceplate in such a way that the anchor tension is converted into a compressive force at the rock face. Spherical washers enable to fix the tendon against the faceplate also in cases when the tendon is not perpendicular to the rock surface (Fig. 5.16). The faceplate also helps to fix the wire mesh (Fig. 5.17).

---

[12] so-called IBO-anchors, e.g. the brands TITAN by Ischebek and MAI by Atlas Copco

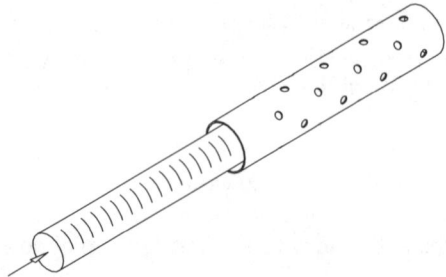

Fig. 5.15. Perfobolt

Fig. 5.16. Spherical washer[13]

Fig. 5.17. The faceplate also serves to fix the wire mesh.

5.4.2 Tensioning

Anchors can be tensioned using torque wrenches, air impact wrenches or hydraulic tensioners. The applied tensile force for short-term applications

---
[13]DYWIDAG

amounts to ca 70 % of the yield force. Alternatively, the tensile force is gradually applied by the expansion of the surrounding rock ('self-tensioning').

5.4.3 Testing

Quality control of anchors comprises checking the following items: depth and diameter of the drillhole, cleaning the drillhole from muck spoil, clean surface of the steel rod, appropriate grouting, tensioning and pull out tests. The latter are applied for

Initial evaluation, i.e. evaluation of the suitability of one or several alternative rockbolt systems for a particular project. Pull out force and displacement are recorded at each test. 10 - 20 bolts are tested in typical rocks of the project (Fig. 5.18).

Proof testing (quality control) is done on a specified percentage of installed bolts.

5.4.4 Application

There are two types of rockbolt applications in tunnelling:

Spot bolting: Individual rockbolts are placed to stabilise isolated blocks

Pattern bolting: Systematic installation of a more or less regular array of rockbolts. Despite some rational approaches (see Section 15.1), the design of pattern bolting is empirical. For coal mines BIENIAWSKI suggested the following approach:[14] Based on the RMR-value of the rock mass and the span s, one obtains the 'rock load height' h_t as

$$h_t = s(100 - RMR)/100 \ .$$

The bolt length is then determined as $l = \text{Min}\{s/3; h_t/2\}$ and the spacing of the bolts is taken as $(0.65\text{-}0.85)l$.

In unstable rock, bolts should be installed immediately after drilling of the hole. In so doing, the long established safe practise of working forward from solid or secured ground towards unsupported or suspect ground must be adhered to.

Apart from tunnelling, anchors are also applied in other fields of civil engineering and mining, e.g. to secure tied-back retaining walls and slopes or to prevent uplift by hydrostatic pressure.

---

[14]J.A. Franklin, M.B. Dusseault, Rock Engineering, McGraw-Hill 1989, Section 16.3.2

[15]DYWIDAG DSI (Info 10)

Fig. 5.18. Pull-out test of an anchor[15]

5.5 Timbering

In the early days of tunnelling, timbering was the only means for temporary support (Fig. 5.21). Nowadays it is mainly used for the support of small and/or irregular cavities (e.g. resulting from inrushes). Timbering has been systematically used (according to the old Belgian tunnelling method) during the recent construction of the Madrid metro (Fig. 5.19 und 5.20).

Fig. 5.19. Timbering in the metro of Madrid (2000)

Wood is easy to handle and transport and indicates imminent collapse by cracking. On the other hand the discontinuous contact with the rock is problematic.

The spacing of the timber frames is usually 1 - 1.5 m. Care must be taken for a sufficient longitudinal bracing.

---

[16]Historische Alpendurchstiche in der Schweiz, Gesellschaft für Ingenieurbaukunst, Band 2, 1996, ISBN 3-7266-0029-9

Fig. 5.20. Timbering in the metro of Madrid (2000)

Fig. 5.21. Building the bricked lining, Lötschberg tunnel[16]

5.6 Support arches

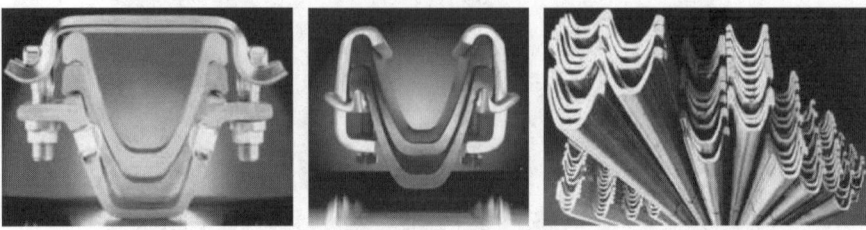

Fig. 5.22. Rolled steel profiles

Fig. 5.23. Lattice girder (Pantex-3-arch), connecting the segments, mounting with rock bolts

Support arches are composed of segments of rolled steel profiles or lattice girders (Fig. 5.22, Fig. 5.23 and Fig. 5.24). The arch segments are placed and mounted together with fixed or compliant joints (to accommodate for large convergences). The contact with the adjacent rock is achieved with wooden wedges or with bagged packing, i.e. bags filled with (initially) soft mortar. Usually, the arches are subsequently covered with shotcrete. This leads to a garland-shaped shotcrete surface, which protrudes to the cavity at the locations of the arches. To achieve a good contact between the shotcrete surface and a geosynthetic sealing membrane, the sag between two adjacent arches should not exceed 1/20 of their spacing. Together with their contribution to support, arches also help to check the excavated profile. They can also serve to mount forepoling spiles in longitudinal direction. Clearly, ʊ-shaped rolled steel profiles have a much higher bearing capacity than lattice girders.

---

[17]Tunnelling Switzerland, Swiss Tunnelling Society, Bertelsmann 2001

Fig. 5.24. Mounting of girder arches, Zürich-Thalwil tunnel[17]

5.7 Forepoling

If the strength of the ground is so low that the excavated space is unstable even for a short time, a pre-driven support is applied in such a way that an excavation increment occurs under the protection of a previously driven canopy.

The traditional method of forepoling was to drive 5 to 7 mm thick steel sheets up to 4 m beyond the face into the ground or 1.5 to 6 m long steel rods (so-

Fig. 5.25. Special forepoling rig, *Rotex*

---

[18]source: http://www.rotex.fi

Fig. 5.26. Forepoling, schematically[18]

called spiles) with a spacing of 30 to 50 cm. Nowadays, forepoling is achieved by spiling, pipe roof, grouting and freezing.

Spiling: This method consists of drilling a canopy of spiles, i.e. steel rods or pipes into the face (Fig. 5.26). A typical length is 4 m. To give an idea, 40-45 tubes, ⌀ 80-200 mm, each 14 m long, enable a total advance of 11-12 m (the last 2-3 metres serve as abutment of the canopy). In order for the spiles to act not only as beams (i.e. in longitudinal direction) but also to form a protective arch over the excavated space, the surrounding soil is grouted through the steel pipes or sealed with shotcrete. Thus, a connected canopy is formed that consists of grouted soil reinforced with spiles. Drilling 40 tubes takes ca 10-12 hours, grouting another 10-12 h. Spile rods can also be placed into drillholes. The remaining annular gap is filled with mortar, whose setting however may prove to be too slow. Alternatively, 'self-drilling' rods are used.

Pipe roof: This method is similar to spiling with the only difference that large diameter (> 200 mm) steel or concrete tubes are jacked into the soil above the space to be excavated. The larger diameter provides a larger bearing capacity. Sometimes, the tubes are filled with concrete. The steel tubes only act as beams and do not form an arch. Pipe roofs do not protect the overburden soil from considerable settlements.

Perforex-method: This method is also called 'peripheral slot pre-cutting method' or 'sciage' (=sawing). A peripheral slot is cut using a movable chainsaw (slot cutter) mounted on a rig (Fig. 5.27). The individual slots have a depth up to 5 m and a thickness between 19 and 35 cm. These are filled with shotcrete, thus forming a vault that protects the space to be subsequently excavated. Immediately after completion the slot is shotcreted while the next one is being cut. The slots are staggered in such a way that consecutive canopies overlap by 0.5 to 2 m. This method allows large advance steps. The resulting canopy is relatively rigid and, therefore, does not induce stress relief by yielding. This effect combined with a possible incomplete setting of shotcrete may possibly cause collapse. Peripheral slots are also applied in hard rock in combination with drill

& blast. There, the slot protects the surrounding rock from explosion damage.[19]

Grouting: Grouting has multiple applications in tunnelling. Therefore, it is presented separately in Chapter 6

Soil freezing: see Chapter 6.

Fig. 5.27. Perforex forepoling

5.8 Face support

Unstable faces can be backed (buttressed) with a heap of muck or reinforced with fiberglass rods. Both methods are temporary and have to be removed before or during the next excavation step. In soft underground and for tunnel diameters larger than 4 m the face should not be vertical but inclined with ca 60 - 70°.

5.9 Sealing

Surface support, such as shotcrete, may act in two distinct ways (Fig. 5.28). With uniform convergence of the rock, the support responds with arching, i.e. mobilization of axial thrust within the lining.

At local spots of weakness or at keystones (in case of jointed rock), shear and tensile stresses are mobilised within the lining. At an initial stage of these deformations, a thin lining is sufficient to resist loosening of the rock and the

---

[19] see also: S. Morgan, Prevaulting success at Ramsgate Harbour, *Tunnels & Tunnelling International* July 1999, 31-34; A. Rozsypal, From the New Austrian Method to the peripheral slot pre-cutting method, *Tunel* Vol. 9, No. 1/2000, ISSN 1211-0728,6-15; P. Lunardi, Pretunnel advance system. *Tunnels & Tunnelling International*, October 1997, 35-38

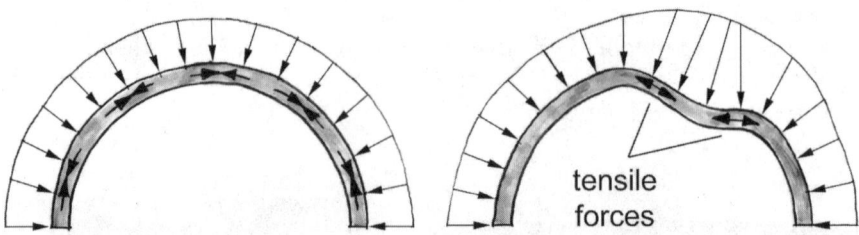

Fig. 5.28. Axial thrust within the lining at uniform radial pressure (left) and bending with non-uniform pressure (right)

related increase of loads. This supporting action is called sealing and can be obtained with thin layers of shotcrete. A recent development is to seal with 3-6 mm thick spray-on polymer liners. They have good adhesive bond, when applied to clean rock, and develop a good performance in tension and shear. It should be mentioned, though, that creep is still an open question. In contrast to shotcrete, the compliant nature of synthetic liners allows them to continue to function over a wide displacement range.[20]

5.10 Recommendations for support

Originally, the support measures were determined in an empirical way, following a sort of trial and error procedure, which is often (but falsely) attributed as 'observational method'. Later on, empirical rules have been based upon rock classification schemes. In contrast, rational analysis seeks to design the support on the basis of the interplay between ground and the several support elements.

The rational approach, based on computations, is increasingly applied. However, there are still important gaps in knowledge. To give an example, the mechanical behaviour of green shotcrete is poorly known (see Section 22.3) and also the loads exerted by the ground upon the lining cannot be exactly determined. Therefore, computations are often biased and recommendations based on rock mass rating are welcome to somehow fill the gap.

Clearly, the required support depends not only on the quality of the ground but also on the size and depth of the cavity and on the allowed deformations. When combining two or more types of support (e.g. shotcrete and rockbolts), attention should be paid to the fact that they may have different compliances, i.e. their resistance is mobilised at different deformations.

---

[20] D. D. Tannant, Development of thin spray-on liners for underground rock support – an alternative to shotcrete? In: Spritzbeton Technologie 2002, published by W. Kusterle, University of Innsbruck, Institut für Betonbau, Baustoffe und Bauphysik, 141-153.

Support recommendations based on RMR:

BIENIAWSKI recommends the support measures shown in table 5.2. They are based on RMR[21] (see Section 3.6.1) and refer to a tunnel of 10 m diameter.

RMR	Heading	Anchoring ⌀ 20 mm, fully bonded	Shotcrete	Ribs
81-100	full face, advance 3 m	–	–	–
61-80	full face, advance 1-1.5 m, complete support 20 m from face	locally bolts in crown, 3 m long, spaced 2.5 m, with occasional wire mesh	5 cm in crown where required	–
41-60	top heading and bench: 1.5-3 m advance in top heading, commence support after each blast, complete support 10 m from face	systematic bolts 4 m long, spaced 1.5-2 m in crown and walls with wire mesh in crown	5-10 cm in crown, 3 cm in sides	–
21-40	top heading and bench: 1-1.5 m advance in top heading, install support concurrently with excavation −10 m from face	systematic bolts 4-5 m long, spaced 1-1.5 m in crown and walls with wire mesh	10-15 cm in crown and 10 cm in sides	light ribs spaced 1.5 m where required
≤ 20	multiple drifts: 0.5-1.5 m advance in top heading, install support concurrently with excavation	systematic bolts 5-6 m long, spaced 1-1.5 m in crown and walls with wire mesh. Bolt invert.	15-20 cm in crown, 15 cm in sides and 5 cm in face	medium to heavy ribs spaced 0.75 m with steel lagging and forepoling if required. Close invert

Table 5.2. Support measures based on RMR (according to BIENIAWSKI)

---

[21]Bieniawski, Z.T., Rock Mechanics Design in Mining and Tunnelling. Balkema, 1984

Support recommendations based on Q-values:

Depending on the rock quality and on the size of the cavity (expressed by its span s or height) the recommended support is indicated in a Q-s-diagram (Fig. 5.29).

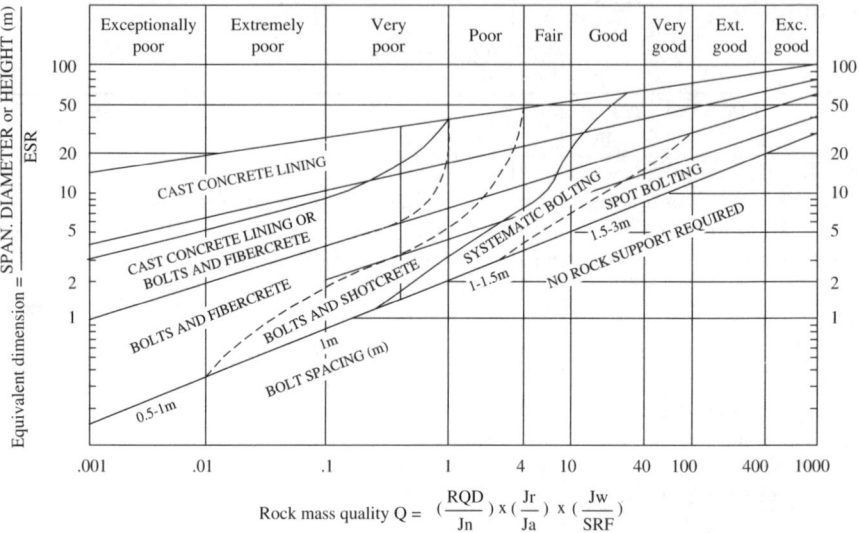

Fig. 5.29. Recommended types of support[23]. ESR is the so-called excavation support ratio. Its values range between 0.5 and 5 and are given in a table of the cited paper for various types of excavations.

5.11 Temporary and permanent linings

In sufficiently strong rock (as often encountered e.g. in Scandinavia) permanent lining is not provided for.[24] Usually, however, conventionally driven tunnels obtain a permanent lining (inner lining) of cast concrete in addition to the temporary lining of shotcrete (outer lining). The prevailing idea is that the loads exerted on the shotcrete lining are initially reduced due to arching but then slowly increase. It is also believed that the shotcrete lining decays with time so that an inner lining of cast concrete becomes necessary. These ideas have never been confirmed. Of course, there is no doubt that the inner lining increases safety. There are also some other benefits from the inner

---

[23]Barton, N., Grimstad, E.: The Q-System following 20 years of application, *Felsbau* 12, No. 6 (1994), 428-436

[24]E.g. the Gjøvik Olympic Cavern Hall in Norway with 91 m span, 24 m height and a capacity of 5,800 persons

lining: A sealing membrane (if necessary) can be mounted between the outer and inner linings. In addition, a smooth surface of the tunnel wall (as is the case with a cast concrete lining) is advantageous from the points of view of aerodynamics (ventilation) and illumination.

In the case of segmental linings (as used in shield driven tunnels) there is usually no inner lining of cast concrete. Such linings are watertight up to water pressures of 6 bar.

5.12 Permanent lining

The usual thickness of a permanent lining is at least 25 cm. For reinforced and watertight linings a minimum thickness of 35 cm is recommended[25]. Blocks of 8 to 12 m in length are separated with extension joints. Usually concrete C20/25 is used. Concretes of higher strengths develop higher temperatures during setting (fissures!) and are more brittle.

Fig. 5.30. Rolling formwork (Engelberg base tunnel)[26]

The concrete is poured into rolling formworks (Fig. 5.30, 5.31) and compacted with vibrators in the invert and with external vibrators in the crown (one

---
[25] Concrete Linings for Mined Tunnels, Recommendations by DAUB, Dec. 2000, *Tunnel*, 3/2001, 27-43
[26] *Tunnel*, 3/2001, p. 30
[27] *Tunnel*, 3/2001, p. 31

154   5 Support

Fig. 5.31. Rolling formwork (Nebenwegtunnel, Vaihingen/Enz) [27]

vibrator for 3 to 4 m^2). It is difficult to achieve complete filling of the crown space with concrete: The pumping pressure should be limited, otherwise the rolling formwork can be destroyed. Possibly unfilled parts should be regrouted with pressures of ≤2 bar 56 days after concreting. Usage of 'self-compacting concrete'[28] can possibly help to avoid incomplete filling of the formwork.
Within 8 hours the concrete should attain a sufficient strength, so that the formwork can be removed. However, there are cases reported where the setting was insufficient and the lining collapsed after early removal of the formwork.

5.12.1 Reinforcement of the permanent lining

Since the loads acting upon the inner lining are not exactly known, the requirement for its reinforcement is an open question.[29] In France and Austria, for instance, inner linings are usually not reinforced. The German Rail, on the other hand, decided to reinforce the inner linings, based on its experiences from the new Hannover-Würzburg line. One should also consider the hindrance to traffic due to repair works of defective linings. Apart from the

---

[28]This is a concrete of high flowability (spread > 70 cm)

[29]One of the greatest figures in contemporary tunnelling, LEOPOLD MÜLLER-SALZBURG writes: "Experience teaches that our inability to design tunnels realistically leads to considerable overdesign and "fear-reinforcement", without any additional safety despite substantial additional costs."

forces exerted by the surrounding ground, the permanent lining is exposed to a series of other loads:

- Own weight
- Shrinkage
- Temperature differences
- Aerodynamic pressure (see Section 1.4.2). Trains which are moving faster than 200 km/h may cause considerable longitudinal fissures in the crown. Their width was reported to be <1.0 mm in not-reinforced and <0.3 mm in reinforced linings.

The reinforcement cage is fabricated in situ. This is troublesome especially in the crown part, because the workers have to construct the cage overhead. In the Engelberg tunnel formworks have been applied that made it possible to fabricate the cage on form panels and then heave it to its final position.
In tunnels with electrically driven trains, the reinforcement may corrode because of creep currents and should therefore be grounded.
The following hints from the Guidelines for the Design of Tunnels (International Tunnelling Association, ITA)[30] should be taken into account:

Minimum thickness of the shotcrete lining:
    15 cm   for cross-overs and shafts
    20 cm   for excavation cross sections ≤ 50 m^2
    25 cm   for larger cross sections
Minimum thickness of cast concrete lining:
    20 cm   for not reinforced lining
    25 cm   for reinforced lining
    30 cm   for watertight lining
Minimum cover of reinforcement:
    3 cm   at the outer surface (extrados), if it is not protected with a membrane
    5-6 cm   at the outer surface, if it is adjacent to ground water
    4-5 cm   at the inner surface (intrados).

5.12.2 Quality assessment of the lining

As already mentioned, one of the aims of the lining is to back the sealing membrane. If, however, the lining has deficient thickness in some places, then the membrane can be damaged by the naked reinforcement. It is therefore important to assure that the lining has a sufficient thickness everywhere. This can be checked by analysing the travel time of a sonic wave created by the

---

[30] Working Group on General Approaches to the Design of Tunnels. *Tunnelling and Underground Space Technology*, Vol. 3, No. 3, 237-249, 1988

[31] *Tunnel*, 8/2001, p. 42

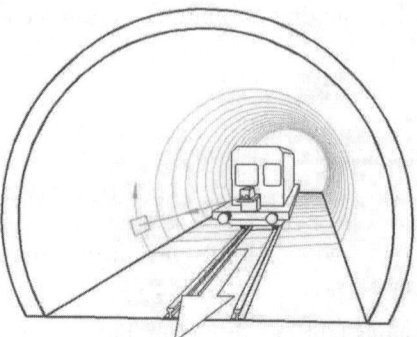

Fig. 5.32. Tunnel-Scanner[31]

impact of a small steel ball.[32] The wave is reflected at the boundary of the lining, and, if the propagation velocity of the wave is known, the thickness of the lining can be inferred from the travel time. The measurements are executed with a spacing of, say, 40 cm. Detected cavities in the intrados can then be grouted. Furthermore, a visual inspection of the state of the lining should be carried out in regular intervals (e.g. every 6 years). This inspection can be facilitated by the use of the 'tunnel scanner' (Fig. 5.32) which records continuous images of the tunnel wall in the visible and in the infrared range. Spallings and fissures > 0.3 mm can thus be easily recognised.

5.13 Single-shell (monocoque) lining

As mentioned before, the shotcrete lining is seen as a temporary measure. In the long term the loads exerted by the surrounding ground are expected to be carried by the inner lining. However, the complete depreciation of the shotcrete lining has never been proved. There is now a trend to design and construct the shotcrete lining in such a way that it can serve permanently or, at least, that it can be integrated into the permanent one. By doing so, the total lining has a reduced thickness and is called single-shell or monocoque lining.[33] The main problem with monocoque lining is the sealing against pressurized groundwater. Shotcrete lining is usually fissured and thus water permeable.

A new development is to construct the inner lining with steel fibre reinforced concrete so that inner and outer lining constitute a composite lining. This

---

[32] W.D. Friebel, J. Krieger, Quality Assurance and Assessing the State of Road Tunnels Using Non-Destructive Test Methods. *Tunnel* 8/2001, p. 38-46. W. Brameshuber, Qualitätskontrolle von Tunnelinnenschalen mit zerstörungsfreien Prüfmethoden. STUVA Tagung 1997, (Vol. 37) Berlin, 126-129.

[33] J. Schreyer: Constructional and economic solutions for monocoque tunnel lining, *Tunnel* 2/96, 14-28.

method was first applied in the metro of Bielefeld.[34] On a 15 cm thick shotcrete lining a 10 cm thick layer of steel fibre reinforced concrete was applied, using the wet mix method. The resulting concrete corresponded to a C20/25 and contained 70 kg steel fibres per m^3. Its permeability was reduced by a factor of 10 - 100 with microsilica.

Monocoque lining of not reinforced shotcrete is only recommended above the groundwater table and only, if considerable fissures (e.g. due to asymmetric loading or bad geological conditions) are not anticipated. In this case the thickness of the shotcrete lining should be at least 30 cm. Furthermore, extension joints should not be used.

Compound monocoque lining (i.e. shotcrete and cast concrete linings acting together) should be considered only for water pressures up to 1.5 bar. In this constellation the thickness of the cast inner lining should be at least 25 cm. The reinforcement should be distributed uniformly and its concrete cover should be 5 cm at the extrados and 4 cm at the intrados. The cast concrete lining should have extension joints (sealed with gaskets) every 8 to 10 m.

---

[34] M. Ziegler: U-Bahn Tunnel in Verbundbauweise mit Innenschale aus Stahlfaserspritzbeton. Berichte des 7. Internationalen Kongresses über Felsmechanik, Aachen 1991, p. 1399-1403

6
Grouting and freezing

Grouting is the introduction of a hardening fluid or mortar into the ground to improve its stiffness, strength and/or impermeability. There are various patterns of the propagation of the grout within the ground:

Low pressure grouting (permeation grouting): The grout propagates into the pores of the soil but leaves the grain skeleton unchanged. The resulting grouted regions are spherical, if the soil is homogeneous and isotropic and if the source can be considered as a point. If the pore fluid, which initially fills the voids, has a higher viscosity than the grout (as is e.g. the case when water is pumped in into a porous rock filled with oil) then the so-called fingering is observed. The resulting boundary of the grouted region is fractal shaped.[1]

Compensation grouting: When the applied grouting pressure is too high, the grout does not propagate into the pores of the ground. Instead, the ground is cracked and the grout propagates into the created cracks (or in case of soft soil the grout pushes the soil ahead). This type of grouting is applied to reverse (compensate) surface settlements (e.g. due to tunnelling).

Jet grouting: A grout jet protrudes from a nozzle into the surrounding soil. With an initial pressure between 300 and 600 bar it completely remoulds the soil and gets mixed with it.

6.1 Low pressure grouting

In most cases the grout is introduced into the ground with a double packer movable within a tube à manchette (also called 'sleeve pipe', Fig. 6.1). The tube à manchette is fixed within a borehole, the annular gap between the tube and the borehole wall being filled with a hardening bentonite-cement slurry.

---

[1] J. Feder: 'Fractals', Plenum Press, New York and London, 1989.

In fine sand, the tube à manchette can be vibrated into the ground, in which case a borehole is not needed.

The double packer is brought down to the depth of the manchettes and the grout is pumped in. It cracks the annular cement body and enters into the ground. The grouting pressure is recorded. The plot exhibits an initial peak, which shows the cracking of the annular cement. Subsequently, the pressure is reduced to the value required to push the grout into the pores (or joints) of the ground. This pressure must not be too high, otherwise the ground is cracked and then high amounts of grout can be introduced and propagate in an uncontrollable manner. To avoid this, the pressure and the discharge of grout must be continuously recorded and controlled. The pressure must not exceed the value $\alpha\gamma h$, where γh is the overburden pressure and α an empirical factor (usually $\alpha \approx 1$). For grounds with very high or very low strength α may vary between the values of ca 0.3 and 3. Furthermore, it must be taken into account that the pressure measured at the pump is not identical with the pressure at the manchette (the pressure loss in the pipe may amount from 2 to 6 bar per 100 m).

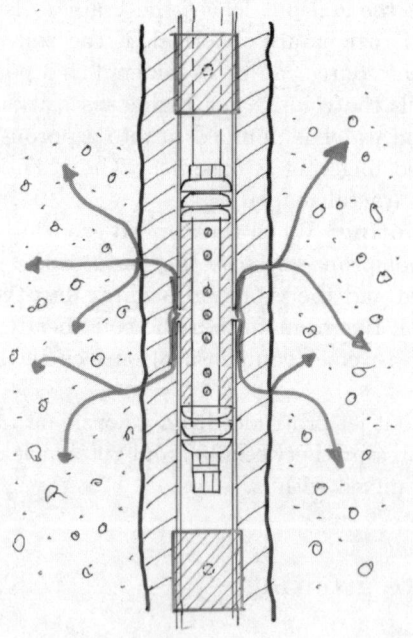

Fig. 6.1. Tube à manchette and double packer. The pressurized grout opens the manchette, cracks the annular cement ring and enters into the soil.

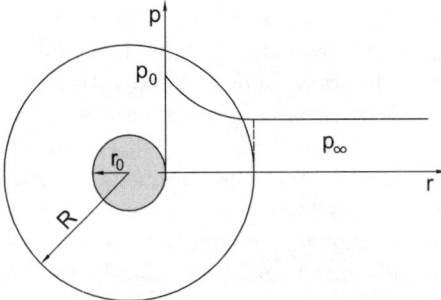

Fig. 6.2. Distribution of pressure around a spherical grout source in homogeneous and isotropic soil.

If the manchette is idealised as a spherical source of radius r_0, then the pressure p_0 needed to push the grout discharge Q into the pores of the ground can be estimated (for the case of isotropic permeability) by the following equation:

$$p_0 - p_\infty = -\int_{r_0}^{R} dp = \frac{\gamma_g}{k_g} \cdot \frac{Q}{4\pi} \int_{r_0}^{R} \frac{dr}{r^2} = \frac{\gamma_g}{k_g} \cdot \frac{Q}{4\pi} \left(\frac{1}{r_0} - \frac{1}{R} \right) \approx \frac{Q\gamma_g}{4\pi k_g r_0} \quad (6.1)$$

The radial velocity v of the grout in a distance r is obtained from $Q = 4\pi r^2 v$. The discharge Q results from the required volume V of grouted soil and the hardening time t_G of the grout ($Q > V/t_G$), with $v = k_g i = -k_g/\gamma_g \cdot dp/dr$. γ_g is the specific weight of the grout and k_g is the permeability of the soil with respect to the grout. With μ_g and μ_w being the viscosities of the grout and water, respectively, k_g can be obtained as:

$$k_g = \frac{\mu_w \gamma_g}{\mu_g \gamma_w} k \quad , \quad (6.2)$$

where k is the permeability with respect to water. Note that the grout viscosity μ_g increases with time, a fact which is not taken into account in this simplified analysis. p_∞ is the pressure of the surrounding groundwater. If the groundwater flows with the superficial velocity v_∞, then the grout will be carried away if $Q < 4\pi r_0^2 v_\infty$. If the ground is inhomogeneous, the grout may escape along coarse grained permeable layers.

6.2 Soil fracturing, compensation grouting

As mentioned above, increased grouting pressure fractures the ground. If the ground has an isotropic strength, the cracks are oriented perpendicular to the minimum principal stress. In a first grouting stage ('conditioning') such cracks

162   6 Grouting and freezing

are opened and filled with grout. In doing so, the minimum principal stress is increased and a hydrostatic stress state results. In a subsequent grouting stage, cracks open in random directions and are filled with grout. As a consequence, the ground 'swells' and the ground surface can be heaved. Previous settlements can thus be reversed (hence the name 'compensation grouting').[2] The upheavals of buildings must be recorded on-line. This is usually achieved with water levels. The elevation of each sensor is measured with accurate pressure transducers. The water must be de-aired and a temperature compensation must be provided for. Setting of the grout must be taken into account: If the grout remains fluid for too long a time, then it will be squeezed out as soon as pumping stops. Compensation grouting is also called 'grout jacking'.[3] The application of compensation grouting to reverse settlements due to tunnelling should be very cautious, because the applied pressure can severely load the tunnel lining (Fig. 6.3). At the tunnel collapse of Heathrow Airport, forces due to grout jacking caused excessive movements of the lining.

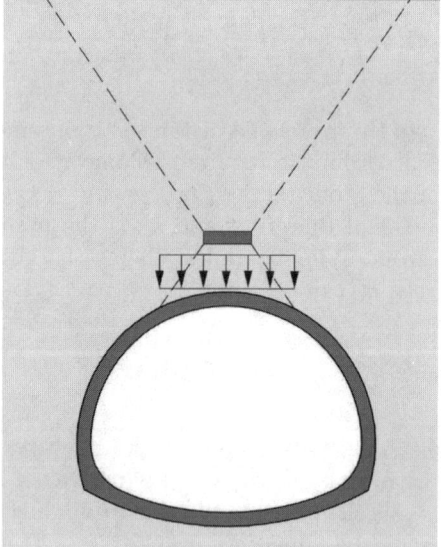

Fig. 6.3. Loading of the tunnel lining by compensation grouting. Assuming a simplified model of stress propagation within the dashed cone, we obtain that the load applied upon the lining can correspond to the weight of the upper cone.

---

[2] E.W. Raabe and K. Esters: Injektionstechniken zur Stillsetzung und zum Rückstellen von Bauwerkssetzungen. In: Baugrundtagung 1986, 337-366

[3] Some authors differentiate between these two types of grouting. This differentiation is, however, not comprehensible.

6.3 Jet grouting

A high pressure (300 to 600 bar) is applied to a cement suspension which is pumped through a pipe with a lateral nozzle at its bottom end (Fig. 6.4). The jet erodes the surrounding soil. When the pipe is pulled out and rotated simultaneously, a cylindrical body, composed of soil and cement, is formed (Fig. 6.5). The diameter of the cylinder depends on many factors, e.g. on the speed of rotation. Recently, diameters of 5 m have been achieved. A part of the suspension escapes to the ground surface along the pipe. With the so-called duplex method, the suspension jet is surrounded by an air jet and is thus more focused. With the triplex method, the soil is pre-cut with a water jet, the cement suspension is subsequently grouted into the created cavity.

For horizontal columns (i.e. for forepoling), the 'simplex' method is applied. The consistence of the cement suspension is important. If it is too liquid, it can easily escape and settlements can occur. If it is too thick, it can cause upheavals of the ground surface. It should also be taken into account that the position accuracy of the grouting pipes is limited. Therefore, the length of the columns should not exceed ca 20 m.

Fig. 6.4. Grout jet

6.4 Grouts

Considering low pressure grouting into soil, the grout has to be selected according to the grain size distribution of the surrounding soil (Fig. 6.6)[5]. Rock fissures can be grouted if their thickness exceeds the maximum particle diameter of the grout by a factor of 3. Thin grouts can be considered as Newtonian fluids and characterised by their viscosity μ. In contrast, thick grouts can be considered as BINGHAM fluids, i.e. they do not flow unless the shear stress exceeds a yield limit τ_f (which is a sort of undrained cohesion). μ controls

---

[4] Bilfinger und Berger company
[5] C. Kutzner, Injektionen im Baugrund, Ferdinand Enke Verlag, Stuttgart 1991

164   6 Grouting and freezing

Fig. 6.5. Uncovered jet-grout columns produced in layered soil[4]

the discharge Q of grouting at a specific grouting pressure (Equ. 6.1 and 6.2), whereas τ_f controls the range l of coverage. This can be easily shown if one considers an idealised pore in the form of a cylinder of length l and diameter d (neglecting thus its tortuosity). The driving pressure p exerts the force $p\pi d^2/4$ on the grout inside the pore. This force has to overcome the flow resistance $\pi dl\tau_f$. Hence, $l = pd/(4\tau_f)$.

The following types of grout can be used:

Cement grouts: The cement content varies between 100 and 500 kg per m^3 mixture. To avoid sedimentation during transport, bentonite is added (10 to 60 kg/m^3). Bentonite reduces not only the permeability of the grouted soil but also its strength (by 50% and more). To achieve groutability into finer soils, ultra-fine cements are used with grain diameters between 1 and 20 μm. These are roughly 3-10 times as expensive as normal cement, but allow to grout medium sand with up to 30% fine sand content. Ultra-fine cements need more water, more intense mixing (which may cause increased heat), but have a quicker hydration and obtain higher strengths than usual cement. No bentonite is used with ultra-fine cements. Additives may accelerate setting. To grout into flowing groundwater (e.g. in karst cavities), up to 10% sodium silicate can be added. Attention should be paid if the grout contacts chlorides, sulfates and lignite. In this case, appropriate cement must be used. The properties of the grout may vary with time not only due to setting. It should also be taken into account that

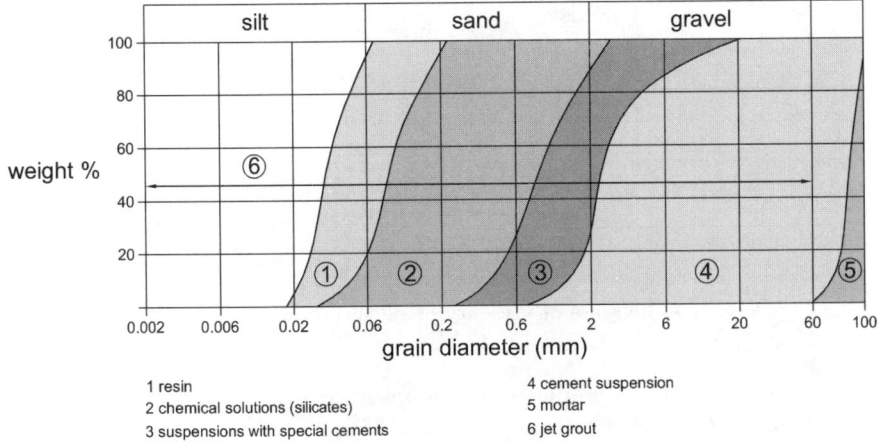

Fig. 6.6. Ranges of application (injectability limits) of several grouts, according to KUTZNER

its water content (and, as a result, the viscosity) can be altered either due to convection of silt particles or due to squeezing of water ('filtration').
The latter effect refers to the so-called pressure stability of cement grouts.[6] Squeezing out of water reduces the flowability of a grout and leads to plugs ('filter cakes') that can form in openings much larger that 3 times the maximum particle diameter. Therefore, the pressure stability of grouts is very important for permeation.

After grouting, a sufficient time of several hours must be awaited for setting before any blasting and drilling into the grouted area.

For advance grouting of tunnels the cement grout consumption varies between 15 and 500 kg/m tunnel.

Chemical grouts:

Silicates: The basic material is sodium silicate ('waterglas'). The method of JOOSTEN has been widely used for grouting into fine grained soils: Concentrated sodium silicate is grouted first. In a subsequent step calcium chloride is injected into the ground which leads to an instantaneous setting. There are also one-component grouts, where the sodium silicate is already mixed with a reactive substance (ester) in such a way that the setting occurs gradually. This can be seen as increase of viscosity with time (Fig. 6.7). The time for setting (also called 'gelatinisation') depends on the temperature and ranges from 30 to 60 minutes. Of course, grouting has to be completed within this time lapse.

---

[6]K.F. Garshol, Pre-Excavation Grouting in Rock Tunnelling, MBT International Underground Construction Group, Division of MBT (Switzerland) Ltd., 2003

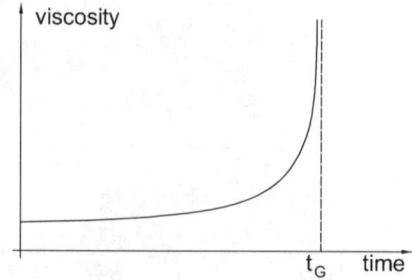

Fig. 6.7. Increase of viscosity of silicate solutions with time

The mechanical properties of the resulting gel can be tailored according to the individual requirements. If only sealing is to be achieved, then the gel may be soft. The gel weeps a fluid (sodium hydroxide) and, in doing so, reduces its volume ('syneresis'). This fluid induces precipitation of iron initially dissolved in the groundwater. As a result, the groundwater obtains a brown colour, a fact which may concern the people. Soil that has been solidified with chemical grout exhibits creep and its strength depends on the rate of deformation. Silicates should not be used for permanent water control.

Polyurethanes: Polyurethanes react with water and produce CO_2, thus causing the formation of foam. One litre polyurethane produces 12 litres of foam which sets very quickly (within 30 seconds to 3 minutes). The created pressure up to 50 bar drives the foam into small fissures. The foam remains ductile after hardening.

Acrylic grouts: Acrylic monomers are liquids of low viscosity until the polymerisation sets on. This occurs rather suddenly with gel-times of up to one hour. Acrylic grouts based on acrylamide should not be used, because they are toxic.

Epoxy resins are of less importance in tunnelling because of difficult handling.

Thermoplastic materials such as bitumen (asphalt) or polyamides melt at approx. 200 °C and can be pumped into cavities filled with fast flowing ground water. They can be effective in plugging off the water flow even of they are grouted with a discharge rate of only 1 % of the water discharge rate.

6.5 Rock grouting

Rock has a much smaller pore volume than soil (e.g. 1 m^3 of soil can have 300 l volume of voids, whereas 1 m^3 rock can have 0.1 to 0.4 l volume of voids). It is, therefore, difficult to uniformly grout all voids (joints) of rock.

Grout can easily escape through large joints leaving smaller joints aside. This can be avoided by

- thicker grouts[7]
- limiting the grout volume V
- limiting the grouting pressure p.[8]

LOMBARDI[9] recommends the use of relatively thick grouts and the addition of concrete liquefiers. Furthermore, he recommends limiting V in cases where large masses of grout can be pumped in at low pressure, and limiting p where it is difficult to grout rock. If high grouting pressures are applied, the rock can be hydraulically fractured. Hydraulic fracture is, however, unlikely to occur if the aperture of the joints is small and the overburden larger than 5-10 m because, in this case, the pressure is rapidly attenuated. Thus, in such cases (i.e. for low acceptance of grout) the grouting pressure can be increased up to 4 MPa. For $p < p_{max}$ and $V < V_{max}$ LOMBARDI recommends keeping the so-called *Grout Intensity Number GIN*, i.e. the product pV, constant, see Fig. 6.8. Typical GIN values vary between 500 and 2,500 bar·l/min.

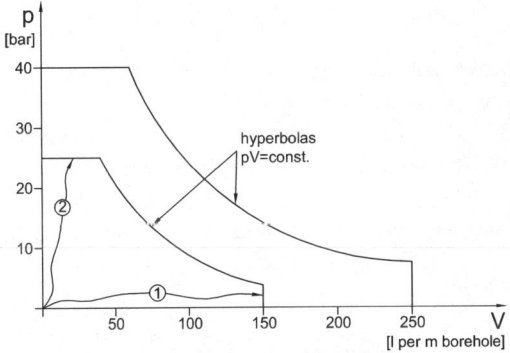

Fig. 6.8. LOMBARDI's GIN-concept. Grouting path 1 corresponds to large joint apertures, path 2 corresponds to small joints

---

[7]It is common to start grouting with a high w/c-ratio (e.g. w/c=3.0) and reduce it in steps whenever the pressure limit is reached.

[8]This procedure is also called 'grout to refusal'.

[9]G. Lombardi and D. Deere, Grouting design and control using the GIN principle. *Intern. Water Power & Dam Construction*, June 1993, 6.H1. ISRM Commission on Rock Grouting. *Int. J. Rock Mech. Min. Sci. & Geomech. Abstracts* Vol. 33, No. **8**, 803-847, 1996

6.6 Advance grouting

Advance grouting is used to seal tunnels against groundwater and thus prevent heading inrushes. Usually, the water inflow has to be limited to an acceptable value, say 1-5 litres per minute and 100 m tunnel length. Staggered boreholes with lengths of ca 20 m are driven from the face and grouted with ultra-fine cements or with chemicals using pressures of 50 to 60 bar. If this procedure is repeated every 10 m of advance, a good overlapping of the grout umbrella is obtained. Each borehole is grouted until a specified pressure (e.g. 60 bar) or a specified grout volume (e.g. 500 l) is achieved. It must be added that the success of this measure cannot be guaranteed. Whenever grouting is applied to confine water flow, it should be taken into account that incomplete waterproofing implies increased flow velocity and, consequently, erosion.

6.7 Soil freezing

The groundwater freezes if a sufficient amount of heat is extracted. The frozen ground temporarily attains a strength which stabilises the cavity until a support is installed. Attention should be paid to the following issues:

- The groundwater velocity must not exceed ca 2 m/s, otherwise heat is permanently supplied and freezing is prevented.
- Minerals dissolved in the groundwater may lower the freezing temperature
- Some fine grained soils may suffer upheavals when freezing (see Section 6.7.1).
- A saturation degree of at least 0.50-0.70 is required. This can be achieved by adding water, e.g. by sprinkling.

Fig. 6.9. Soil freezing: main collector Mitte Düsseldorf, Germany

The common cooling fluids are salt solutions that remain fluid up to temperatures of $-35°C$ and liquid nitrogen with a temperature of $-196°C$. The cooling fluid circulates within pipes that are driven into the soil. The precise placement of these pipes is crucial for the success. Frozen soil is a creeping material. Therefore, its stiffness and its strength (given e.g. in terms of friction angle and cohesion) cannot be specified independently of the rate of deformation. For rough estimations some approximate values are given in Tables 6.1 and 6.2, according to JESSBERGER. To avoid large creep deformations (and, hence, possible breakage of freezing pipes), the applied stresses must be considerably lower than the strength of frozen soil.

Soil	q_u (MN/m^2)	φ	c (MN/m^2)	YOUNG's modulus (MN/m^2)
non-cohesive medium density	4,3	20°-25°	1,5	500
cohesive stiff	2,2	15°-20°	0,8	300

Table 6.1. Short-term properties of frozen soils (for durations up to one week)

Soil	q_u (MN/m^2)	φ	c (MN/m^2)	YOUNG's modulus (MN/m^2)
non-cohesive medium density	3,6	20°-25°	1,2	250
cohesive stiff	1,6	15°-20°	0,6	120

Table 6.2. Long-term properties of frozen soil (for durations up to one year), q_u is the unconfined (uniaxial) strength

6.7.1 Frost heaves

The attraction forces acting on a mineral surface lower the freezing temperature. Therefore, the freezing of the porewater in fine grained soils is less uniform. Ice aggregations ('lenses') can form that grow by attracting water from the surrounding pores. Such ice lenses may cause upheavals of the ground surface. Upon thawing, the ice lenses collapse and chuckholes are created. There are several criteria for the susceptibility of a soil to formation of ice lenses at freezing (e.g. Fig. 6.10)[10].

---

[10] see also A. Kézdi: Handbuch der Bodenmechanik, Band 2, 238 ff, VEB Verlag für Bauwesen, Berlin 1970

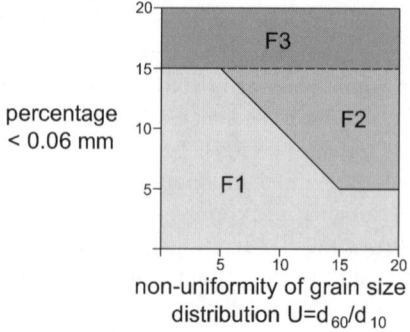

Fig. 6.10. Sensitivity to freezing according to the German road standard ZTVE-StB94. F1: non-sensitive, F2: low to medium sensitivity, F3: very sensitive

6.8 Propagation of frost

The following problem is relevant to the construction of tunnels and shafts using the ground freezing method: How fast does the region of frozen soil surrounding the freezing pipe expand? To answer this question, one has to resort to complicated numerical codes which are not commonly available. Therefore, a simple analytical approximation is

$$t_s \approx \frac{1}{3} \cdot \frac{A}{Br_0} \cdot \left(\frac{a}{2}\right)^3 \quad . \tag{6.3}$$

Herein, t_s is the closure time, i.e. the time needed for two adjacent cylindrical freezing fronts with a distance a to get in touch. The derivation of equation 6.3 and the definition of the quantities A and B can be found in appendix C.

7
The New Austrian Tunnelling Method

The New Austrian Tunnelling Method (NATM, in German: NÖT) emerged in the years 1957 to 1965[1] and was entitled in this way to be distinguished from the Old Austrian Tunnelling Method. The NATM was developed by Austrian tunnelling specialists (VON RABCEWICZ, PACHER, MÜLLER-SALZBURG). Its main idea is to head the tunnel conventionally, to apply support (mainly shotcrete) sparingly and to follow the principles of the observational method. The NATM requires the distortion of the ground to be kept to a minimum (in order to avoid softening and thus loss of strength). But at the same time sufficient ground deformations should be allowed in order to mobilise the strength of the ground. Consequently, thick and stiff linings which do not completely abut on the rock, are no longer in use. According to MÜLLER-SALZBURG[2] the main principles of the NATM were guesstimates (mainly by RŽIHA, HEIM, ANDREAE), which could not be applied until the techniques for shotcrete and rock monitoring had been developed. As many of the NATM's recommendations were already in use, it is not easy to differentiate NATM against other tunnelling methods. This has led to a lengthy controversy, which is still underway. The debate does not refer to the content but rather to the name of the NATM because the lack of an exact definition makes it unclear in which cases this name should be used.

One attempt to define NATM was made by the Research Society for Road Engineering of the Austrian Union of Engineers and Architects,[3] who published a complex and not very illuminating definition composed of not less than 5 basic principles and explanations, 3 general principles and 8 specific

---

[1] L. Müller-Salzburg und E. Fecker: Grundgedanken und Grundsätze der 'Neuen Österreichischen Tunnelbauweise', in Felsmechanik Kolloquium Karlsruhe 1978, Trans Tech Publications, Clausthal 1978, 247-262

[2] L. Müller-Salzburg, Der Felsbau, dritter Band: Tunnelbau, p. 562, Enke-Verlag 1978

[3] Schriftenreihe der Forschungsgesellschaft für das Straßenwesen im Österreichischen Ingenieur- und Architektenverein, Heft 74, 1980

principles.[4] The definition states that the NATM activates a bearing ring in the surrounding ground. However, this statement has been widely criticised, because any cavity in the rock is — at least partially — supported by the rock itself, no matter if this cavity has been headed according to NATM or not.[5]

Another debate surrounding the NATM is the Austrian designation,[6] which has been questioned with regard to the first application of shotcrete. But VON RABCEWICZ himself confirms that shotcrete was first applied during the construction of the Swiss Lodagno-Losogno-tunnels between 1951 and 1955. The Austrians were quick to follow. Shotcrete and rockbolts were systematically applied in the pressure tunnel Wenns of the Prutz-Imst water power plant from 1953 onwards.[7] There is no doubt that Austrian engineers contributed with courageous and pioneering applications to the propagation and foundation of NATM. The most widely held view is that NATM has come to represent conventional heading with shotcrete support.[8]

Confusion over the designation 'NATM' was also apparent in the *HSE Review*[9] which emerged after two inrushes of NATM headings in London clay in 1994. Two notations were used: 'N.A.T.M.' for the method according to the Austrian Union of Engineers and Architects, and *'NATM'* for tunnels headed with open face excavation and supported as soon as possible with shotcrete, anchors, nails and bolts.

NATM's recommendations were launched originally as empirical guidelines which can be interpreted today in terms of theoretical analysis. The theoretical foundations were always missing.[10] But even without such scrutiny, NATM has achieved remarkable successes (Tauern tunnel, Arlberg tunnel, Inntal tunnel, metro Frankfurt, Schweikheim tunnel, Tarbela caverns).

When applying the NATM in urban areas with a soft ground, one of its rules has to be relaxed: deformations to mobilise rock strength should be limited as otherwise the surface settlements can become excessive. Nevertheless, NATM was successfully applied in the metro construction in Frankfurt

---

[4] The perception of this definition as 'impenetrable shroud of complexity' (A. Muir-Wood, Tunnelling: Management by design. Spon, London 2000) appears thus understandable.

[5] K. Kovári: Gibt es eine NÖT?, XLII. Geomechanik Kolloquium 1993, Salzburg

[6] Note that in Norway the NATM is known as 'Norwegian Method of Tunnelling'.

[7] In US mining, shotcrete has been used since about 1925.

[8] see *Tunnels & Tunnelling*, September 1995, p. 5, for the history of NATM see also J. Spang: Die Geschichte des Spritzbetons und seiner Anwendung beim untertägigen Hohlraumbau. Taschenbuch für den Tunnelbau 1996, p. 321 ff, Verlag Glückauf.

[9] Health and Safety Executive: Safety of New Austrian Tunnelling Method (NATM) tunnels. A review of sprayed concrete lined tunnels with particular reference to London clay. HMSO, 1996

[10] 'We need a scientific foundation, otherwise we have to disappear', F. Laabmayr in *a3BAU* 12/1994, p. 88

clay between 1969 and 1971 and proved itself economical when compared with shield heading. Successful applications followed in Nürnberg, Bochum, Bonn, Stuttgart, Vienna. Since 1992 NATM has also been applied in London clay, which has similarities with Frankfurt clay. In October 1994 two inrushes occurred which gave rise to detailed investigations to NATM (see *HSE Review* and *ICE design and practice guide*[11]). The successful completion of the Heathrow baggage transfer tunnel and the Römerberg tunnel[12] are further vindications of NATM. There have also been some drawbacks (München-Orleansplatz, München-Trudering, Heathrow and the new rail track from Hannover to Würzburg), which should be attributed not to the method itself but rather to its improper application or other reasons. A list of 39 inrushes or daylight collapses of NATM headings is given in the *HSE Review*.

If we look at NATM in the wider sense — i.e. as the Austrian tunnelling school, we can appreciate all the more the globally respected expertise of Austrian tunnelling specialists. Their philosophy has been sharpened by the highly variable geology of the Alps and relies not so much on previous site investigations as on the flexibility to find the correct support measures on the spot.

In conclusion, probably the best definition of NATM belongs to H. LAUFFER:[13]

> NATM is a tunnelling method in which excavation and support procedures, as well as measures to improve the ground — which should be distorted as low as possible, — depend on observations of deformation and are continuously adjusted to the encountered conditions.

Consequently, NATM contrasts the *design and construct* principle, where the construction has to proceed as originally designed.

7.1 HSE Review

The HSE Review mainly addresses the NATM applications in urban areas with soft ground. Its conclusion is that NATM is indeed a safe construction method, as long as some principles are taken into account. The increasing number of accidents on NATM construction sites can be attributed to several reasons: application in difficult ground, improper application, shortcomings of the NATM, insufficient control, over-confidence in the method and more open reporting of failures. However, it cannot be deduced that NATM is less safe compared with other methods. Tunnel construction sites can hardly be

---

[11] ICE design and practice guide. Sprayed concrete linings (NATM) for tunnels in soft ground. Thomas Telford, London, 1996

[12] see *Tunnels & Tunnelling*, July 1995, 17-18, and H. Lutz: Driving the Römerberg tunnel given slight Overburden, *Tunnel* 4, 1995, 18-21

[13] personal communication

compared with each other as each has its own typical conditions. Limited existing investigations reveal a comparable frequency of accidents in NATM and non-NATM sites.

Among others, the following conclusions are drawn:

Reduction of risk: Daylight collapses in urban areas can have grave consequences. To minimise the risk one should, if necessary, change the tunnel alignment to avoid sensitive ground. Other measures are to relocate bus stops and traffic lights and hindering access to endangered sites.[14]

Qualified personnel: NATM requires the in situ fabrication of shotcrete, which is a highly complex procedure. To assure safety, only trained personnel should be employed, no training phase during construction must be allowed for.

Management: Poor site management is a main reason for accidents. Good management contributes to:
- coping with unforeseen events
- proper application of the observational method
- elimination of human errors.

It is important to inform the involved persons of hazards. Compatibility and cooperation of the teams should be assured.

Collapses: Most of the inrushes in NATM headings occur in the unsupported face. Therefore, a sufficient stand-up time is necessary. However, observations are of little use if the collapse is unannounced.[15] A feature of the NATM is that it is unable to provide a support for a suddenly destabilised face. The probability of a daylight collapse in shallow tunnels is particularly high if watersaturated permeable layers of low strength are encountered. In absence of groundwater, the inrushing earth masses form a heap that prevents further soil inrush. In most cases, there is sufficient warning for the personnel to escape. The largest hazard for persons is due to blocks falling from the unsupported face. Reports on such accidents are rare, but there is some indication that on average one in 15 inrushes causes injury.

Geology: Bad geologic conditions are often blamed for inrushes. In particular, unexpected erosion structures, such as lenses or old wells filled with watersaturated cohesionless material, can be troublesome. Therefore, a forwards exploration and a thorough geological record of the face are recommended.

Observational method: This method requires:

---

[14] see also W. Schiele: Findings from the Underground Shield Drive for the Munich Underground Lot 1 West 5, *Tunnel* 6/1996, 23-30

[15] In this context the law of the Japanese seismologist K. Mogi should be mentioned, according to which the fracture process strongly depends on the degree of heterogeneity of materials: the more heterogeneous a material is, the more warnings one gets before collapse (cited in D. SORNETTE: Critical Phenomena in Natural Sciences, Springer, 2000)

1. Determination of acceptable limits for the behaviour of a construction
2. Verification that these limits will (with sufficient probability) not be exceeded
3. Establishing a monitoring programme that gives sufficient warning of whether these limits are kept
4. Providing measures for the case that these limits are exceeded.

To date, items No. 1 and 2 have not been considered in a convincing way by the NATM literature, i.e. the decision of which deformations are acceptable is left to the experience and intuition of the engineer in charge.[16] Contemporary measuring programmes usually produce an overwhelming amount of data which are very difficult to grasp. An appropriate processing and graphic representation of the data is therefore highly advisable.

Protection measures: The following measures can be applied to increase the stability of the excavated cavity:

- heap
- sidewall drift
- elephant foots
- anchors in the crown
- forepoling
- drainage and pressure relief
- temporary ring closure
- thicker shotcrete lining
- larger crown curvature
- additional ribs
- compressed air support
- reduction of advance step
- reduction of partial face cross sections
- earlier construction of the cast concrete lining.

---

[16]K. Kovári and P. Lunardi point to this shortage of NATM (On the observational method in tunnelling. Proceedings of GeoEng 2000, Melbourne, Australia, 2000, Vol. 1, 692-707). However, their explanatory statement *"NATM can be disregarded as an observational method"* because *"Pacher's concept violates the fundamental principles of the conservation of energy"* is not tractable.

8

Management of groundwater

In this chapter particular attention is paid to the groundwater flow within rock, as the percolation of soil is rather well-known from textbooks on soil mechanics.

8.1 Flow within rock

8.1.1 Porosity of rock

Unjointed rock is porous, in exactly the same way as soil. But usually it has a much smaller porosity (ratio of the volume of voids to the total volume). The porosity of magmatic and metamorphic rocks is rarely larger than 2%. For sandstones the porosity is much smaller than for sand and usually has values between 1 and 5%. For sediment slates it varies between 5 and 20%, and for soft limestones between 20 and 50%. The porosity of unjointed rock is called primary porosity. In jointed rocks also the secondary porosity, which represents the volume of the open joints, must be considered.

The voids (pores) of a rock contain various fluids: Gases (like air, methane etc.) and liquids (water, oil). The pore water is also called groundwater. One part of the pore water is electro-chemically bound to the minerals. The remainder is mobile. Pore fluids can be important economic goods, oil and drinking water come to mind. The Karwendel mountain (northern Tyrol), for instance, is one of the most important reservoirs for drinking water in Central Europe.

The pore water can dissolve minerals of the adjacent rock. If it is not flowing, then the dissolution process stops as soon as the saturation concentration is reached. However, if fresh unsaturated water is constantly supplied, then the dissolution does not stop. Thus, cavities (so-called karst) can develop, which can endanger the stability of buildings[1] and cause water inrushes in

---

[1]In 1969 three pillars of a bridge in Florida disappeared into karst cavities, the bridge collapsed and there was one casualty. In 1970 a 12,000 m^3 cavity has

178   8 Management of groundwater

tunnelling, Fig. 8.1. The largest solubility is exhibited by the evaporites[2] rock salt (NaCl), gypsum (CaSO$_4$· 2H$_2$O) and anhydrite (CaSO$_4$). These are followed by limestone (CaCO$_3$) and dolomite, which is a mixture of calcium and magnesium carbonates. Even quartz is water soluble: A gap in quartzite can grow by ca 0.4 mm in 100,000 years due to dissolution.

Fig. 8.1. Karst cavity encountered at the excavation of the Rollenberg tunnel[3]

Pore water can physico-chemically affect the rock and can cause some minerals to swell. By altering the surface tension it can favour the propagation of cracks. In particular, slates and shales can disintegrate (slake) at contact with water. This is either due to an increase of the pressure of enclosed air, caused by the penetration of water, or to osmotic swelling of the clay minerals.[4]

8.1.2 Pore pressure

The pressure p of groundwater is also expressed as pressure height p/γ_w. It can be determined by standpipes, also called piezometers (open standpipe piezometers). These are placed into boreholes. At the depth where the measurement is taken, they must be permeable (slit or porous stone, surrounded by filter sand). The annular gaps above and below the location of measuring are sealed with clay or cement. The water rises to a height p/γ_w over the measurement location, and the water level in the pipe can be determined

---

been discovered in 2 m depth underneath the runway of the airport of Palermo (R. E. Goodman, Engineering Geology, John Wiley & Sons, 1993)

[2]These are rocks formed by evaporation.

[3]IBW – Engineering structures, No. 3, 12/86, edited by G. Prommersberger

[4]R.W. Seedsman, Characterizing Clay Shales. *In:* Comprehensive Rock Engineering, Volume 3, 131-165, Pergamon, 1993

with a water level probe. Since the water volume in the standpipe is considerable, this measuring procedure takes longer in impermeable ground. The water pressure p can be measured faster with electrical pressure transducers or with pneumatic sensors.[5]

For water-saturated pores, another issue is whether and how the pore pressure p impairs the rock strength. As in soil mechanics, the so-called effective stress σ' (or σ'_{ij}) can be defined with the help of the total stress σ (resp. σ_{ij}) and the pore pressure p:

$$\sigma' := \sigma - p \; ; \qquad \sigma'_{ij} := \sigma_{ij} - p\delta_{ij} \tag{8.1}$$

The principle of effective stresses, according to which deformation and strength of soil are governed by the effective stresses exclusively, also applies to rock.[6] This can be examined, e.g., with the triaxial test conducted with unjacketed rock samples: The curve of $(\sigma_1 - \sigma_2)$ plotted over the strain ε_1 remains the same under various cell pressures. A condition for this behaviour is that the pores are connected and that the deformation rate is sufficiently small, so that the sample can drain.

It should be taken into account that the position of the ground water table can range within wide limits, especially in jointed rock. Therefore, related statements should be treated with caution.

8.1.3 Permeability of rock

When the gradient of energy head $h = p/\gamma_w + z$ does not vanish, i.e. if grad $h \neq \mathbf{0}$, the groundwater flows. Considering one spatial direction x only, the superficial velocity v (i.e. the flow per unit of total cross-section area) amounts to $v = -Ki$ with $i := \frac{\partial h}{\partial x}$ (DARCY's law). K is the so-called hydraulic conductivity and is sometimes also called permeability. However, since K depends on the viscosity μ and density ρ of the fluid as well as on the gravity g, it is advisable to introduce the material-independent quantity

$$k := K \frac{\mu}{\rho g}$$

and call *this one* permeability. k only depends on the geometry of the pore system (size, tortuosity). The unit of k is m^2.[7] The product Kd, where d is the thickness of the aquifer (=permeable layer) is called transmissivity.

With regard to the primary porosity the hydraulic conductivity of rock is quite small:

---

[5] J. A. Franklin, M.B. Dusseault, *Rock Engineering*, Mc. Graw-Hill, 1990.

[6] S. K. Garg, A. Nur: Effective Stress Laws for Fluid-Saturated Porous Rock. *Journal of Geophysical Research*, Volume 78, No. 26, 1973, 5911-5921

[7] Petroleum engineers use the unit "darcy" ($\approx 10^{-8}$ cm^2).

Rock	K (m/s)
Slates	$10^{-12} - 10^{-9}$
Limestone	$< 10^{-7}$
Hard coal	$10^{-6} - 10^{-4}$
Magmatic and metamorphic rocks	$10^{-12} - 10^{-11}$

Laminar flow within the joints can be regarded as COUETTE-flow. If the joint system consists of parallel joints with a spacing s and an opening width b, then the seepage parallel to the joins is governed by the permeability

$$k = b^3/(12s)$$

E.g., for $s = 1$ m and $b = 0.1$ mm is $K \approx 10^{-6}$ m/s; for $s = 1$ m and $b = 1$ mm is $K \approx 10^{-3}$ m/s. The cubic law ($K \propto b^3$) applies down to joint apertures of 10 μm. If the rock mass is crossed by joints, one can regard them as 'smeared' (lumped) and use the tensorial relationship

$$v_i = -K_{ij}\frac{\partial h}{\partial x_j}$$

where

$$K_{ij} = \frac{\rho g}{\mu} \frac{b^3}{12s}(\delta_{ij} - n_i n_j) \ .$$

Herein, δ_{ij} is the KRONECKER-symbol and n_i is the unit vector normal to the joints.

The hydraulic conductivity is determined on rock samples in the laboratory with the same methods as for soil, i.e. with tests of constant (for $K \geq 10^{-6}$ m/s) and falling pressure height. Pressure pulses can be applied to the one side of impermeable ($K \leq 10^{-10}$ m/s) samples whilst on the other side of the sample the decrease of pressure is being measured. One can also let gas (e.g. nitrogen) flow through the rock sample, which permits a fast determination of k.

In the field, the hydraulic conductivity can be estimated with various methods:

Standpipe tests: The water level is measured in a standpipe ($\varnothing\, d$) which has hydraulic contact to the rock over a length l. Switching the feeding pump on and off, one observes how the water level in the standpipe settles down to its original location. If h_1 and h_2 are the heights at the beginning and at the end of the time interval t, then K can be obtained as

$$K \approx M\frac{d^2}{lt}\ln\left(\frac{h_1}{h_2}\right)$$

The coefficient M depends, among other quantities, on l/d (d is the borehole diameter) and is nearly equal to 1.

Packer tests:[8] In a borehole ($\varnothing\,d$) a test section of the length l (about 3 to 6 m long) is delimited by 2 inflatable packers. In order to keep the hydraulic height in the test section constant at a value h_1, water must be pumped into the borehole at a flow rate Q. At a horizontal distance r from the borehole the hydraulic height h_2 is measured in another borehole. Then the hydraulic conductivity results to

$$K = \frac{Q}{2\pi l(h_1 - h_2)} \ln\left(\frac{2r}{d}\right) \qquad (8.2)$$

Usually, a test is carried out with only one borehole. Water is pumped in with a given pressure p. For each p-value[9] a steady discharge Q has to be reached (which cannot always be achieved). Thus, p-Q-diagrams are obtained, which may have various forms and, therefore, need appropriate interpretation.[10]

In the so-called LUGEON-test, water is pumped in into the entrance range between two packers with a pressure of 1 MPa (= 10 bar). Q (in litres per minute and per meter of the test section) is called the Lugeon value. 1 Lugeon corresponds to a hydraulic conductivity of ca 10^{-7} m/s. The LUGEON-test was originally conceived as a criterion for groutability (with is considered as given for $Q > 1$ Lugeon). Pressures higher than 1 MPa are critical because they can cause expansion of joints.

Strictly speaking, the standpipe and packer tests do not allow the determination of K, because the hydraulic gradient is not known (except from Equ. 8.2). Their main purpose is to indicate 'impermeable' rock (for $Q < 1$ Lugeon). They can also be used to assess the efficiency of grouting (by comparing the flow rates before and after grouting).

8.2 Inflow in the construction phase

Heading inflows occur when a water-bearing zone is penetrated during tunnelling. Appropriate resources (pumps etc.) must be available on site, because inflows are very difficult to predict. Water inrush can be critical, especially if the tunnel is headed downhill or starting from a shaft. The flow rate can be very high (cases with more than 1,000 l/s are reported), but usually slows down quickly as the water stored in the rock is depleted. If not, one possible countermeasure is to grout the water carrying joints using cement mortars

---

[8]ISRM, Commission on Rock Grouting, Final report 1995

[9]p is not the pressure at the pump but the over-pressure applied at the entrance into the rock.

[10]F.K. Ewert, 70 Jahre Erfahrungen mit WD-Versuchen – wozu sind sie nützlich? *Geotechnik* **27** (2004), Nr. 1, 13-23

with up to 10% sodium silicate or polyurethan foams. A successful grouting may require the reduction of the water inrush by pressure relief (achieved with draining boreholes or with diaphragm walls).

During heading of the Tecolate pressure tunnel in California water inrushes at the face of up to 580 l/s were encountered. The water had a temperature of 40° C. A source with a supply of 180 l/s could not be plugged in with grouting and the heading had to be stopped for 16 months. Also the construction of the road tunnel at Füssen was considerably delayed in 1996/97 by water inrush.

During construction of the 8.6 km long twin tube Hallandsås tunnel of the Swedish rail, water inrushes up to $15\,\mathrm{m}^3/\mathrm{min}$ were encountered.[11] Grouting with cement did not help, therefore a chemical product based on acrylamide was grouted. Usually, this toxic substance is bonded and therefore harmless. In this case, however, the ground water velocity was too high and the bonding was incomplete. As a result, the environment was contaminated. Within an area of 10 km^2 all agricultural products were spoilt and 24 wells of drinking water had to be abandoned.

Fig. 8.2. Water inrush. Left: Füssen tunnel[12]; right: Simplon tunnel[13]

Clearly, groundwater and other waters must be withdrawn from the tunnel. In case of water-sensitive rock (e.g. rock containing dispersive clay minerals), the contact with water should be minimised.

The water ingress into tunnels can be measured by dams in the tunnel floor and an overflow V-notch. The acceptable limits of water ingress depend on the excavation method (e.g. 2.0-2.5 m^3/m for TBM and ≤ 0.5 m^3/m for drill & blast).

---

[11] *Tunnels and Tunnelling*, November 1997

[12] ÖSTU Stettin Hoch- und Tiefbau GmbH

[13] Historische Alpendurchstiche in der Schweiz, Gesellschaft für Ingenieurbaukunst, Band 2, 1996, ISBN 3-7266-0029-9

8.3 To drain or to seal?

Tunnels below the groundwater table can be either sealed or drained. Sealed tunnels do not influence the groundwater but their lining has to support the full water pressure. This is technically possible down to depths of 60 m below the groundwater table. The hydrostatic pressure can be relieved if the tunnel is drained. Note, however, that also a drained tunnel can be sealed (waterproofed) in the sense that the water is guided to the longitudinal drainage pipes but is not allowed to enter the tunnel interior.

Drainage can influence the surrounding groundwater considerably. An intermediate solution is the so-called partial pressure relief: The pressure is limited by special valves. This solution reduces the disturbance of the groundwater. The condition 'sealed' can be easily changed to 'drained' by the appearance of cracks in the lining or by opening a valve in the circumferential drainage (Fig. 8.7). Note, however, that a redistribution of pore pressure and the corresponding discharge of pore water will need some time, which can be considerable in case of soil/rock with low permeability.

8.4 Drainage

Drainage affects the distribution of hydraulic head by attracting groundwater and relieving the lining from hydrostatic pressure. The groundwater is then collected and appropriately discharged. This needs to be achieved in a permanent way and maintenance must always be possible. It should be added that there are tunnels, exclusively devoted to drainage, e.g. to stabilise a slope. The drainage path of the groundwater is as follows:

1. The groundwater penetrates the shotcrete shell through fissures and ad hoc bored holes (to enhance the mobility of groundwater towards the drainage system, radial boreholes may be drilled into the ground).
2. The interface drainage systems consist either of fleece (for low discharge) or of composite geosynthetics or air-gap membranes (for high discharge) and are placed in the interspace between shotcrete and concrete lining. There are many types of geosynthetics (so-called geospacers), designed to provide a stable interspace for water discharge. Higher water discharges emanating from local sources are caught with separate pipes and guided to the drainage pipes (Fig. 8.3).

184   8 Management of groundwater

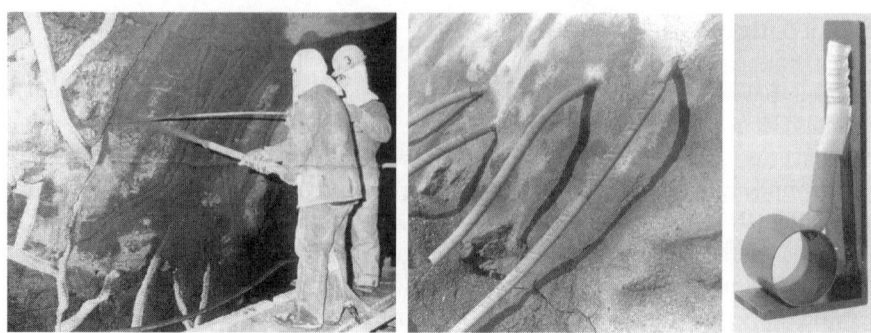

Fig. 8.3. Draining pipes in shotcrete, connection to main drainage

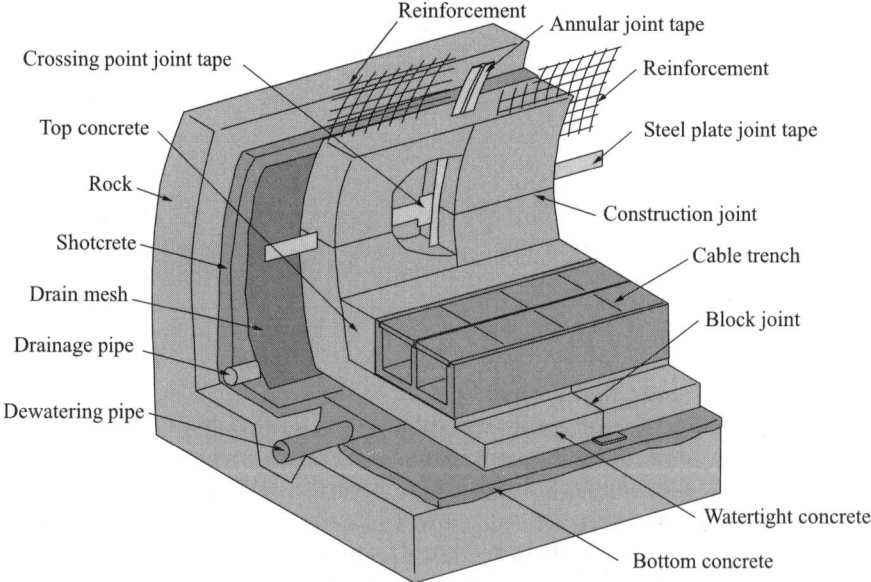

Fig. 8.4. Drainage of a rail tunnel[14]

This area drainage receives the groundwater flowing to the crown and the sides of the tunnel and guides it to the longitudinal drainage pipes, which are installed at the merges of the sides with the invert (Fig. 8.4). The interface drainage and the drainage pipes are embedded within granular filters ('dry pack'), the pipes are perforated in their upper parts (Fig. 8.5).

---

[14]G. Prommersberger, H. Schmidt, Planning and execution of the tunnels Markstein, Nebenweg and Pulverdinger. In: Engineering Structures, DB New Railway Line Mannheim-Stuttgart, No. 5, 9/89, 46-80

3. Transversal slots (Fig. 8.11) guide the groundwater from the side pipes to the main collector which is placed underneath the carriageway.
4. The groundwater flowing into the tunnel invert is collected in a similar way, i.e. with a granular filter and a perforated pipe in longitudinal direction, placed in the deepest part of the invert.

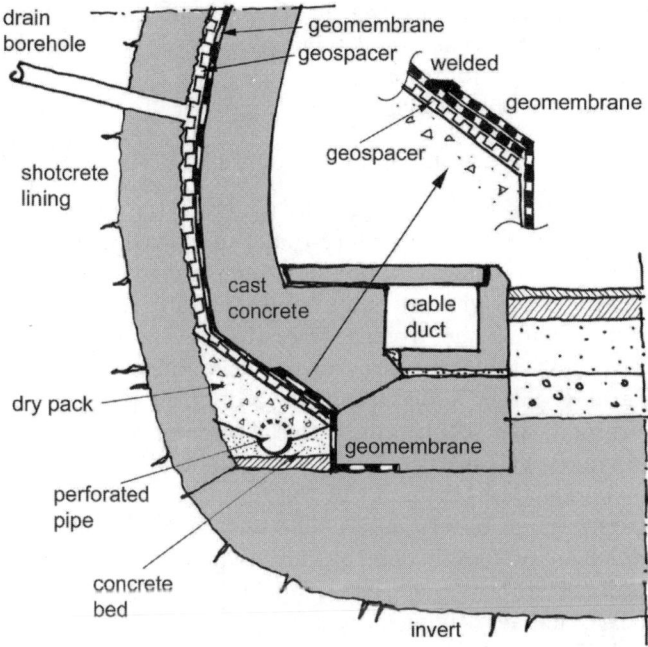

Fig. 8.5. Example for drainage into the side pipe

Polluted water and other liquids are drained from the carriageway into longitudinal pipes. The access to these pipes is given either via gullies or slots. An aim of this drainage system is to collect leaking inflammable fluids in case of accidents. The polluted water is temporarily piped into reservoirs outside the tunnel to be treated elsewhere later.

The longitudinal pipes underneath the carriageway are accessible via shafts (manholes) for inspection and cleaning. The longitudinal pipes along the tunnel sides are accessible via niches. The cleaning is performed with pressurised water (up to 150 bar at the nozzle) at a discharge of up to 500 l/min. The details of cleaning should be contained in a manual of maintenance.

8 Management of groundwater

The following figures may serve as examples for the design of a tunnel drainage:[15]

- Drainage fleece for the shotcrete-concrete interface, drainage capacity 7 - 14 l/(m·h) at $p = 200$ kPa, $i = 1$.
- Composite geosynthetics, drainage capacity > 14 l/(m·h) at $p = 200$ kPa, $i = 1$.
- Air gap membranes, array of 6 - 20 mm high naps, membrane thickness $\geq$ 1.2 mm, compressive strength ≥ 200 kN/m^2 (for $\varepsilon < 20\%$) .
- Drainage pipes, external $\varnothing \geq 200$ mm
- Pipe perforation, $\varnothing$ 10...15 mm
- Longitudinal slots 5...10 mm

The following items are of interest with respect to drainage:

Clogging: Drainage can be clogged by sintering (i.e. precipitation of carbonates) and/or fouling. Calcium (lime), dissolved in groundwater, can precipitate due to changes of pressure, temperature, pH, entrance of oxygen and interaction with cement. Countermeasures are (apart from regular cleaning) siphoning the drainage lines (to prevent contact with air) and use of hardness stabilisers. Asparangine acid impedes the precipitation of carbonates. It can be added as a fluid to drainage pipes (if the water discharge is more than 1-2 l/s) or it can be placed into the drainage system as solid cubes. This additive reduces the amount of sinter and renders it softer. Maintenance (cleaning) of the drainage pipes is very costly and inhibits the operation of tunnels. The same holds for the related inspections which are preferably done by video scanning. The cleaning of clogged pipes is done with water jets, as explained before. Alternatively, chain and rope flails or impact drilling cutters can be used.[16] The German Rail (DB) had to pay up to 60 € per m tunnel drainage annually for cleaning the drainage. Therefore, German Rail tunnels are now generally sealed.

Temporary Drainage: Localised outflow of water impedes the application of shotcrete or waterproofing geomembranes. Therefore, a temporary drainage must be provided for to catch and divert the water. This can be achieved, e.g., with semi-cylindrical flexible pipes or with strips (Fig. 8.6).[17]

Porous concrete: It has a void ratio ≥ 15 % and, consequently, a high permeability and a reduced strength. To obtain such a high porosity, gap graded aggregates, recommendably siliceous ones, should be used. The

---

[15] Austrian standard 'Richtlinie Ausbildung von Tunnelentwässerungen', 19.12.2002

[16] G. Girmscheid et al, Versinterung von Tunneldrainagen – Empfehlungen für die Instandhaltung von Tunneln. *Bauingenieur*, **78**, Dez. 2003, 562-570

[17] AFTES Guidelines on waterproofing and drainage of underground structures, May 2000 (http://www.aftes.asso.fr)

cement content should be ≥ 350 kg/m^3 and its chemical composition should not enhance clogging.

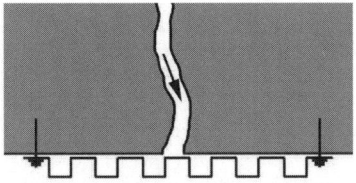

Fig. 8.6. Temporary drainage with panel strips

8.5 Water ingress into a drained circular tunnel

Let us consider a drained tunnel with a circular cross section. The surrounding ground has the isotropic permeability K. If the inflow into the tunnel is fed by a large reservoir, then drainage does not affect the position of the ground water table and a steady groundwater flow is achieved (Fig. 8.7).

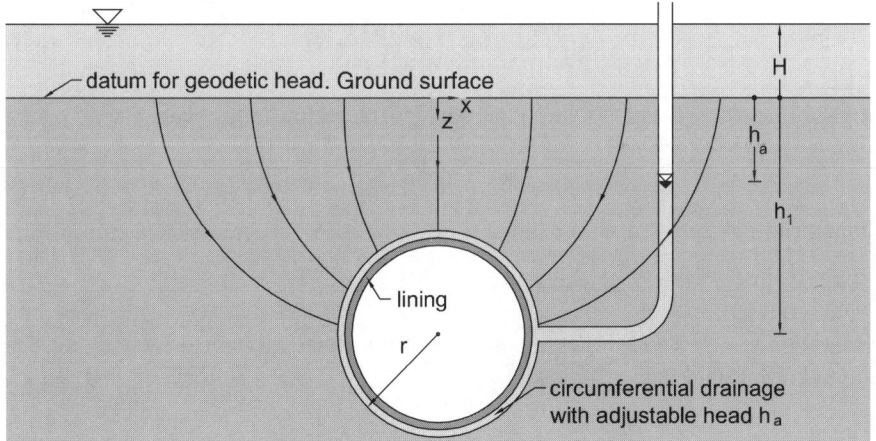

Fig. 8.7. Ground water inflow into a drained tunnel (radius r)

To obtain computational estimations for the two-dimensional problem we need to know the energy head h as function of x and z. The function $h(x,z)$ must have constant values (i) at the ground surface $z = 0$, and (ii) along the tunnel circumference.[18] There is a widespread solution which is based on a function

---

[18]We assume that the circumferential drainage of the tunnel is a closed hydraulic system, the energy head of which is kept to the constant value h_a. If, however, in

that fulfils only the first requirement, whereas the second one is only approximately fulfilled.[19] Thereby, the degree of approximation increases with increasing depth of the tunnel, i.e. for $h_1 \gg r$. This function reads:

$$h(x,z) = c \log \frac{x^2 + (z+h_1)^2}{x^2 + (z-h_1)^2} + H$$

with c = const. On the tunnel circumference $x^2 + (z-h_1)^2 = r^2$ we have $h(x,z) = c \log(1 + 4(zh_1)/r^2) + H$. Instead of $h_1 - r \leq z \leq h_1 + r$ we can also write $z = h_1 - r + 2\xi r$ mit $0 \leq \xi \leq 1$. Thus, for the tunnel circumference it is obtained:

$$h(x,z) = c \log \left(1 + 4 \frac{h_1^2}{r^2} + (2\xi - 1)\frac{h_1}{r}\right) + H \ .$$

and for deep tunnels ($h_1 \gg r$) is approximately obtained

$$h \approx c \log \left(4 \frac{h_1^2}{r_0^2}\right) + H = \text{const}$$

For a close neighbourhood $\rho > r$ around the tunnel we can write (with ρ being the radial coordinate):

$$h(\rho) \approx c \log \left(4 \frac{h_1^2}{\rho^2}\right) + H \quad (8.3)$$

and obtain with DARCY's law $\mathbf{v} = -K\nabla h$ the radial inflow velocity

$$v_r = -2Kc/r.$$

Thus, the volume q of water flowing into one meter of tunnel length reads:

$$q = -2\pi r v_r = -4\pi K c.$$

Hence $c = -q/(4\pi K)$. Taking into account the energy head h_a along the tunnel circumference, it follows from (8.3)

$$h_a = -\frac{q}{4\pi K} \log \left(4 \frac{h_1^2}{r^2}\right) + H$$

or

$$q = \frac{2\pi K}{\log(2h_1/r)}(H - h_a) \ . \quad (8.4)$$

---

the circumferential tunnel drainage prevails atmospheric pressure, then this circumference is not an equipotential curve with constant energy head.

[19] P. Ya. Polubarinova-Kochina: Theory of Ground Water Movement. Princeton University Press 1962, p. 374

Despite the existence of a rigorous solution (Equ. 8.5), the approximate solution given by Equ. 8.4 is widespread in tunnelling. To take into account the irregular and anisotropic permeability of hard rock, the obtained value has to be reduced by HEUER's empirical factor 1/8.[20] The inflow q depends linearly on the permeability K. In evaluating q, using field measurements of the permeability (obtained e.g. with packer tests), one should take into account the exponential scatter of the measured K-values. Following RAYMER, a log-normal distribution applies to the statistics. The K values at the high end tail of the distribution are decisive for the water inflow into the tunnel. The rigorous solution (Appendix D) is not restricted to deep tunnels and reads

$$q = \frac{\pi K(H - h_a)}{\log\left(\dfrac{r}{h - \sqrt{h^2 - r^2}}\right)} \quad . \tag{8.5}$$

There is also a solution for the case of permeability decreasing with depth.[21] The resulting expression is, however, quite complicated. In view of the always existing inhomogeneity and the large scattering of measurements of the permeability, the application of this solution does not appear to be justified.

8.5.1 Seepage force

An undrained lining is exposed to the full hydrostatic water pressure, whereas for a drained one the pressure is obtained from the constant energy head h_a. The ground pressure upon the lining is obtained using the buoyant (submerged) unit weight γ'. For the drained case, however, the seepage force should be taken into account. It acts (as a volume force) upon the ground and thus increases the rock/earth pressure on the lining.

To demonstrate the effect of the seepage force, the approximate solution for the pressure upon the crown, presented in chapter 16, should be consulted. According to Equ. 16.11 the pressure exerted by the ground upon the crown of an undrained tunnel (Fig. 8.8) reads:

$$p_c = h \frac{\gamma' - \dfrac{c}{r} \cdot \dfrac{\cos\varphi}{1 - \sin\varphi}}{1 + \dfrac{h}{r} \cdot \dfrac{\sin\varphi}{1 - \sin\varphi}} + \gamma_w(H + h) \quad . \tag{8.6}$$

---

[20] cited in J.H. Raymer, Predicting groundwater inflow into hard-rock tunnels: Estimating the high-end of the premeability distribution. Proceedings of the Rapid Excavation and Tunnelling Conference, Society for Mining, Metallurgy and Exploration, Inc., 2001

[21] L. Zhang, J. A. Franklin, Prediction of Water Flow into Rock Tunnels: an Analytical Solution Assuming a Hydraulic Conductivity Gradient. *Int. J. Rock Mech. Min. Sci. & Geomech. Abstr.*, 30, No 1, 37-46, 1993

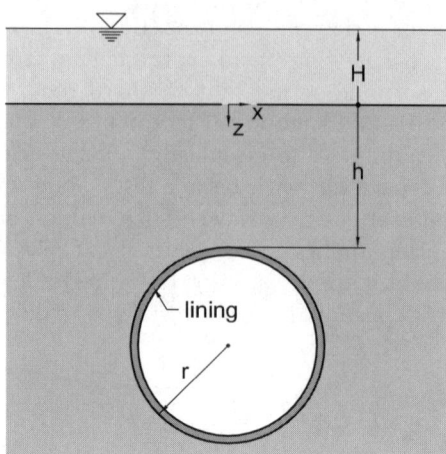

Fig. 8.8. Tunnel below groundwater level

The second term in Equ. 8.6 represents the water pressure upon the crown. For a drained tunnel, this term drops out and the unit weight γ' of the ground has to be increased by the seepage force $i\gamma_w$. Assuming a linear distribution of water pressure in the vicinity of the crown, we obtain the hydraulic gradient head

$$i = \frac{H+h}{h}$$

Consequently, the ground pressure acting upon the crown of a drained tunnel reads

$$p_c = h \frac{\gamma' - \dfrac{c}{r}\dfrac{\cos\varphi}{1-\sin\varphi}}{1 + \dfrac{h}{r}\dfrac{\sin\varphi}{1-\sin\varphi}} + \gamma_w \frac{H+h}{1 + \dfrac{h}{r}\dfrac{\sin\varphi}{1-\sin\varphi}} \; .$$

8.6 Influence of drainage

Drainage changes the initial distribution of groundwater and may, thus, have a severe environmental impact. Sources may run dry and settlements may occur. E.g., the drainage of the Rawil tunnel in Switzerland caused a settlement of 12 cm of the Zeuzier arch dam. To avoid failure of the dam, the water table in the reservoir had to be lowered considerably. Therefore, drainage may be prohibited by environmental protection agencies. Even drainage during construction can be prohibited, in which case the inrushing water has to be fed back into the ground. A previous treatment may be necessary if the water has been polluted in the meanwhile.

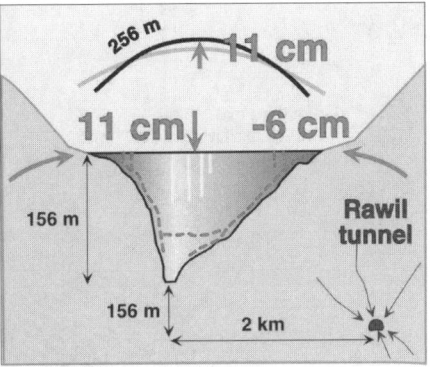

Fig. 8.9. Zeuzier arch dam (Switzerland)

Since heading and operation of (drained) tunnels may influence water sources in the surrounding area, a perpetuation of evidence should be provided for. This should be done well in advance of the heading, as it is time consuming.[22] To this end, repeated observations of the discharges of sources have to be carried out and electric conductivity, chemical and bacteriological composition of the source waters have to be recorded. These results are then compared with the corresponding values of the water rushing into the tunnel. E.g. the Tauernkraftwerke (Tauern water power plants) have observed 724 sources over 40 years. Only 6% of them were influenced by tunnelling.

8.7 Sealing (waterproofing)

Above the groundwater table, a tunnel has to be protected against downwards percolating water. This is achieved with a so-called umbrella waterproofing (Fig. 8.12). Below the groundwater table, groundwater is pressurized so that an all-embracing waterproofing must be applied.
Up to a water pressure of 3 bar, water-tight concrete can be used,[23] above 3 bar, and up to approx. 15 bar, watertight membranes should be used in addition (Fig. 8.13).[24] The sealing membrane is fixed between the outer and

---

[22]P. Steyrer, P. Ganahl, R. Gerstner, Stollenvortrieb und Quellbeeinflussung, *Felsbau* **12** (1994), Nr. 6, 474-480
[23]RVS 9.32, Blatt 4, Abs. 3.6.3.2

192   8 Management of groundwater

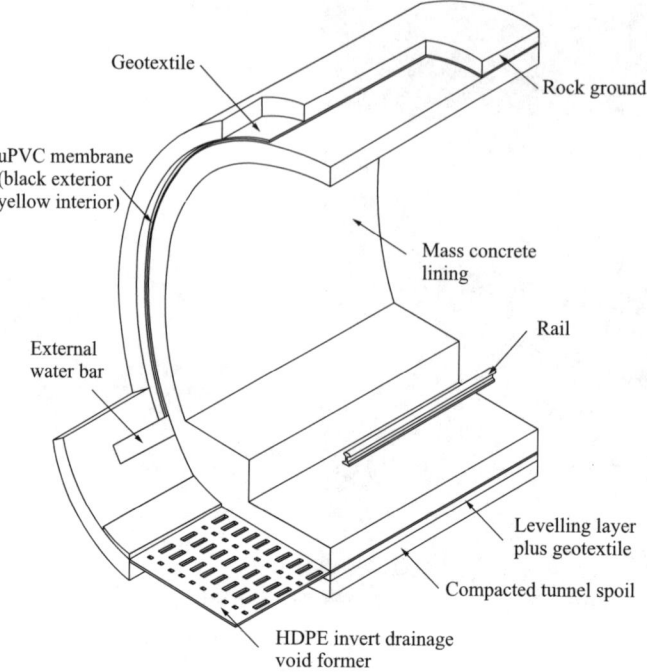

Fig. 8.10. Example for waterproofing a rail tunnel

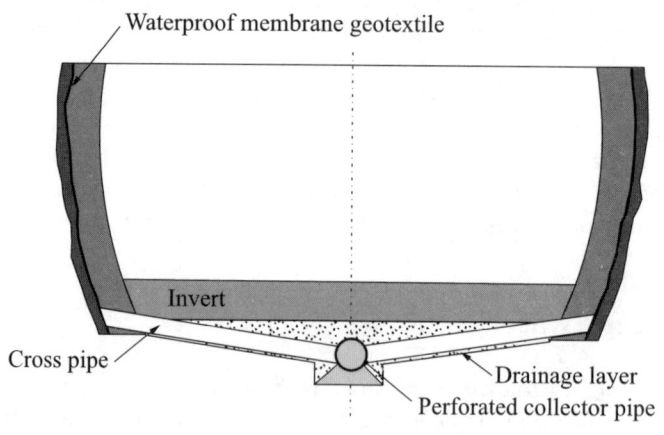

Fig. 8.11. Invert drainage

---

[24]Empfehlungen Doppeldichtung Tunnel – EDT, Deutsche Gesellschaft für Geotechnik, Ernst & Sohn, 1997

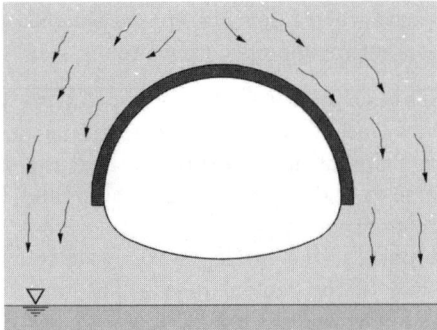

Fig. 8.12. Umbrella waterproofing

inner linings. For pressures higher than 15 bar in permeable rock, a sealing ring around the tunnel has to be obtained with advance grouting (see Sect. 6.6). When advance grouting proves to be insufficient, post-grouting can be applied. At any rate, advance grouting should be tried first, as it is much more efficient and economic than post-grouting.

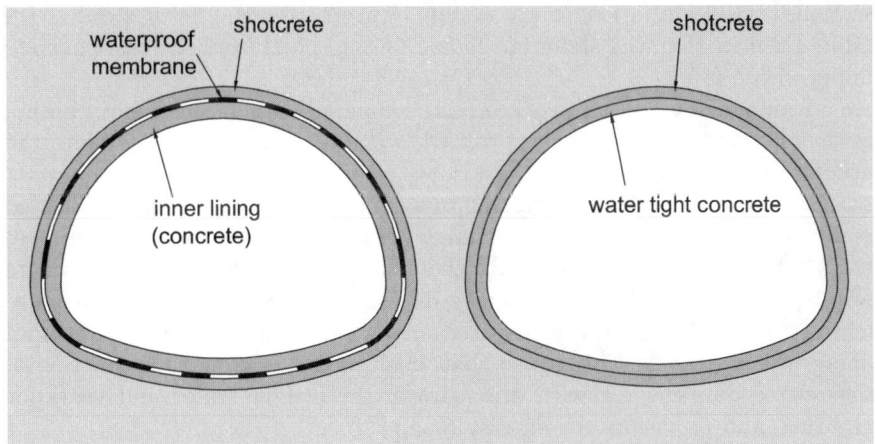

Fig. 8.13. Principles of tunnel sealing

A properly fabricated cast concrete is watertight if some conditions, regarding the grain size distribution of the aggregates and the water content, are adhered to. In this case, its pores are not interconnected and the concrete can be considered as impermeable (for thicknesses ≥ 30 cm). Its impermeability can only be reduced by fissures, which may appear due to too large tensile stresses, temperature gradient, creep and shrinkage.[25] With respect to watertight concrete attention should be paid to:

Hydration heat: This is produced during the setting process. The subsequent cooling may lead to incompatible stressing and, thus, to fissures. Remedies are late dismantling of formwork, heat isolating formwork, cooling of the aggregates and of the mixing water and long (10 - 14 days) moistening of the concrete.

Shrinkage: Only a part of the mixing water is chemically bonded (corresponding to ca 25% of the cement mass). The remaining water occupies the pores. Thus, a low water content helps to keep the porosity small. Shrinkage is due to the evaporation of the free water. Thin parts are more prone to shrinkage, their length reduction corresponds to the one caused by a temperature reduction of 15 - 20°C.

Cement: Portland cement is preferably used for watertight concrete, while blast furnace slag cement produces less heat but needs longer time for setting.

Longitudinal reinforcement reduces the spacing and the width of the fissures but does not avoid them. The only remedy is to keep short sections of concreting. On the other hand, too many joints should be avoided. As a compromise, sections of ca 12-20 m length are usually applied. To avoid fissures due to inhibited contraction, the shotcrete should be separated from the cast concrete lining with a foil.

An advantage of watertight concrete (as compared with synthetic membrane) is that leakages can be easily localised, whereas in case of membranes the leaking water is spread.

It should be noted that the watertightness of tunnels is often over-emphasised and the related measures are overdone. Some droplets of water are in most cases tolerable.[26] What should be avoided (especially in road tunnels) are water puddles, reduction of visibility due to inrushing water, black ice and icicles.[27] It should also be taken into account that, in the end, every waterproof lining will have some leakages so that what counts more is the reduction of the related damage (e.g. with drainages of the leaking water and vents for grouting) and provisions for an easy repair.[28]

---

[25] RVS 9.32.

[26] Leakage rates of, say, 2-40 litres per minute and 100 m tunnel can be permitted, dependent on the usage of the tunnel.

[27] In cold climates and ventilated tunnels even small drips of less than 1 l per minute and 100 m tunnel can be problematic because of freezing.

[28] D. Kirschke: Neue Tendenzen bei der Dränage und Abdichtung bergmännisch aufgefahrener Tunnel. *Bautechnik* 74 (1997), Heft 1, 11-20

8.8 Geosynthetics in tunnelling

Geosynthetics are applied for drainage (geospacers, geocomposite drains) and for waterproofing (geomembranes). Polyester should not be used, as it can be destroyed by hydrolysis in alkaline environment, such as concrete. PVC (polyvinyl chloride) produces hydrochloric acid in case of fire. It may, however, be used if it is covered by concrete lining.

Geomembranes are not mounted directly on the rough shotcrete lining. To protect them against puncturing or tearing, a geotextile or geocomposite protective barrier of 600 to 1,200 g/m^2 is interposed. Geomembrane strips are mounted with synthetic disks and welded with each other. This is done with hot air of 200-300° C, which, however, must not deteriorate the underlying protective barrier.

9

Application of compressed air

The idea to apply compressed air to prevent groundwater from entering into excavated spaces goes back to Sir THOMAS COCHRANE, who obtained a patent in 1830.[1]

Tunnelling under compressed air is connected with the following hazards:[2]

1. Health problems
2. Fire due to increased oxygen concentration (fires ignite more easily, burn more rigorously and are more difficult to extinguish)
3. Blow outs.

Compressed air is mainly applied in loose sandy or silty soils which are headed conventionally or with shield and is also applicable in the cover and cut method:

Shield heading: The face is often supported with pressurised slurry or earth spoil. However, for maintenance one has to enter the excavation chamber. On this, slurry is removed and the support is accomplished by air pressure. Air pressure support can also be permanent within an appropriately closed part of the tunnel. Of course, locks must be provided for. The same applies to pipejacking (Fig. 9.1).

Conventional heading: The tunnel is sealed by a bulkhead and pressurized with air. Air locks permit access through the bulkhead. At the tunnel face, hydrostatic groundwater pressure increases linearly with depth, whereas the air pressure is approximately constant. Consequently, the air pressure can only balance the water pressure at some level, above which the air pressure exceeds the water pressure. Air escapes through the unprotected face and through fissures and weak spots of the shotcrete lining. Of course, the proportion of the latter losses increases with the length of pressurized tunnel. It is very difficult to predict the air losses, which are reported to

---

[1]R. Glossop, The invention and early use of compressed air to exclude water from shafts and tunnels during construction. *Géotechnique* **26**, No. 2, 253-280 (1976)

[2]Changes in the air. *Tunnels & Tunnelling International*, January 2002, 26-29

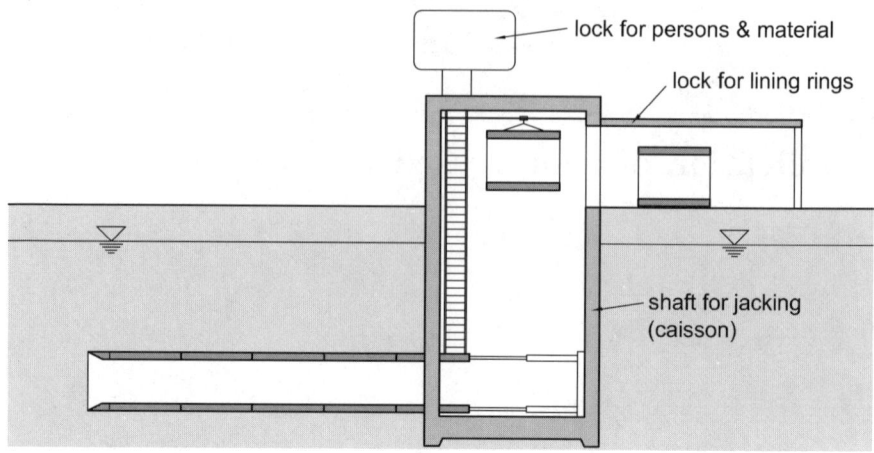

Fig. 9.1. Pipejacking with air pressure and locks at the start shaft.

amount to between 20 and 700 m^3/min. Table 9.1 shows some specific examples. The corresponding costs for compressors range from 10,000 to 200,000 € per month.[3] It should be taken into account that the permeability of partially saturated soil with respect to air increases with time. Compressed air is effective in all ground conditions (including fissured rock), provided the air losses can be controlled. For the U2-subway in Munich, the permeability of the ground was extremely high so that the overburden had to be grouted to avoid extreme air loss.

City	Length (m) driven under compressed air	Air pressure (bar)	Air loss (m^3/min)
Munich	6,961	0.3 - 1.1	25 - 580
Essen	1,330	0.4 - 1.2	52 - 250
Taipei	400	0.8 - 1.4	50 - 180
Siegburg	240	0.6 - 1.2	50 - 450

Table 9.1. Examples of application of compressed air in combination with NATM[4]

Cover and cut: The application of air pressure in the cover and cut tunnel construction in groundwater is shown in Fig. 9.2: The cover is buttressed

---

[3]S. Semprich, Tunnelbau unter Druckluft – ein immer wiederkehrendes Bauverfahren zur Verdrängung des Grundwassers. TA Esslingen, Kolloquium 'Bauen in Boden und Fels', Januar 2002.

[4]S. Semprich and Y. Scheid, Unsaturated flow in a laboratory test for tunnelling under compressed air. 15th Int. Conf. Soil Mech. and Geot. Eng., Istanbul, Balkema, 2001, Vol. 2, 1413-1417.

on two diaphragm walls, subsequently the soil is removed whereas groundwater is kept off by means or air pressure. The needed air supply must be empirically estimated for the proper choice of compressors. The air leakage is composed of the losses at the face, along the tunnel wall and at the locks.

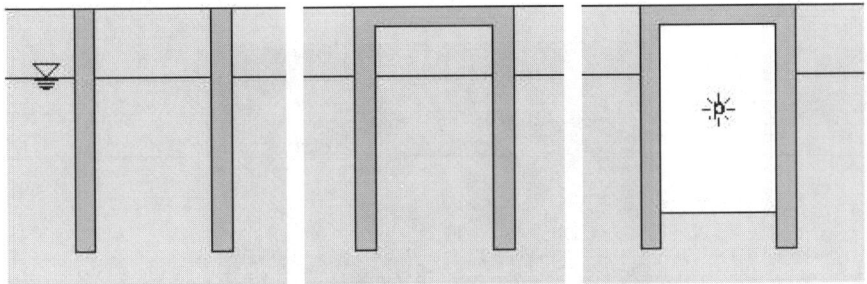

Fig. 9.2. Application of compressed air in cover and cut tunnelling

9.1 Health problems

Increasing the air pressure implies that more air is dissolved into the blood. The surplus oxygen is supplied to the cells, whereas nitrogen remains dissolved and drops out in case of decompression. If the decompression occurs too fast, then bubbles can appear in the blood, the joints and the tissue causing decompression illness, which is accompanied with pain in the joints and can lead to embolism. Therefore, decompression has to occur gradually. The required time increases with pressure and retention period.

In the German Standard[5] tables ('diving tables') indicate the required decompression time. Persons are allowed stay in pressures up to 3.6 atmospheres, their age is limited between 21 and 50 years. Complaints can appear even 12 hours after decompression. The only reasonable treatment is to put the person again into a pressurized chamber. In construction sites with more than 1 atmosphere air pressure a special recompression chamber for ill persons must be provided for. Persons working within pressurized air must always carry a red card with appropriate hints that help to avoid unneeded medical treatments in case of sudden illness. According to Table 9.2, in cases I and II

---

[5]Druckluftverordnung of 4th October 1972 (BGBl. I p. 1909) last change of 19th June 1997 (BGBl. I p. 1384). See also: Work in Compressed Air Regulations 1996 and accompanying guidance document L96 (UK), BS 6164: 2001 'Code of practice for safety in tunnelling in the construction industry', and the draft CEN standard prEN12110 'Airlocks-safety requirements'.

the person must be withdrawn from the lock, whereas in case III a recompression is needed. The decompression time can be reduced by ca 40% with respiration of pure oxygen (with a mask). Note, however, that pure oxygen is toxic in pressures above 1 bar. Another disease, osteonecrosis, is manifested as corroboration of the joints. It can appear many years after the work under compressed air.

Fig. 9.3. Tunnelers in decompression chamber[6]

In the Netherlands exposures to air pressures up to 4.5 bar (in some cases 7 bar) are allowed for maintenance works in slurry shields.[7] At pressures above 3.6 bar, nitrogen narcosis was observed: divers worked slower and made more mistakes. Special gas mixtures had to be inhaled via helmets.
The following air pressure diseases are distinguished:

---

[6] *Österreichische Ingenieur- und Architekten-Zeitschrift (ÖIAZ)*, **142**, 4/1997, p. 247

[7] J. Heijbor, J. van der Hoonaard, F.W.J. van de Linde, The Westerschelde Tunnel, Balkema 2004

	Symptom	appears at
I	pain in the tympanum	increasing pressure
II	rapture of the deep	constant pressure
III	pain in sinus and joints	decreasing pressure

Table 9.2. Symptoms due to compressed air

9.2 Influence on shotcrete

The increased moisture of compressed air accelerates setting. If the adjacent rock is permeable, the shotcrete will be percolated by large amounts of air, which lead to drying and shrinkage and, thus, to reduced strength. Countermeasures are: moistening of fresh shotcrete, additives to reduce its permeability and early sealing the shotcrete surface. If the adjacent rock is relatively impermeable, compressed air is favourable for the shotcrete.

9.3 Blow-outs

If the pressurized air pipes through the soil, it can abruptly escape, thus causing a sudden pressure drop in the tunnel. The pressure drop is accompanied with a bang and the formation of mist and can lead to collapse of the tunnel. It is in most cases announced with increasing air leakage. To avoid such blow-outs it is necessary that the total primary stress at the tunnel crown exceeds the air pressure by 10%. To achieve this, it is sometimes needed to bring in an embankment at the ground surface. Its width should be 6 times the diameter of the tunnel.

10

Subaqueous tunnels

The following possibilities are available for the crossing of water ways[1]:

Ferries: These are slow and have a reduced transport capacity when compared with bridges and tunnels (e.g. the Channel tunnel has shortened the drive from Paris to London from 6 hours to 2 hours and 40 minutes). In addition they are dangerous to other ships and are affected by bad weather.[2]

Stationary or mobile bridges: The following types can be considered:
- floating pontons with movable sections in the centre to allow the passage of ships
- mobile bridges.

Opening and closing of the bridges is time consuming and hinders the traffic.

High bridges: These must be sufficiently high to allow the passage of ships:
10 m for inland navigation,
25 m for coastal navigation,
70 m for open sea navigation.
With a longitudinal slope of 4% for roads and 1.25% for rail, long ramps may be necessary. High bridges may affect the scenery and traffic may be impaired by bad weather (e.g. hurricanes).

Subaqueous tunnels: They can be up to 50% more expensive than bridges but they help to avoid the disadvantages cited above. The following variants are available:
- anchored floating tunnels
- subaqueous bridges erected on piers

---

[1] M. Kretschmer und E. Fliegner: Unterwassertunnel, Ernst und Sohn, 1987

[2] In 1954 a storm caused 5 ferryboats to sink in Japan, 1,430 persons died. Thereupon, the 54 km long Seikan tunnel was built (from 1971 to 1988), that connects the islands Honshu and Hokkaido.

- cut and cover below water level, either within a sheet- or bored pile wall, or constructed with towed and lowered parts, or built as caissons or mined.

Mined tunnels have covers between 30 and 100 m and are, therefore, deeper than tunnels constructed with cut and cover. Therefore, they need longer ramps. Navigation requires a free water depth between 50% (for inland navigation) and 10% (for open sea navigation) of the clearance above sea level.

10.1 Towing and lowering method

Pre-fabricated blocks (Fig. 10.1, 10.2) with lengths between 40 and 160 m are towed and subsequently lowered to their final position. To render them floatable, their front parts are temporarily bulkheaded off. They are designed to withstand loads between 10 und 80 kN/m^2 due to wracks as well as an impact of anchor of 1,300 kN distributed over a surface of 1.5×3 m. A pressure drop of 7.5 bar due to passage of ships should also be taken into account.

Tunnel elements of concrete are waterproofed by means of steel sheets, bitumen or synthetic membranes. Water-tight concrete is increasingly used. If steel is used, corrosion should be taken into account. If the water contains dissolved carbonate, a protective layer is created which inhibits corrosion. Otherwise losses of 0.01 to 0.02 mm steel annually have to be taken into account due to corrosion. The so-called cathodic corrosion protection is based on a galvanic voltage applied between the steel to be protected and an anode made of zinc, magnesium or aluminium.

The stability of the floating tunnel elements has to be checked according to the rules of ship-building. A resonance of the tunnel element and of the water swapping within the ballast tanks should be avoided.

The subsoil investigation is based on sampling from boreholes and sounding. This can be carried out from ships, if the water depths do not exceed 15 m. For depths up to 25 m the investigations can be operated from leg jack-up drilling rigs. Remote controlled vehicles that are capable of sampling and sounding are also applicable on the sea ground.

The depth of the basement must comply with the requirements of navigation, i.e. a minimum free water depth of 15 m must be allowed for. Taking into account a tunnel height of 8 m and the corresponding cover, depths of ca 30 m must be dredged. The surface of the dredged channel should be controlled e.g. with ultrasonic sounding. Before touching the ground the lowered element causes the water to escape from the gap. The accelerated water flow may induce flutes. Therefore, sandy ground should be protected with a gravel cover.

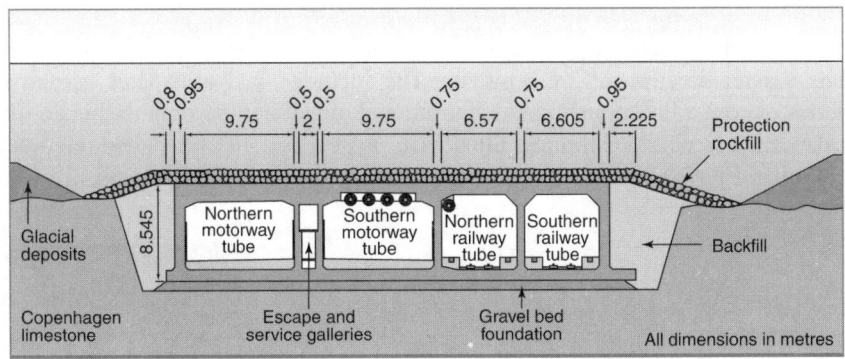

Fig. 10.1. Cross section of the Øresund tunnel[3]

Fig. 10.2. Pre-fabricated tunnel blocks for Øresund tunnel[4]

Due to buoyancy the bottom contact pressure is low (usually between 5 and 10 kN/m^2). A safety factor of 1.1 with respect to buoyancy has to be assured. The following types of bedding can be applied:

Bedding upon gravel: The gravel bed has to be bulldozed.

Underfilled basement: The tunnel element is deposed upon temporary footings. The remaining space is subsequently filled with hydrauliking material (mixture of sand and water). The sand bed obtained is rather loose and, thus, prone to settlement and liquefaction. A remedy is to add cement to the hydrauliking material.

Pile foundation is to be applied in case of weak underground.

---

[3] Busby, J., Marshall, C., Design and construction of the Øresund tunnel, Proceedings of the ICE, Civil Engineering 138, November 2000, 157-166

[4] www.thornvig.dk

10.2 Caissons

Submerged tunnels can be constructed as arrays of caissons. A caisson is a box driven downwards by removing the included soil and, thus, evoking a series of controlled punchings. The removal of soil can be done either in open caissons or in closed chambers filled with pressurized air[5] (so-called pneumatic caissons, Fig. 10.3).

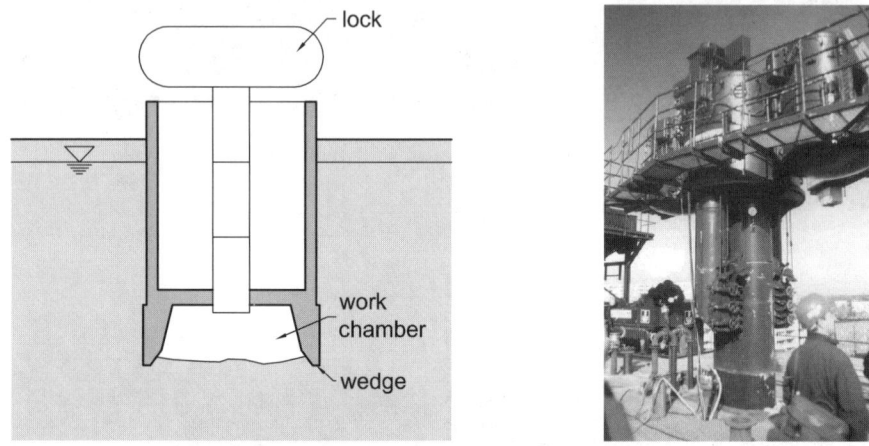

Fig. 10.3. Pneumatic caisson, lock (right)

Fig. 10.4. Work chamber of a pneumatic caisson (Metro Amsterdam)

---

[5]H. Lingenfelser: Senkkästen, Grundbau-Taschenbuch, 4. Auflage, Teil 3, Ernst und Sohn, Berlin 1992

Fig. 10.5. Caisson for Metro Amsterdam: mould for concrete

The working chamber of pneumatic caissons is filled with concrete after arrival at the final depth. Pneumatic caissons have a series of advantages:

- The penetrated soil can be inspected during lowering and obstacles can easily be removed.
- The groundwater is not disturbed
- No vibrations are produced
- Loading the space above the working chamber with soil or water renders lowering always possible.
- The caisson can be built elsewhere (e.g. on an artificial island) and towed to its final position.

An offset of 3 - 10 cm width is provided 3 m above the cutting edge (Fig. 10.6). The resulting gap is filled (or grouted) with bentonite slurry to reduce wall friction (Fig. 10.6). Rough external walls cause an increased wall friction.

The inner wall of the cutting edge should be neither too flat (to enable access to the soil) nor too steep (to avoid a too deep penetration into the soil). The tip of the cutting edge is reinforced with a steel shoe appropriately anchored within the concrete. The clearance within the working chamber should be ca 2.0 - 2.5 m. The ceiling of the working chamber should have a thickness of at least 60 cm to support the loads exerted by the pressurized air and the ballast. An appropriately formed sand heap can serve as formwork for the inner wall (Fig. 10.5). The excavation within the working chamber can be done with the help of water jets. Muck can be pumped away (Fig. 10.4).

By means of a lowering plan it should be assessed that the ballast load is sufficient to punch the caisson (Fig. 10.7).

On this, the driving forces $G+B$ are balanced to the resisting forces $P+R+V$ for every intermediate state (Fig. 10.7). Herein are

P: Air pressure resultant $= p_l \cdot A$, where p_l is the air pressure and A the loaded surface.

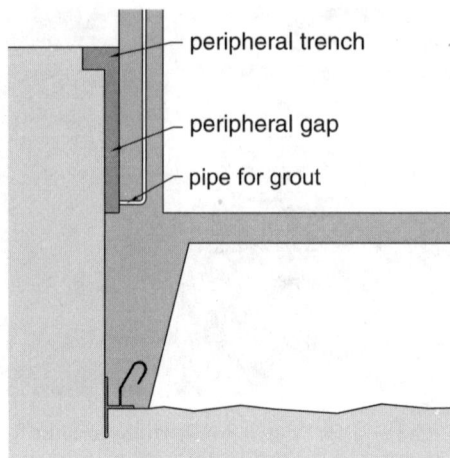

Fig. 10.6. Cutting edge and wall gap filled with bentonite suspension

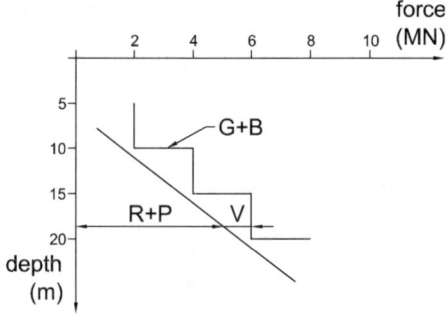

Fig. 10.7. Lowering plan

R: Wall friction resulting from the horizontal earth pressure[6] multiplied with $\tan\delta$. δ is the wall friction angle and can be estimated to $\frac{2}{3}\varphi$. Adjacent to the bentonite lubrication is $\delta = 5°$, or a wall shear stress of 5 - 10 kN/m² can be assumed.

V: Vertical component of the edge force.[7] V is obtained from the limit load p_g, the distribution of which is assumed as shown in Fig. 10.8.

---

[6] although not theoretically justified, usually *active* earth pressure is assumed.

[7] P. Arz, H.G. Schmidt, J. Seitz, S. Semprich: Grundbau, Abschnitt 5: Senkkästen. In: Beton-Kalender 1994, Ernst und Sohn, Berlin

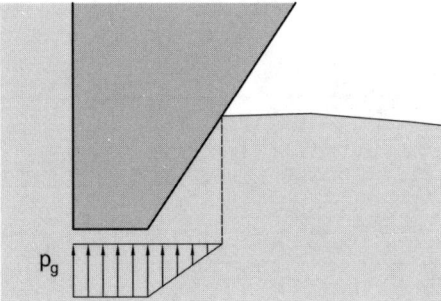

Fig. 10.8. Assumed distribution of stress at the cutting edge

p_g can be estimated as follows:

$$1.2 - 1.6 \text{ MN/m}^2 \text{ (gravel)}$$
$$0.9 - 1.3 \text{ MN/m}^2 \text{ (sand)} \quad .$$

11
Shafts

11.1 Driving of shafts

Shafts are vertical or inclined tunnels. They are either driven downwards or upwards. In the former case, muck and inrushing groundwater has to be conveyed upwards, i.e. against gravity. This can be done, for example, with the help of air lift. If the shaft is driven towards a pre-existing tunnel (so-called bottom access), then muck and water can be conveyed downwards through a pilot borehole. Reaming of the pilot borehole occurs either downwards ('downreaming') or upwards ('upreaming' or 'raise boring', see fig. 11.1). For downreaming with machines the following two methods apply:[1] A vertical TBM with conical cutting wheel is braced with grippers against the shaft wall and pushes downwards (Wirth). Alternatively the downwards thrust is provided by the own weight of the TBM (weight stack downreaming of Robbins, Fig. 11.2). For large diameters upreaming is risky because of possible instability of the face and the wall, difficult exchange of the disk rolls, limited thrust against the face and water inrush. Note that countermeasures can only be taken after the completion of the shaft.

Raise boring is the most economic method for shafts up to 6 m diameter and 1,000 m depth[3] provided that the rock has sufficient strength, bottom access and a precise borehole.

Shafts are lined with bored pile walls, diaphragm walls or with frozen soil.[4] The latter is mainly applied for deep shafts, e.g. for mining purposes. The world's deepest shaft (gold mine in South Africa) lies at 2.7 km depth.

---

[1] R.J. Robbins: Recent Experience with a Mechanically Excavated Shaft. Vorträge STUVA-Tagung 1995, Stuttgart, 111-118

[2] Atlas Copco prospectus

[3] A. Wagner: Raise-boring, heute und morgen. 38. Salzburger Kolloquium für Geomechanik, 1989

[4] J. Klein (ed.), Gefrierschachtbau. Glückauf Verlag, Essen, 1985

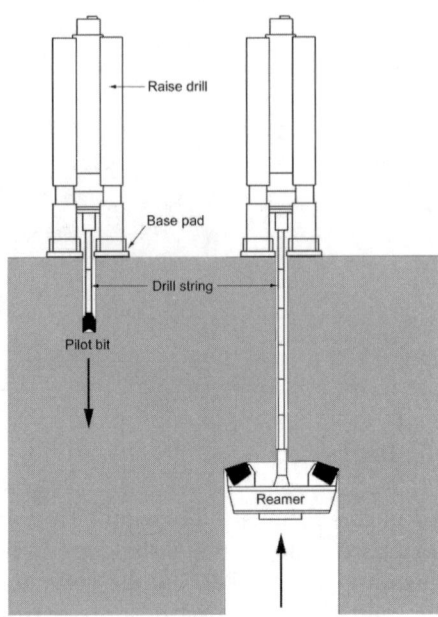

Fig. 11.1. Shaft sinking with raise-boring. Left: driving pilot borehole, right: up-reaming[2]

The bases of shafts are to be ventilated by an exhaust ventilation system.[5]
Drilling rods are equipped with built-in inclinometers. The measured signals are transmitted as pressure pulses through the flushing fluid. Changes of the drilling direction are accomplished by remote controlled vats in the cutterhead. For the shaft of the power plant Uttendorf II, a 600 m deep borehole was drilled with an accuracy of 0.17 %.

For upraised shafts, the ALIMAK excavation platform can be applied (driven with pressurized air). This platform climbs along a rail that is mounted to the rock with spreading bolts. A canopy of steel mesh protects the workers from falling stones. Blastholes are bored and charged from this platform, which is removed for the explosions. Subsequently the platform is raised to install the support.

---

[5]Model specification for tunnelling. The British Tunnelling Society and the Institution of Civil Engineers. Thomas Telford, London, 1997

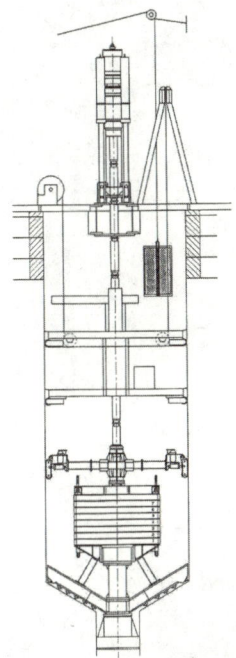 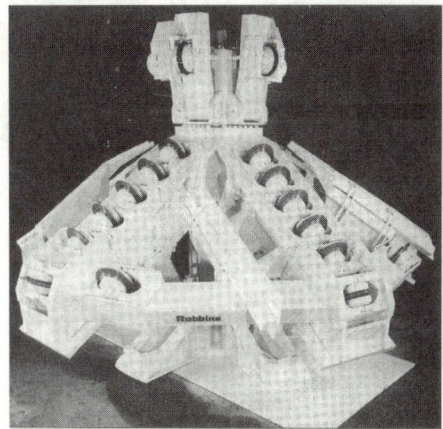

Fig. 11.2. Left: Weight Stack Downreamer with 83RM Raise Drill, Right: Downreamer Cutterhead

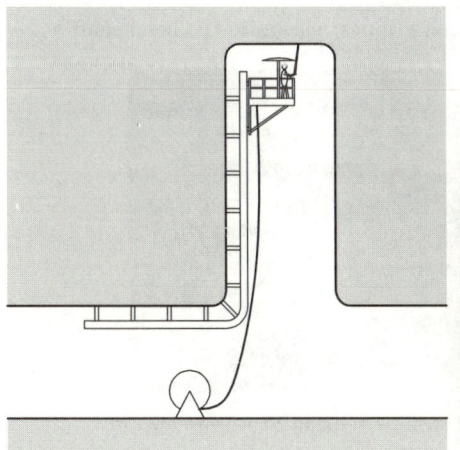 

Fig. 11.3. *Alimak* - lift platform

Fig. 11.4. Shaft Sedrun, muck hoppit and cable winch[6]

Fig. 11.5. Shaft Sedrun, excavation gripper; view into the lined shaft[6]

Fig. 11.6. Launch shaft, city rail extension Stuttgart airport

11.2 Earth pressure on shafts

Shafts can be regarded as vertical tunnels. We can apply the same solutions as we can for tunnels as long as the increase of stress with depth can be neglected.

---

[6]source: http://www.tunnel.ch/ass/

11.2 Earth pressure on shafts

The determination of the earth pressure on a shaft is a 3D problem, for which solutions exist that are hard to understand.[7]
One obtains a simple estimation of the earth pressure p acting upon the shaft lining, if one assumes full mobilization of the rock strength and furthermore that $\sigma_r (\equiv p), \sigma_\theta$ and σ_z are principal stresses. It then follows $\sigma_r = \gamma z$. Hence,

$$p = K_a \gamma z - 2c \frac{\cos \varphi}{1 + \sin \varphi} \quad .$$

In mining, p is usually set to $0.3\, \sigma'_{z,\infty} + u$. Herein, $\sigma'_{z,\infty} = \gamma' z$ is the effective vertical stress in the far field and u is the porewater pressure.

---

[7] A. Kézdi : Erddrucktheorien (p. 291), Springer Verlag, Berlin, 1962, as well as K. Széchy: Tunnelbau (p. 732), Springer Verlag, Wien, 1969, and B. Walz, K. Hock: Berechnung des räumlich aktiven Erddrucks mit der modifizierten Elementscheibenmethode. Bericht Nr. **6** (March 1987), Bergische Universität-GH Wuppertal. In some of these works a 3D gliding wedge is regarded, which is kinematically not possible.

12

Safety during construction

12.1 Health hazards

In early days of tunnelling the toll in accidents was very high. It could, however, be gradually reduced, as inferred from the following table:

Tunnel	Length (km)	Construction time	Casualties
Gotthard rail	15.0	1873-1882	177
Simplon I rail	19.8	1898-1906	67
Lötschberg rail	14.6	1907-1913	64
Gotthard road	16.3	1970-1980	17

Traffic accidents are common hazards faced by tunnelers, apart from hardness of hearing caused by noise and silicosis caused by quartz dust.[1] Besides of accidents due to cave-ins and machines, further sources of hazard are:

Gas intrusions: The following gases may occur in the subsurface:

> Methane and higher hydrocarbon gases: Methane can form an inflammable mixture together with air. At atmospheric pressure there is an acute danger of explosion if the methane concentration is between 5 and 14 %. These limits are reduced for higher hydrocarbons. Methane is lighter than air and, therefore, can build up in high locations.[2,3,4] Good ventilation and gas detectors are the best precaution.

---

[1] H.D. Serwas, Gefährdungen der Gesundheit bei Tunnelvortrieben, *Bauingenieur* **75**, März 2000, 119-122

[2] R. Wyss, Danger Caused by Gas Intrusions on Tunnelling Sites, *Tunnel*, 4/2002, 85-90

[3] During the construction of the metro in Taegu (South Korea) a gas pipe was accidentally tapped. The resulting explosion cost 100 lifes.

[4] The 7 m ⌀ San Francisco water tunnel was driven by TBM in the area of Los Angeles. In 1971 a gas explosion took the lives of 17 workers. The underground contained oil and gas.

Hydrogen sulphide (H_2S) is an extremely toxic gas. It is perceptible even at very low concentrations of a few ppm (cm^3 gas in one m^3 air). Symptoms of poisoning can result after a few minutes within a concentration of 100 ppm, a concentration of 1,500 ppm is lethal. H_2S is slightly denser than air, therefore it builds up close to the ground.

Carbon dioxide (CO_2) is dangerous on account of its suffocating effect: 8-10 vol % lead to headaches and lack of breath, a 20 vol % concentration can be fatal. CO_2 is heavier than air.

Radon: Outdoor radon never reaches dangerous concentrations because of diffusion. Radon is hazard indoor because of its accumulation in enclosed spaces that are not ventilated. This is the case when the surrounding ground contains sufficient uranium, is sufficiently permeable and has a pore pressure higher than within the adjacent space. The radioactivity of radon may lead to lung cancer.

Noise can cause irreparable damage of hearing. Moreover, continuing and loud noise is a risk factor for heart attacks.

Examples of noise	
Moving electric locomotive	95 dB(A)
Moving diesel locomotive	100 dB(A)
Pressurized excavation chamber	105 dB(A)
Air pressure hammer	115-120 dB(A)
Pain threshold	120 dB(A)
Explosion	>140 dB(A)

Beyond 85 dB(A) ear protection should be supplied and beyond 90 dB(A) this protection must be used.

Ear protection		
Type	Average quietising	Range of application
Cotton	20 dB(A)	up to 105 dB(A)
Plugs	25 dB(A)	up to 110 dB(A)
Capsules	30 dB(A)	up to 115 dB(A)

Dust: Nearly every rock contains quartz and quartz dust is produced with drilling, blasting, loading and mucking. It penetrates into the deepest branches of the lungs (alveoli) and releases a chronic inflammation (silicosis) which leads to malfunctions and overload of the right chamber of the heart. In grave cases it may cause invalidity. At that, silicosis doubles the probability of lung cancer.

The following measures can reduce the production of dust:
- drilling equipment with dust suction
- spraying water (water droplets can bind large quantities of dust)
- accurate blasting makes dust-producing after-profiling dispensable
- moistening and keeping small filling degrees when loading trucks

- maintenance of carriageways and use of bonding agents.
- use of the minimum required amount of explosives.

If the MAC limits are exceeded, a filtering facepiece P2 according to Euronorm EN 149 should be worn.

Asbestos fibres: Respired fibres can stay in the lung and cause inflammation which may lead to tumours.[5] Such fibres have lengths $\geq 5\mu$m and diameters $\leq 3\mu$m. The minerals crocidolite (blue asbestos) and amosite (brown asbestos) are particularly dangerous; actinolite, anthophyllite, chrysotile and tremolite are also potential threats to health. The related illnesses are asbestosis and chest or abdominal cancer. Protective measures are:

- protective clothing and mask of class FFP3
- washing exposed parts of the body after work
- spraying excavated material with water and installation of water mist dischargers to bind fibres
- cleaning exposed machines and vehicles
- extract ventilation system.

Information of the personnel about the dangers and possibly also financial gratification for working with the protective breathing mask are additional preventive measures.

Explosion fumes: A detonation transforms the explosive into a mixture of gases containing the nontoxic components water vapour (H_2O), carbon dioxide (CO_2) und nitrogen (N_2), and the toxic gases carbon monoxide (CO), nitrogen monoxide (NO) and nitrogen dioxide (NO_2). Note also that the nontoxic gases can cause asphyxia if the oxygen concentration is too low.

Fraction of CO and NO_x in explosion fumes		
Explosive	CO (l/kg)	NO_x (l/kg)
Ammon-Gelite 2	24.0	3.5
Nobelite 310	7.5	1.1
Nobelite 216	4.1	0.1
Emulgite	1.1-4.6	0.1-0.2

CO inhibits the oxygen transport within the blood and causes feebleness, headaches, tinnitus, vomiting, loss of decision capacity, unconsciousness and breathing paralysis. The MAC value is 30 ppm (30 cm^3/m^3).

Nitrose gases[6] (in particular NO_2, which results from NO and oxygen) are malignant toxic, as the damage appears with a time delay. NO_2 together with humidity of the transpiration paths forms nitrogen acid (HNO_3 and HNO_2), which etches the bronchi and may lead to lung oedema. In fresh

---

[5]M. Aeschbach, Asbestos in the Lötschberg Base Tunnel Drive. *Tunnel*, **73**/2004, 34-43

[6]NO and NO_2 are denoted as "nitrose gases" (NO_x).

air the discomforts (coughing, throat scrape, headaches, vertigo) soon disappear, but after two or three hours the symptoms of the second step of intoxication appear, which can be lethal for concentrations >150 ppm NO_2.

The MAC values are exceeded shortly after blasting. A blowing ventilation pushes the toxic fumes along the entire tunnel. Therefore, workers should either leave the tunnel or go into an appropriately ventilated container before the explosion.

Nitroglycerin and nitroglycol are oils contained within the explosives. They can enter the human body by skin contact or breathing and cause headache and sickness due to vascular dilation and corresponding shortage of oxygen in the brain.

Diesel combustion products (DCP): These comprise the tumorigenic carbon black and connected carbohydrates such as fuel, lubricants etc. The maximum permissible concentration for tunnelling is 0.3 mg/m^3 DCP. Diesel motors in tunnelling must be equipped with particle filters (except for motors with very short operation time) and maintained regularly.

For accidents, especially fire, a safety container (Fig. 12.1) must be provided for near to the face.

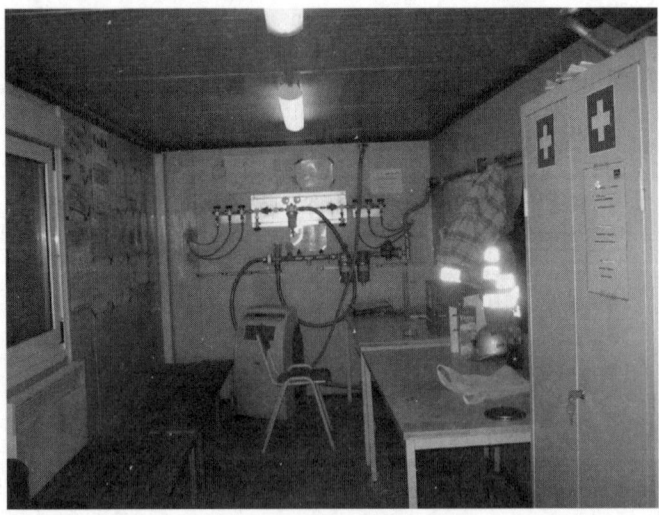

Fig. 12.1. Safety container in Lötschberg base tunnel

12.2 Electrical installations in tunnelling

Electrical installations for underground work need additional provisions.[7] The considerable amounts of fine dust combined with humidity cause creep currents. Therefore, electrical devices need protection against dust and also against sprayed water and flooding. They must also be protected against falling stones and earth inrushes. Explosive gas mixtures can emanate from rock, gas bottles, contamination and accumulators. They must be detected by appropriate monitors. If they should appear, the following measures must be taken:

- shutdown of electric devices
- removal of personnel
- ventilation.

12.2.1 Hazards due to failure of vital installations

The following hazards may occur with blackouts of electricity supply:

- failure of pumps can cause flooding
- failure of compressors can cause pressure reduction
- failure of ventilation.

To meet these hazards, hot standby electricity supply must be provided.

12.2.2 Special provisions

Transformers: The cooling medium must be neither inflammable nor toxic. Its leakage should not endanger the groundwater (collect tray should be provided for).

Accumulator recharge stations: During recharging a gas is created which is explosive if mixed with air. Therefore, the following precautions must be taken:
  - special containments with doors opening outwards
  - special designation
  - placement of accumulators upon timber grids
  - sufficient ventilation (per box: $Q[l/h]=55 \cdot I[A]$)

Electric potential equalisation: An equalisation conductor is strictly required in tunnelling.

Emergency shut down: Machines and installations underground must be provided with a red emergency shut down button on a yellow background.

---

[7] "Elektrische Einrichtungen im Tunnelbau", Tiefbauberufsgenossenschaft, Am Knie 6, D-81241 München

Cables must be laid outside the carriageways and outside the access of vehicles and construction machines. It is recommended that they are suspended together with water and air pipelines. Suspension from wires or rod hooks could damage their isolation. Flat hooks with widths of 4 to 5 cm and maximum spacing of 5 m should be used. In exposed positions cables can be protected with tubes or laths. However, a permanent access for inspection should be preserved.

12.2.3 Energy supply of excavation machines

Modern excavation machines have power consumptions of between 50 and 1.000 kVA. Therefore, only high voltage supplies are relevant. The corresponding transformers are either built in or carried along. The cables are either suspended (Fig. 12.2a) or carried along rolled in bobbins (Fig. 12.2b) or just pulled along, in which case a traction limiter must be provided for (Fig. 12.2c).

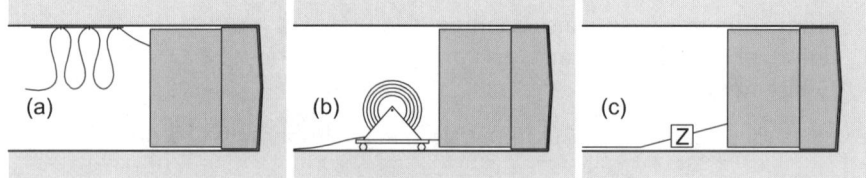

Fig. 12.2. TBM power supply cable

Fig. 12.3. Rolled cables for TBM power supply

12.2.4 Illumination during construction

Sufficient illumination assures quality, improves response times, prevents fatigue and accidents.[8] A tunnelling construction site may only be accessed if there is a general illumination and a safety illumination. The latter must immediately set on in case of a black out of the general illumination. The safety illumination is dispensable if each worker is equipped with a lamp. Table 12.1 indicates the required illumination intensity.

General illumination	
Traffic routes	10 Lux
Workplaces	60 Lux
Permanent installations	120 Lux

Safety illumination	
Escape and rescue routes	1 Lux (at least 1 h)
Workplaces	15 Lux (at least 10 min)

Table 12.1. Required illumination during construction

In escape routes and bifurcation points the escape direction must be indicated with illuminated or luminescent signs. Spotlights have a long range but they may dazzle. Tunnelling dirt can reduce the illumination intensity by 70% within a few weeks. Therefore, a regular maintenance must be provided for.

12.3 Controls

The observational method is of high importance in tunnelling. During and after heading a tunnel must be observed via measurements. If the measured values exceed some pre-specified thresholds, then pre-defined measures must be taken. To this end a control plan (also called 'inspection and test plan') is of high value. It is part of the Method Statement and contains specifications of the following items:

- responsibility (who?)
- topic of control (what and where?)
- time of control (when?)
- control procedure (how? e.g. measuring device, check list)
- limit values, tolerances, rules for rating

---

[8] It should be mentioned, however, that according to experience in mining, the accident rate was higher in excellent illuminated mines than in less well illuminated ones, where the staff was more alert.

- prescriptions for documentation
- procedures in case of deviation from limit values.

Controls are also a matter of quality assurance, which requires inspection and testing work.

The checks can be carried out as spot checks. The percentage of the checks depends on the importance of activity. Some checks can be executed by the contractor (so-called self-certification). An appropriate training of the involved personnel is necessary. Everyone should have an understanding of what is required.

The 'required excavation and support sheet' (RESS) is important for quality control. It is a one-sheet method-statement summarising the construction details. Every working day, a formal meeting to produce the RESS is to be held. The RESS is a form containing specifications and it may look as follows (as an example):

Advance length:	*1.00 m*
Overbreak:	*left side: 6 cm, right side: 6 cm, crown: 8 cm*
Spiles:	*none*
Elephant foots:	*left and right = 50 cm*
Face sealing:	*where required*
Arches:	*lattice girder $W_x > 65\ cm^3$*
Shotcrete thickness:	*35 cm*
Quality:	*J3*
Wire mesh:	*2 layers (inner/outer) AQ 50*
Top invert:	*thickness = 10 cm, 1 × AQ 50 outside*
Advance length of top invert:	*max. 4.0 m*
Supporting core (Stützkern):	*none*
Bolts:	*SN-bolts 250 kN*
	6 m long with fast setting mortar
	spacing: transversal = 1.2 m, longitudinal = 1.0 m
Limitation of advance:	*max. 7 m in 24 h*
Forerun of top heading to top-invert:	*max. 8.5 m*

12.4 Risk management

Tunnelling is risky. Hazards are multiple and can never be excluded. However, elements of uncertainty need to be understood and controlled. Industry cannot simply rely on good fortune. The absence of a previous incident does not denote that a process is without risk. Risk control and management comprise all provisions aiming at minimising risk,[9] i.e. measures to identify potential

---
[9] A *hazard* is a possible and adverse or dangerous situation. The related *risk* is the product of the probability of hazard with its magnitude, expressed e.g. by the costs of repair.

risks, to reduce uncertainty, to mitigate risks and assess residual risk. A basic idea (Fig. 12.4) is to record as quickly as possible all the signals that warn of forthcoming damage, and provide for appropriate response. Notabene, in cases where there is no warning (e.g. brittle collapse) it is impossible to prevent damage.

Risks can be reduced, as can uncertainty in apportioning responsibility, by accurate and comprehensive site investigation, and by documents being explicit as to the presence of and the provision for risk.[10] A pervasive risk results from pre-existing pipes, cables and (in some places) also from undetonated buried bombs.

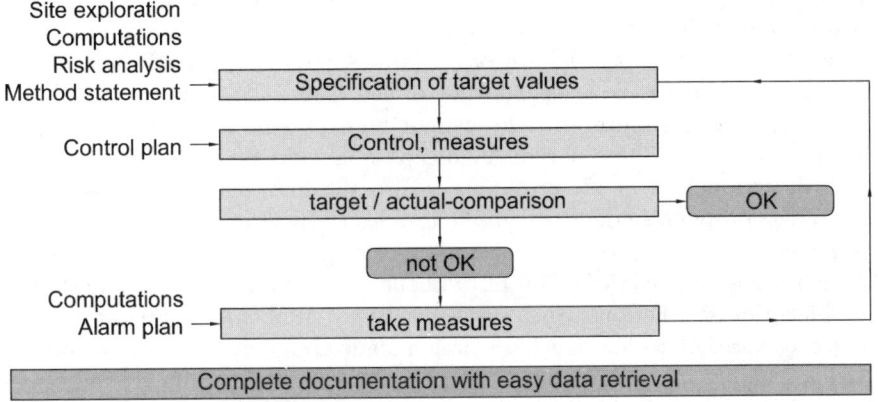

Fig. 12.4. Sequences of risk management

12.5 Emergency plan and rescue concept

The emergency plan[11] comprises all precautions aiming to speed up the response in case of emergency, i.e.

- Telephonic access of construction manager, fire brigade, rescue and police
- Description of locations that are important in case of emergency (access, management offices, first aid facilities, emergency supplies, keys)
- Possibilities to rescue and lift injured persons
- Provisions for evacuation (when? whom? whereto? how?)

A list of items and events for the rescue concept of underground work has been published by the Swiss Accident Insurance Institute.[12]

---
[10] CIRIA report 79
[11] see RVS 9.32
[12] *Tunnel* 2/2002, 49-57

12.6 Quantification of safety

Civil engineers operate since long with the so-called factor of safety. This factor is considered to quantify the safety of a construction or a part of it. Virtually, however, the factor of safety is not an objective physical quantity, as it depends on a series of conventions, the most pronounced one being the calculation method. The only objective measure of safety is the so-called reliability, defined as $1 - P_f$, where P_f is the probability of failure (collapse). Despite many expectations, however, P_f cannot be quantified in geotechnical engineering.[13]

The evaluation of P_f presupposes the knowledge of the limit equation, i.e. an algebraic equation that comprises all significant variables and attains the value of 0 at failure. This equation is however hardly known, there are only very rough approaches to the failure based on assumed stress and displacement fields. Moreover, the path dependence of the material behaviour of soil and rock implies that failure virtually cannot be expressed with algebraic equations. Apart from these severe handicaps it should be taken into account that the evaluation of P_f presupposes also the knowledge of the statistics (i.e. of the probability densities and joint density functions) of the underlying variables, which have to be considered as random variables. The science of statistics offers a rich selection of methods to determine or to estimate the aforementioned functions. Such methods, however, should not be overtaxed. The properties of ground (such as friction angle etc.) constitute random fields that are highly inhomogeneous. At that, sampling is very difficult and alters the addressed properties. Taking into account that the related laboratory tests are also burdened with errors and that our mechanical models are rough approximations, one can easily see that P_f can hardly be estimated. Fig. 12.5 shows the experimental outcomes for the friction angle of a particular soil and some overlaid probability density distributions, all of which have passed the known statistical tests. Nevertheless, the resulting failure probabilities exhibit a large scatter and range between 10^{-11} and 10^{-3}.[14]

In this context it should be mentioned that the so-called krigging is often but misleadingly attributed as geostatistics, while virtually it is a method for spatial interpolation.

12.7 Collapses

Small inrushes can be seen as overexcavation, but major inrushes (cave-ins, or daylight collapses) are catastrophes frequently resulting in high costs, down-

---

[13] W. Fellin et al (eds.), Analyzing Uncertainty in Civil Engineering, Springer 2005

[14] Oberguggenberger, M. and Fellin, W. (2002): From probability to fuzzy sets: The struggle for meaning in geotechnical risk assessment. Proceedings of Probabilistics in GeoTechnics: Technical and Economic Risk Estimation September 15-19, 2002, Graz, Austria, Verlag Glückauf Essen

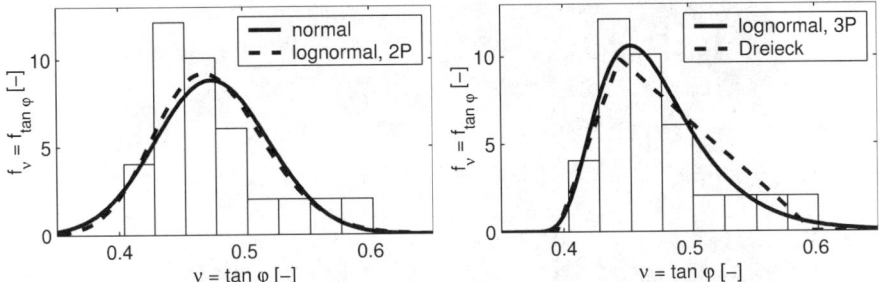

Fig. 12.5. Several probability density functions adapted to the same histogram of measured φ-values

times and even death. These may be caused by inadequate planning or execution, but in many cases also from unforeseen underground conditions. If we take as an example the new Hannover-Würzburg line of German Rail, with approx. 120 km tunnels, whose cover was on average only about 30 m, we can see that many collapses occurred especially following lengthy work stoppages of heading during Christmas or Easter.[15] On average, one collapse occurred every 10 km of tunnel. Fortunately, there were no serious injuries and the related costs only amounted to 1% of the construction costs. To give an example, during driving the Witzelshöhe tunnel a daylight collapse occurred at a location with 14 m cover. The resulting crater had a diameter of 15 m. Luckily, there were no victims, as the night shift was taking its break.

Often, a collapse is announced by cracks and spalling of the shotcrete lining. So, there is time to withdraw machines and personnel. Measurements are expected to warn against an imminent collapse, which (often, though not always) announces itself with increasing deformations. In such cases it is tried first to install additional support (e.g. rockbolts). If this does not help, the collapse occurs. Collapses were also caused by slotting the invert or after-profiling.

Yet another cause of collapse is the crossing of undetected weak zones, e.g. shafts (natural or artificial ones, such as old wells) filled with loose, watersaturated cohesionless material (Fig. 12.7).

For instance, in the course of the construction of the Lötschberg rail tunnel in the years 1906 and 1913, the pilot tunnel unexpectedly ran into an overdeepened valley, filled with saturated gravel. Nearly one kilometre of the pilot tunnel was inundated with soil, killing and permanently burying the entire shift of thirty men.[16]

---

[15]Therefore, RVS 9.32, Blatt 6, requires to seal the face with shotcrete, steel mesh and drains for week-ends and other breaks.

[16]P.K. Kaiser, M.S. Diederichs, C.D. Martin, J. Sharp, and W. Steiner, Underground works in hard rock tunnelling and mining, GeoEng 2000, Melbourne 2000, Technomic Lancaster-Basel, Vol. 1, 841-926

Fig. 12.6. Tunnel collapse in München-Trudering (20.09.1994). Three lives were lost and 36 people injured.

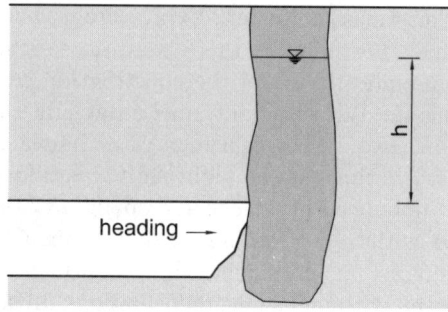

Fig. 12.7. Crossing of a shaft filled with cohesionless material and water.

The seepage force drives the cohesionless material which rushes with high velocity into the excavated space. Neglecting energy losses due to friction, this velocity can be estimated according to TORRICELLI's law $v \approx \sqrt{2gh}$. Such catastrophic events occur very quickly and there is no time for countermeasures. Moreover, this type of catastrophe sets on without announcement. A precautionary measure is pressure relief with advance drillings. In doing so, one should take into account the high flowability of pressurised water-saturated sand and provide special valves ('preventers') at the drillholes that can be closed if the crew looses control of the inflow. During the drive of an exploratory tunnel for the Gotthard base tunnel, exploratory drilling was carried out ahead of the TBM into water-saturated sand pressurised with 1,000 m waterhead. An incident occurred when the blow-out preventer was not properly mounted and operated: Several thousands m³ of sand were washed into the tunnel, the TBM was buried (see also Section 4.5).

12.7 Collapses

The following measures should be taken immediately after the occurrence of a collapse:

- check whether anybody was left in the tunnel
- control access to the site
- removal of water from the crater and avoidance of inflow of surface water
- survey of the crater and (re)establish monitoring instrumentation.

Thereafter, a detailed analysis should be executed to establish the cause of collapse. For remediation, the inrushed masses can be grouted. A crater can be excavated, lined and supported as a shaft. Heading should then be carefully continued (e.g. with sidewall drift method).

12.7.1 Heathrow collapse

The report on the collapse of NATM tunnels at Heathrow Airport[17] is extremely instructive and its reading is highly recommended to anybody working in the field of tunnelling. Here are presented only its main conclusions. The collapse occurred at a construction site for the Heathrow Express Link Project, that connects London (Paddington) with Heathrow Airport. The contract was awarded in the sum of £60.7 million.

The collapses occurred at the Central Terminal Area (CTA) starting at 20 October 1994 and lasting several days. The situation is shown in Fig. 12.8. The underground is London clay. Heading was according to the NATM.

The public and those engaged in the construction work were exposed to grave risk of injury. Workers were evacuated minutes before the first collapse. By remarkable good fortune no one was injured. The cost of the recovery work was of the order of £150 million.

The collapse was an 'organisational accident', i.e. a multiplicity of causes led to failure. These were:

1. Bad quality of the shotcrete lining. Shotcrete was transported by road to the CTA. Although water required to start the cement's hydration was only intended to be added just before use, the inevitable presence of water in the sand and excessively wet aggregates meant that some initial set was bound to occur during transportation and while the shotcrete awaited use at site. Consequently, substantial amount of shotcrete was seriously understrength. Some shotcrete could readily be taken apart by hand due to insufficient cement binder. The thickness of the shotcrete lining was deficient. The repairs revealed shotcrete thicknesses of 100, 150 and, exceptionally, 50 mm in contrast to the 300 mm specified.
2. Lattice girders were not always formed to the correct profile and this slowed progress.

---

[17]The collapse of NATM tunnels at Heathrow Airport, HSE Books, Crown copyright 2000, Her Majesty's Stationery Office, St Clemens House, 2-16 Colegate, Norwich NR3 1BQ, ISBN 0 7176 1792 0

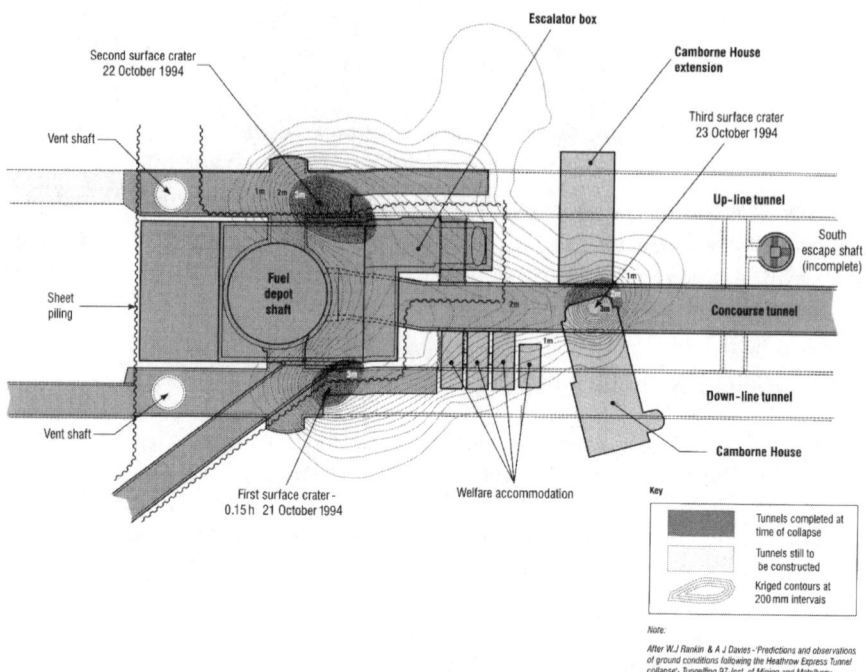

Fig. 12.8. Array of buildings and underground constructions. Surface effects of tunnel collapses.

3. Deficient joints of the several lining segments. The shortages were:
   - insufficient overlap of the wire mesh
   - damage or removal of the exposed wire mesh that projected from completed panels and which was intended to provide continuity in the lining's reinforcement
   - forming the joints with tapered edges rather than right angles
   - rebound was incorporated into bench panels.
4. Lack of robust design, which means that buildability (i.e. ease with which something can be correctly built) issues have been overlooked:
   - Minor variations in shape of the flat invert could result in major adverse effects on the structural performance of the lining.
   - The connection of the temporary wall (of the sidewall drift) to the invert was deficient. The lack of any inspection of the invert joint was a serious omission since the joint was critical to the stability of the tunnel lining.
5. The invert (which was covered by the temporary running fill) was damaged to at least 70 m due to built-in defects or to grout jacking, or to a combination of both. The damage was belatedly detected due to insuffi-

cient monitoring, the subsequent repair did not cover the entire damage length.
6. The formation (profiling) of the temporary central wall was deficient.
7. Three parallel tunnels were constructed at the same time.

The reasons behind the mentioned technical defects were:

1. Poor workmanship
2. Poor engineering control
3. Deficient monitoring. No test coring had been carried out in the invert.
4. NATM design and construction were poorly understood, and past experience has not been adequately taken into account
5. Repairs were permitted without sufficient inspection, engineering analysis and design, and without recording and certificating the repair work that was completed.
6. The numerical model used for design was not correctly calibrated against existing measurements.
7. The management systems and the resources capable of collecting, inputting, processing and interpreting the large amounts of instrumentation data were insufficient. E.g., a surface contour map to indicate distribution of settlements was not provided.
8. The safety plan was deficient in two respects. First, it did not provide guidance on procedures for preventing tunnelling collapses and for recovering from developing adverse events, and, second, it did not fully deal with emergencies.
9. Bad health condition of the personnel. When collapse started at 20th October, the 'NATM engineer' was only at work between 07.00 and 10.00 due to illness. The project director was unwell and was absent from the site. The construction superintendent had just returned from sick leave.

Part II

Tunnelling Mechanics

"We must learn to distinguish between what we understand and what we don't understand."

– Socrates

13
Behaviour of soil and rock

13.1 Soil and rock

Tunnels are driven both in soil and in rock. The transition between soil and rock is not sharp, and there are many kinds of rock, which may be considered either as soft rock or as soil. If one refrains from the separation by arrays of joints, the strength of rock is usually modelled in the same way as the strength of soil. The differences are then quantitative, but not qualitative. On the other hand, if the size of the individual rock blocks is comparable with the tunnel diameter, then continuum-mechanical considerations are inappropriate, and one must regard the individual blocks. This requirement proves often to be rather academic, since we hardly know in advance the position of the individual joints. Thus, in most cases jointed rock is considered as a homogeneous medium with whatsoever assumed mechanical properties.
In order to address geomaterials uniformly, regardless of whether it concerns soil or rock, we use the word "ground". In the following, characteristics of ground are represented, whereby the essentials of notions such as elasticity, plasticity, friction and cohesion are described.

13.2 General notes on material behaviour

It should be clear that problems in tunnelling can only be dealt with, if the mechanical behaviour of the involved materials (such as ground, shotcrete etc.) are sufficiently comprehended. This can only be achieved in the frame of an appropriate constitutive law. Such laws are mathematical expressions that connect stresses and strains. In view of the complexity of the involved materials, constitutive equations can be very complex and, often, confusing. It should be kept in mind that constitutive equations are not only implemented in numerical simulation schemes (cf. Chapter 22) but they also help to detect and describe the material behaviour. E.g., referring to the YOUNG's modulus

E of a particular rock automatically implies that the linear theory of elasticity (i.e. HOOKE's law) is assumed to be valid for the considered rock. The validity of the underlying constitutive law is, however, often overlooked. E.g., the folding of rock strata has been investigated under the assumption that rock is a NEWTONean fluid.[1] Equally, a YOUNG's modulus is often attributed to soft rocks and soils, the mechanical behaviour of which pronouncedly deviates from linear elasticity. Needless to say that such inconsistencies must be avoided. They are not reasonable and, moreover, there is no way to realistically determine the involved material parameters.

The many existing solutions based on the theory of linear elasticity may be questioned in view of the fact that ground does in most cases not comply with the assumptions of this theory (cf. Section 13.3). On the other hand, however, one should take into account that isotropic linear elasticity is the most simple constitutive equation that can be applied to a solid material. Thus, it allows to find analytical solutions which serve as reference ones. They are in themselves consistent, in the sense that they don't violate neither the equilibrium nor the boundary conditions and, in many cases, they help to reveal some basic features of the considered problems.

In selecting one from the many existing constitutive equations proposed for soils and/or rocks, attention should be paid to the fulfilment of the following general requirements:

- It should be complete, i.e. it should provide predictions of the material response for all conceivable stress or strain paths. I.e., given any strain path the corresponding stress path must be predicted and vice versa.
- The pertinent material parameters must be determined on the basis of laboratory or field tests. A procedure on this must be specified and it must be tractable and — as far as possibly — simple.
- The predicted material behaviour must be realistic. It should be noted that constitutive equations are (contrary to the balance equations) always approximative. This means that the predicted outcomes of an experiment will never be identical with the measured ones. The question is how close are predictions and measurements. This must be decided by judgement and it has to be added that there is no objective measure to rate deviations between predictions and observations.

Some additional requirements on constitutive laws, such as objectivity, are explained in the pertinent literature.[2]

---

[1] H. Ramberg, Gravity, Deformation and the Earth's Crust, Academic Press 1981.
[2] See e.g. D. Kolymbas, Introduction to Hypoplasticity, Advances in Geotechnical Engineering and Tunnelling, Balkema, 2000.

13.3 Elasticity

A material is elastic if the stress can be represented as function of the deformation. This means, that the history of the deformation is irrelevant and accordingly, remains undetectable. A material is linear elastic, if the relationship between stress and deformation is linear. A material is isotropic, if its behaviour is independent of preceding rotations.

The stress-strain relationship for isotropic and linear elastic materials is given by HOOKE's law, where two material constants are needed. If one uses the ones given by LAMÉ, namely λ and μ, then HOOKE's law reads:

$$\sigma_{ij} = \lambda \varepsilon_{kk} \delta_{ij} + 2\mu \varepsilon_{ij}$$

resp.

$$\varepsilon_{ij} = -\frac{\lambda \sigma_{kk}}{2\mu(3\lambda + 2\mu)} \delta_{ij} + \frac{1}{2\mu} \sigma_{ij} \quad .$$

Herein δ_{ij} is the KRONECKER-Symbol ($\delta_{ij} = 0$ for $i \neq j$, $\delta_{ij} = 1$ for $i = j$). The index notation and the summation convention have been used here. Written out, HOOKE's law reads

$$\begin{pmatrix} \sigma_{11} & \sigma_{12} & \sigma_{13} \\ \sigma_{21} & \sigma_{22} & \sigma_{23} \\ \sigma_{31} & \sigma_{32} & \sigma_{33} \end{pmatrix} = \lambda(\varepsilon_{11} + \varepsilon_{22} + \varepsilon_{33}) \cdot \begin{pmatrix} 1 & 0 & 0 \\ 0 & 1 & 0 \\ 0 & 0 & 1 \end{pmatrix} + 2\mu \cdot \begin{pmatrix} \varepsilon_{11} & \varepsilon_{12} & \varepsilon_{13} \\ \varepsilon_{21} & \varepsilon_{22} & \varepsilon_{23} \\ \varepsilon_{31} & \varepsilon_{32} & \varepsilon_{33} \end{pmatrix}$$

or, in somewhat shortened way of writing

$$\sigma_{ij} = \lambda \sum_{k=1}^{3} \varepsilon_{kk} \cdot \delta_{ij} + 2\mu \cdot \varepsilon_{ij} \quad .$$

According to the summation convention one omits the sum-sign $\sum$ and sums up automatically, where an index (here: k) occurs twice: $\varepsilon_{kk} = \varepsilon_{11} + \varepsilon_{22} + \varepsilon_{33}$. The constant μ is often designated as shear modulus G ($\mu \equiv G$). One can write HOOKE's law also using the material constants G and ν, whereby ν is the POISSON-ratio:

$$\sigma_{ij} = 2G \left(\varepsilon_{ij} + \frac{\nu}{1 - 2\nu} \varepsilon_{kk} \delta_{ij} \right)$$

resp.

$$\varepsilon_{ij} = \frac{1}{2G} \left(\sigma_{ij} - \frac{\nu}{1 + \nu} \sigma_{kk} \delta_{ij} \right) \quad .$$

HOOKE's law can also be expressed using YOUNG's modulus E and the POISSON-ratio ν:

resp.

$$\sigma_{ij} = \frac{E}{1+\nu}\varepsilon_{ij} + \frac{\nu E}{(1+\nu)\cdot(1-2\nu)}\varepsilon_{kk}\delta_{ij}$$

$$\varepsilon_{ij} = \frac{1}{E}[(1+\nu)\sigma_{ij} - \nu\sigma_{kk}\delta_{ij}] \quad .$$

The following relations apply between the constants:

$$\nu = \frac{\lambda}{2(\lambda+\mu)}$$

$$\lambda = \frac{\nu E}{(1+\nu)(1-2\nu)}$$

$$E = \frac{\mu(2\mu+3\lambda)}{\lambda+\mu}$$

$$\mu \equiv G = \frac{E}{2(1+\nu)} \quad .$$

Also the bulk modulus B, resp. K, is often used as material constant:

$$B \equiv K = \frac{E}{3(1-2\nu)}$$

Some authors prefer to represent stress and strain as vectors of 6 components. Because of symmetry ($\sigma_{ij} = \sigma_{ji}, \varepsilon_{ij} = \varepsilon_{ji}$) the component σ_{21} is omitted, because it is identical to σ_{12}, etc. The stress strain relation of HOOKE then reads:

$$\begin{pmatrix} \varepsilon_{11} \\ \varepsilon_{22} \\ \varepsilon_{33} \\ \varepsilon_{12} \\ \varepsilon_{23} \\ \varepsilon_{13} \end{pmatrix} = \frac{1}{E} \begin{pmatrix} 1 & -\nu & -\nu & 0 & 0 & 0 \\ -\nu & 1 & -\nu & 0 & 0 & 0 \\ -\nu & -\nu & 1 & 0 & 0 & 0 \\ 0 & 0 & 0 & 2(1+\nu) & 0 & 0 \\ 0 & 0 & 0 & 0 & 2(1+\nu) & 0 \\ 0 & 0 & 0 & 0 & 0 & 2(1+\nu) \end{pmatrix} \begin{pmatrix} \sigma_{11} \\ \sigma_{22} \\ \sigma_{33} \\ \sigma_{12} \\ \sigma_{23} \\ \sigma_{13} \end{pmatrix}$$

HOOKE's law is the simplest conceivable constitutive law (i.e. relationship between stress and strain) for solid materials. Therefore it is used occasionally for the computation of deformations in rock. One should realise however that this is allowable only for very small deformations and even then it represents a (possibly rough) approximation of the real material behaviour. It is a widespread opinion that the σ-ε-relation of a solid is linear, at least in its initial part. This is, however, not always true. In such cases, the determination of YOUNG's modulus is not clear and one has to resort to arbitrary definitions, e.g. to take the secant modulus $E \approx \Delta\sigma_{11}/\Delta\varepsilon_{11}$ for some specified values of $\Delta\sigma_{11}$ or $\Delta\varepsilon_{11}$. For some rocks E correlates with the uniaxial compressive strength q_u, e.g. for gneiss E (GPa) $\approx 20 + q_u$ (GPa)/ 7. Note also the difference between intact rock and rock mass. The modulus of rock mass has, very often nothing or little to do with the modulus of intact rock.

13.4 Plasticity

The plastic behaviour is characterised by irreversible deformations, i.e. by deformations, which remain after preceding loading and unloading, Fig. 13.1. The continuum-mechanical presentation of plastic behaviour by tensors is quite complicated, therefore only some fundamentals will be represented here on the basis of one-dimensional stress and strain (e.g. uniaxial compression test).

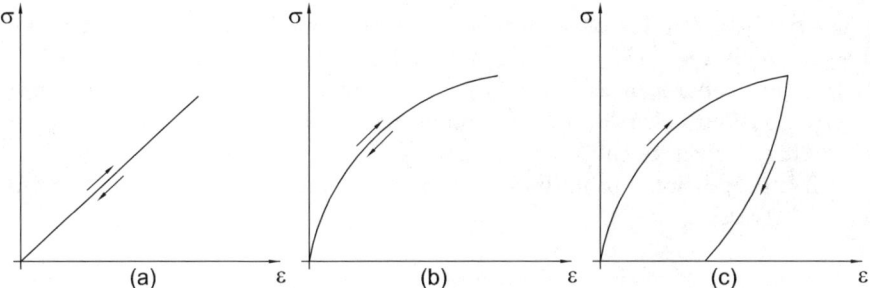

Fig. 13.1. Linear elastic behaviour (a), nonlinear-elastic behaviour (b), plastic behaviour (c)

Following concepts of plastic behaviour can be distinguished:
- rigid - ideally plastic (Fig. 13.2a)
- elastic - ideal plastic (Fig. 13.2b)
- elastic - plastic hardening (Fig. 13.2c)

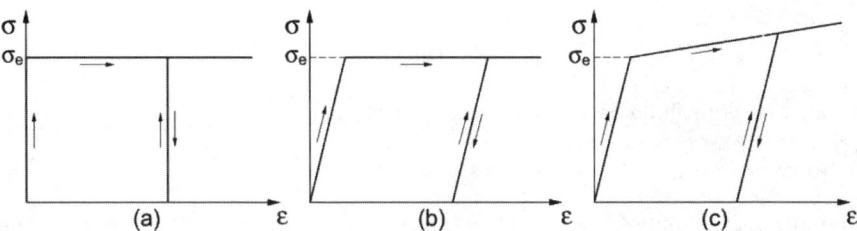

Fig. 13.2. Various concepts of plastic behaviour: (a) rigid - ideal-plastic, (b) elastic - ideal-plastic, (c) elastic - plastic hardening

The basic conception is that as long as $\sigma < \sigma_e$ applies, there is rigid resp. elastic behaviour.[3] The ideal-plastic behaviour (cases (a) and (b) in Fig. 13.2) is marked by the so-called plastic yield or plastic flow, i.e. by the fact, that the deformation ε increases under constant stress.

13.5 Strength

13.5.1 Strength of soil

The strength of soil is characterised by friction and cohesion. Hard rock will be separately regarded, it should be noted, however, that also its strength — if the joint separation is not decisive — is likewise described in most cases by friction and cohesion. In soil, the shear stress τ cannot increase beyond a value τ_f, which is called (shear) strength. τ_f can be split up into a part, which is proportional to normal stress σ, and a part, which is independent of normal stress:

$$\tau_f = \tan\varphi \cdot \sigma + c \quad . \tag{13.1}$$

The first part is due to friction, the second one is the cohesion. The cohesion c is connected with the compression of the soil, caused by a pre-loading[4] σ_v. Even if the load is removed, then its effect remains and manifests itself as cohesion. If the pre-loading is removed (totally or partly), then the current stress σ is smaller than σ_v. Such soils are called overconsolidated:

$\sigma < \sigma_v$: overconsolidated

$\sigma = \sigma_v$: normal consolidated

According to the concept of KREY and TIEDEMANN the cohesion c is set proportional to the pre-loading, and the proportionality constant is called $\tan\varphi_c$:

$$c = \sigma_v \cdot \tan\varphi_c \quad .$$

To normal consolidated soil then applies (with $\sigma_v = \sigma$):

$$\tau_f = (\tan\varphi + \tan\varphi_c)\sigma = \tan\varphi_s \cdot \sigma \tag{13.2}$$

The angle φ_s (defined by $\tan\varphi_s = \tan\varphi + \tan\varphi_c$) is called the angle of the total shear strength (Fig. 13.3).
As mentioned, the cohesion is due to preceding compression. Now, if the soil is loosened, then its cohesion is lost. Then the shear strength τ_f does not

---

[3]Obviously, the criterion $\sigma < \sigma_e$ applies only to uniaxial stress states, since one cannot introduce inequalities for tensors. An appropriate procedure for tensors is introduced by the theory of plasticity.
[4]other sources of cohesion of soils are cementation and capillary effects

13.5 Strength 241

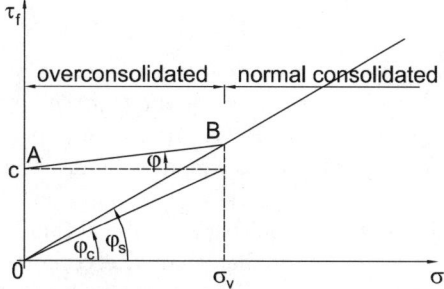

Fig. 13.3. Concept of KREY and TIEDEMANN: For $\sigma < \sigma_v$ (overconsolidated soil) the shear strength is indicated by the line AB ($\tau_f = c + \sigma \tan\varphi$). For $\sigma = \sigma_v$ (normal consolidated soil) the shear strength is indicated by the line $\tau_f = \sigma \tan\varphi_s$

correspond to the straight line AB, but to the straight line OB (Fig. 13.3). The tendency of dense soil, when shearing to increase its volume (and to be accordingly loosened up) is called dilatancy. The loss of cohesion and the associated reduction of the shear strength manifest themselves in the stress-strain curve as softening (Fig. 13.4).

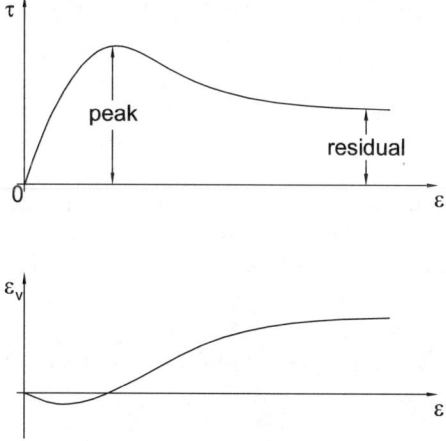

Fig. 13.4. Softening and loosening

Thus, one must distinguish between the peak value of the strength $\tau_{f,peak}$ and the residual value $\tau_{f,residual}$. The peak value is reached after small deformation and decreases with continuation of the deformation to the residual value. If one deals with large deformations, then he should set the shear strength after equation 13.2.

Bound theorems

For a certain group of plastic materials the theory of plasticity permits to estimate collapse loads with the so-called *bound theorems* or collapse theorems. This estimation takes place via the determination of upper resp. lower bounds of the collapse load. An upper bound is a load, which is larger than the collapse load, while a lower bound is smaller than the collapse load. The knowledge of a lower bound permits a safe design, because the actual load can be selected in such a way that it is surely smaller than the collapse load, if it smaller than. The bound theorems apply to those plastic materials, that fulfil the so-called normality rule, which is not further explained here. Materials obeying the normality rule are e.g. cohesive soils ($c > 0$ and $\varphi = 0$).

According to the *lower bound theorem*, collapse does not occur, if a stress field can be found, that fulfils the equilibrium and boundary conditions and does not violate the strength limits of the regarded material. Except from these conditions, this stress field is freely selectable and does not need in particular to be the real stress field.

The *upper bound theorem* considers a possible (though not necessarily the real one) collapse mechanism. The power of dissipation linked with it is calculated on the assumption that the stresses in the regarded body fulfil the limit condition of the shear strength. If this power of dissipation is smaller than the power of the gravity and the other external forces at collapse, then the body will collapse.

Principle of effective stress

For saturated soils, the pressure p of the pore water plays an important role. The principle of the effective stress asserts, that for deformation and strength of soil is responsible only the so-called effective stress σ' (as already stated in Section 8.1.2). It results from the total stress σ by means of the equation $\sigma' = \sigma - u$ whereby the total stress results from the total load, which acts upon a surface. As a consequence, the geostatic effective stress below the groundwater surface results from the so-called buoyant unit weight $\gamma' = \gamma - \gamma_w$ (γ_w = specific gravity of water), cf. Fig. 13.5:

$$d\sigma'_z = \gamma' dz \quad ,$$

Considering the principle of effective stress, the equations 13.1 and 13.2 should be replaced through

$$\tau_f = \tan\varphi \cdot \sigma' + c \quad \text{resp.} \quad \tau_f = \tan\varphi_s \cdot \sigma' \qquad (13.3)$$

The prime is usually omitted where no confusion is to be expected between σ' and σ.

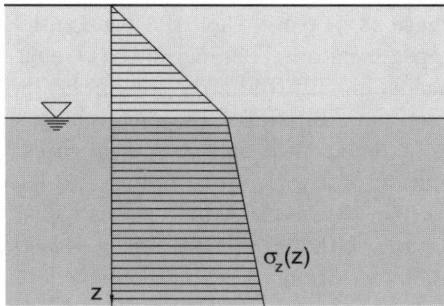

Fig. 13.5. Distribution of the effective vertical stress above and underneath the groundwater surface

Drained and undrained conditions

For problems with water-saturated soil it plays an important role whether the pore water can escape ('drained conditions') or not ('undrained conditions'). Another important role plays thereby the tendency of soil to change its volume at shear deformation: Dense soil behaves dilatant, i.e. it increases its volume (loosening) if normal stresses remain constant. Loose soil, on the other hand, behaves contractant, i.e. it gets compacted (at constant normal stress). The variations in volume addressed here are not possible, if the soil is water-saturated and drainage is impeded.[5] With prevention of contractancy, the pore pressure is increased, and with prevention of dilatancy the pore pressure is decreased. The change of the pore pressure causes substantial changes of the effective stresses, which can lead to collapse of contractant soil. Since the responsible pore water pressure can hardly be computed, strength criteria formulated in effective stresses, e.g. equation 13.3, can hardly be applied to undrained cases. In undrained conditions the strength τ_f is given by

$$\tau_f = c_u \quad . \tag{13.4}$$

c_u is the so-called undrained cohesion (cohesion of the undrained soil) and depends only on the void ratio e. Undrained conditions are given, if the load is applied fast, i.e. within a time, which is small compared to the one needed for the excess pore water pressure to dissipate. The latter is essentially determined by the permeability of the soil. It can be 10^6 times smaller for clay than for sand. Thus, short term stability should be judged by equation 13.4 and long-term stability by equation 13.3.

13.5.2 Strength of rock

With sufficiently small deformations rock behaves elastic, as is manifested by the several elastic waves (e.g. due to earthquakes) that propagate in the

---
[5]The pore water and the individual grains are regarded as incompressible

earth. If the deformations increase, then the (tangential) stiffness decreases, and eventually collapse[6] sets on. This marks the strength of the material. In this section we consider mainly the strength of the intact, unjointed rock. In contrast to it, rock mass separated by joints has a substantially reduced strength, which can be determined only in special cases.

Considering deformation of a rock up to failure, we have to distinguish between ductile and brittle behaviour A material is called *ductile* if it can undergo large deformations without collapse, whereas *brittle* materials collapse after relatively small deformations. The ductile behaviour of rock manifests itself with many geological processes, which are linked with large deformations, without onset of collapse (e.g. folds). The latter is manifested by appearance of discontinuities (e.g. faults). The main characteristic of brittle failure is the onset of collapse without preliminary warning. Therefore, its theoretical forecast is very important. Decisive for ductile or brittle behaviour are pressure, temperature and deformation rate.

In the laboratory, the rock strength is examined on the basis of unconfined or triaxial compression tests. With the triaxial test the sample is subjected to a lateral stress $\sigma_2 = \sigma_3$ and compressed in axial direction. The triaxial test was introduced 1911 by VON KÁRMÁN for the investigation of rock and was later also adopted in soil mechanics. Upon rock samples are applied lateral pressures up to 1000 MPa.[7] They are exerted by a cell fluid on the sample. The sample must be sealed with a diaphragm of rubber or copper so that the cell liquid does not penetrate into its pores. Alternatively, a very viscous cell fluid is used, e.g. kerosene, which develops a high viscosity with high pressure. If the axial stress σ_1 is reduced compared with the lateral stress $\sigma_2 = \sigma_3$, then one speaks of the triaxial extension test (even if all principal stresses are compressive stresses), because the sample is extended in axial direction.

The uniaxial tensile strength of rock is approximately 10 to 20 times smaller than the corresponding compression strength. Its determination on the basis of extension tests is difficult, since a homogeneous distribution of stress and deformation can hardly be achieved. Often the so-called Brazilian test is used (Fig. 13.6), in the case of which (on the assumption of a linear elastic behaviour) an approximately constant tension is obtained in the section between the two force application lines. The tensile strength σ_{xf} results approximately as $F/(\pi r l)$.

Other tests to determine the tensile strength of rock are (i) the 4 point bending test on a beam (sample preparation is very difficult, stress concentration plays a role), (ii) rotating disk (centripetal test, according to MOHR), (iii) direct tension test; the ends of a cylindrical specimen are glued to the testing machine; this is the most reliable test, and (iv) the LUONG-test (Fig. 13.7):

---

[6] Usually, the terms 'collapse', 'failure' and 'plastic flow' are considered as synonymous, although some authors attribute different meanings to each of them.

[7] Referring to aspects of safety at high pressures see: Cox, B.G., Saville, G. (eds.): High Pressure Safety Code. High Pressure Technol. Assoc. U.K., 1975

two concentric circular slots are countersunk from its upper and lower ends, respectively. The tensile stress field in the annular link proves to be inhomogeneous, so that the results depend on the geometry of the sample. It is remarkable that each of the above stated test types gives quite reproducible results which, however, differ considerably among the test types.[8]

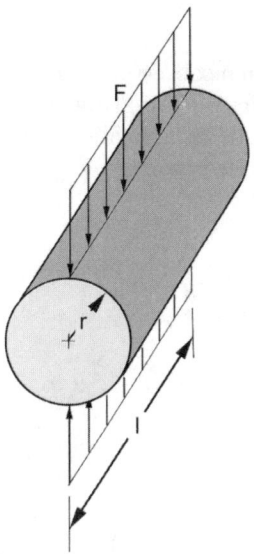

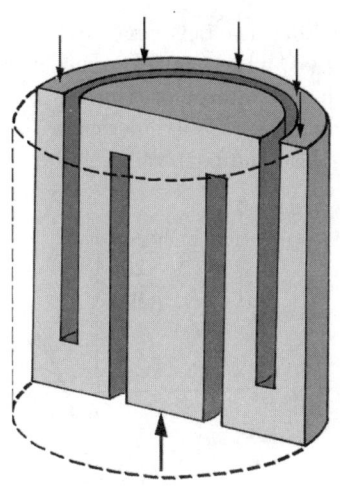

Fig. 13.6. 'Brazilian' test

Fig. 13.7. LUONG-test for the tensile strength of rock[10]

For uniaxial or triaxial compression tests the sample must be provided with parallel ends (necessary accuracy at least 0.02 mm). Due to weathering, the uniaxial compression strength (also called 'unconfined compression strength') q_u of a rock with a particular geological designation (e.g. granite) can vary by orders of magnitude.

In first approximation one can describe the shear strength of rock (in the same way as with weak rock or soil) by the failure criterion of MOHR-COULOMB: Failure occurs, if the shear stress τ reaches the value τ_f:

$$\tau_f = c + \sigma \tan \varphi \quad . \tag{13.5}$$

σ is the normal stress, c the cohesion and φ the friction angle, which varies for rock between 25° and 55°. As for soil, equation 13.5 applies only in first approximation. In reality τ_f increases sub-linear with σ, which means that

---

[8]Oral presentation by R. NOVA in Aussois in 2002

[10]Luong, M.P., 1986. Un nouvel essai pour la mesure de la résistance à la traction. Revue Française de Géotechnique, **34**, 69-74

246   13 Behaviour of soil and rock

the friction angle φ is stress-dependent and decreases with increasing normal stress σ.

Also the shape of a sample influences its strength: the slimmer a sample is (i.e. the larger the ratio height/diameter), the smaller is its strength. This effect is mainly caused by the friction at the sample ends.

13.5.3 Brittle and ductile behaviour

The difference between brittle and ductile behaviour can be observed with rock samples, which are examined with various cell pressures $\sigma_2 = \sigma_3$ (Fig. 13.8). With low lateral pressure is reached a smaller strength at a smaller strain (peak strain). With increasing lateral pressure the sample becomes more ductile. Also the patterns of failure differ (Fig. 13.9).

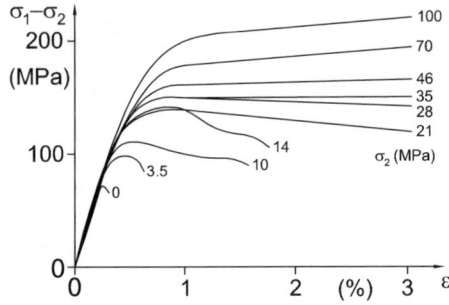

Fig. 13.8. Stress strain curves from triaxial tests with marble with various lateral pressures[11]

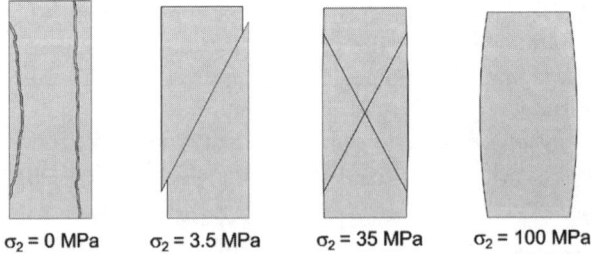

Fig. 13.9. Pattern of failure of marble samples with various lateral pressures[11]

With vanishing lateral pressure occurs the so-called axial splitting tension failure (Fig. 13.9), which appears as paradox, since no macroscopic tensions

---

[11]M.S. Paterson: Experimental rock deformation, the brittle field. Berlin: Springer, 1978

prevail in the sample. For the explanation of this feature, microscopic inhomogeneities are assumed. These are connected with the granular structure of rock. Rock is not a loose granulate, the individual mineral grains are firmly bonded together. By numerical simulation of an agglomerate with the program PFC (*Particle Flow Code*) the important role of the tension fissures in granite[11] could be pointed out (Fig. 13.10).

It was shown that damage starts by occurrence of fissures with a stress, which approximately corresponds to 30% of the peak-value, and it was also shown that 50 times more tension fissures than shear fissures arose. Tension fissures can lead to spalling at the tunnel boundary, in particular if one principal stress is much higher than the other. It appears that the criterion for brittle failure should be expressed in terms of strain rather than in terms of stress.

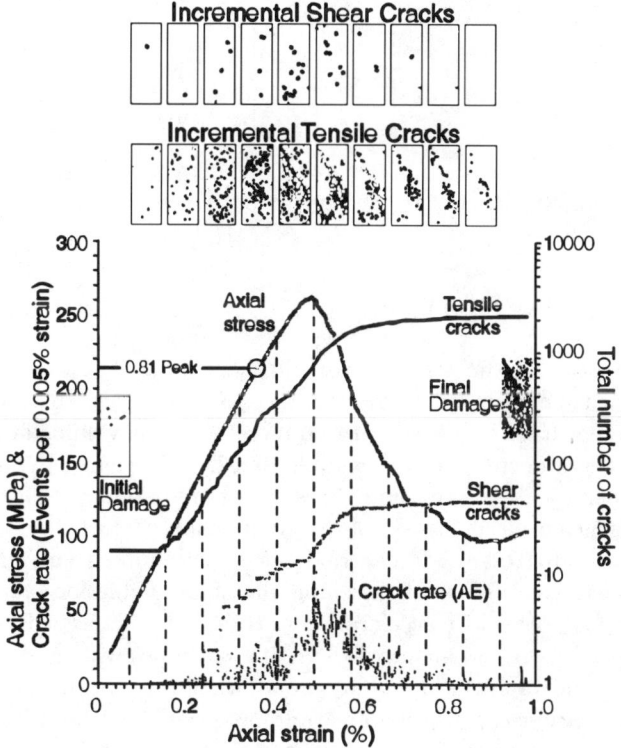

Fig. 13.10. Numerical simulation of the compression of a granite sample with the programme PFC (according to DIEDERICHS)

---

[11]Diederichs, M.S., Instability of hard rock mass: The Role of Tensile Damage and Relaxation. PhD thesis, University of Waterloo, 1999. Quoted in: P.K. Kaiser and others, Underground works in hard rock tunnelling and mining, GeoEng 2000, Melbourne

Core discing, a peculiar phenomenon of disintegration of rock cores extracted from depths where high in situ stress prevails, can be explained with the brittleness of rock: The elastic expansion of the core, which is relieved from high stress causes high expansion strains which cannot be accommodated by the core (Fig. 13.11).

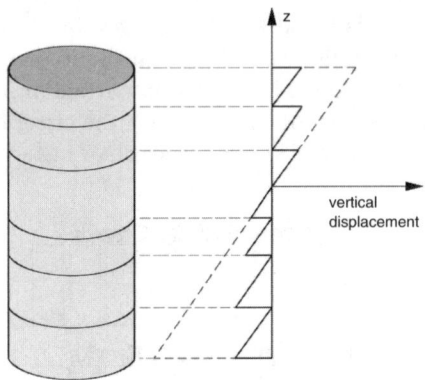

Fig. 13.11. To the explanation of core discing

13.6 Post-peak deformation

Beyond the peak of the stress strain diagram (e.g. Fig. 13.10) the stress decreases with increasing deformation. In soil mechanics, this behaviour is called softening. It is connected with an increase of the volume of the sample (Fig. 13.4), i.e. a decrease of the sample density ('loosening'). The registration of the post-peak deformation poses special requirements on the loading procedure, which is either stress-controlled or strain-controlled. Stress-control means that small stress (or force) increments are imposed and the resulting strain is registered. In case of softening the stress reduction steps must be applied very fast, otherwise acceleration to complete collapse will set on. This can be achieved with servo-controlled testing machines which execute several thousands of control cycles within a second. Strain-control means that small strain (or displacement) increments are imposed and the resulting stress (or force) changes are registered. It should be taken, however, into account that the elongation Δs of the actuator cannot be completely transferred to the rock specimen: A part of Δs is consumed to deform the frame of the testing machine and the load cell, as shown in Fig. 13.12. Let the stiffnesses of the frame and the sample be c_{frame} and c_{sample}, respectively. In the softening regime, c_{sample} is < 0. From $\Delta s = \Delta s_{frame} + \Delta s_{sample}$ and $c_{frame}\, \Delta s_{frame} = c_{sample}\, \Delta s_{sample}$ we obtain

13.6 Post-peak deformation

$$\Delta s_{sample} = \frac{c_{frame}}{c_{sample} + c_{frame}} \Delta s.$$

In order Δs_{sample} to be positive, the stiffness of the frame must be sufficiently high:

$$c_{frame} > -c_{sample}$$

This is the case with so-called hard testing machines. For very brittle rock, i.e. rock with very steep softening, no machine is stiff enough so that servo-controlling has to be applied.

Taking into account that beyond the peak the sample is either disintegrated or inhomogeneously deformed, the sense of stress strain curves in the post-peak regime has to be questioned. Within an inhomogeneously deformed sample the strain varies more or less irregularly from point to point and so does the stress as well. Therefore, it does not make sense to refer to *the* stress or strain in the sample.

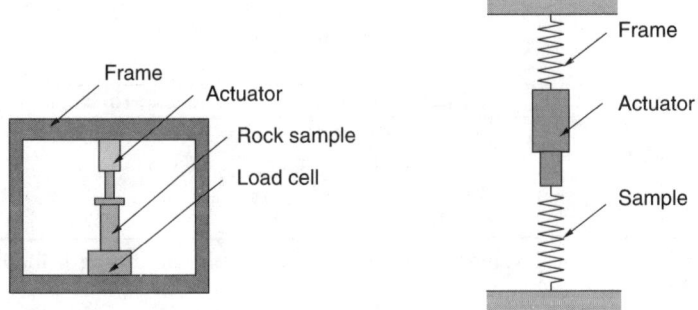

Fig. 13.12. Principal array of a testing machine and its idealisation.

13.6.1 Point load test

As sufficiently large rock samples cannot always be retrieved from exploration drillings in jointed rock, the point load test has been introduced. Rock hand pieces of arbitrary form are compressed to failure by two forces applied to two opposite points of the sample. Let F be the collapse load and a the distance of the two application points. The strength index I_s is defined as

$$I_s := \frac{F}{a^2}$$

and serves for the fast classification of rock. The uniaxial compression strength q_u can be inferred by experience from I_s (Table 13.1). For rocks with $q_u < 25$ MPa the point load test is unsuitable.

---

[13] from E. Hoek, P.K. Kaiser, W.F. Bawden, Support of Underground Excavations in Hard Rock. Balkema, 1995

250   13 Behaviour of soil and rock

Unconfined q_u (MPa)	Strength index, point-loading test I_s (MPa)	Estimation in the field	Examples
> 250	> 10	small fragments can be separated by repeated hammer blows. Hard sounding rock	basalt, diabas, gneiss, granite, quartzite
100 - 250	4 - 10	rock samples can be carved up only by several hammer blows	amphibolite, sandstone, basalt, gabbro, gneiss, granodiorit, limestone, marble, rhyolit, tuff
50 - 100	2 - 4	hand piece can be broken with a hammer blow	limestone, marble, phyllit, sandstone, slate
25 - 50	1 - 2	with hard impact the picke of the geologist hammer penetrates up to 5 mm into the rock. The rock surface can be grazed with a knife	slate, coal
5 - 25	–	cutable with knife	chalk, rock salt
1 - 5	–	disintegrates with hammer blows. Strong weathered rock can be treated.	

Table 13.1. Empirical values for unconfined strength and strength index of the point load test.[13]

13.6.2 Griffith's theory

Fracture mechanics try to explain failure by microscopic considerations of the material structure. GRIFFITH set up in 1920 a theory, according to which the strength of brittle materials is determined by the presence of microscopic fissures. Failure occurs if these fissures can grow. GRIFFITH assumed the fissures as cylindrical cavities with elliptical cross section and computed the stress distribution around the cavities with the help of the elasticity theory. When a fissure grows, the surface energy of the free surface increases. Also the potential energy in the stress field around the fissure is changed. A fissure will grow, if thereby energy can be released. From this energetic criterion GRIFFITH could determine the tensile strength with uniaxial extension tests to

$$\sigma = \sqrt{\beta \frac{E\gamma}{c}} \quad , \tag{13.6}$$

whereby E is YOUNG's modulus, γ the specific surface energy, $2c$ the fissure length and β a numerical constant of the order of magnitude of 1. The theory of GRIFFITH supplies only rarely realistic forecasts and today is not used, it does serve however as starting point for more detailed investigations.

13.6.3 Acoustic emission

During the deformation of rock, in particular in the proximity of failure, acoustic impulses are generated, which can be heard by amplifiers or with naked ear. This phenomenon was examined in detail, in hope that one can develop from it a warning method against forthcoming failure. It was observed that the number of impulses per time unit (frequency) rises with increasing stress. This increasing takes place however continuously, so that no significant threshold value is reached with forthcoming failure. The acoustic emission continues, even if the stress does not increase any longer but remains constant. The larger the energy of the sound impulses is, the smaller is their frequency. If the stress increases beyond the value of a possible pre-loading, then one registers a significant increase of the acoustic emission (so-called KAISER-effect).

13.6.4 Friction of joints

The maximum shear force T_f applicable on an plane joint is proportional to the normal force N, $T_f = \mu N$, where, according to the law of AMONTON, the coefficient of friction μ is independent of N and from the macroscopic contact area. Strictly speaking T_f grows sub-linear with N, which is approximated (with consideration of the associated shear- and normal stresses) by relations of the form $\tau_f = c + \mu\sigma$ or $\tau_f = \mu\sigma^n$. This non-linearity is not considered further here. For rock the coefficient of friction μ usually amounts between 0.4 and 0.7. The individual mineral constituent parts of the rock can have substantially smaller coefficients of friction, e.g. 0.1 to 0.2 with dry quartz or calcite.

The joint roughness affects the coefficient of friction,[14] one must however consider that it is altered by abrasion due to relative displacement. The resulting powdered mineral can increase the coefficient of friction. On freshly developed shear failure planes the coefficient of friction μ lies between 0.6 and 1.0.

Moistening the joint surface with water can increase or decrease the friction depending upon the rock. If the water in the joints is at a pressure p, then it is the effective normal stress that is determining the friction force.

A remarkable phenomenon with the friction between two rock blocks is the so-called *stick-slip*, a consecutive increase and decrease of friction (Fig. 13.13). Stick-slip is to be observed with all rocks, whereby quartz content increases the tendency to stick-slip. It is also favoured by increasing the normal stress.

---

[14]M.S. Paterson: Experimental rock deformation, the brittle field. Berlin: Springer, 1978

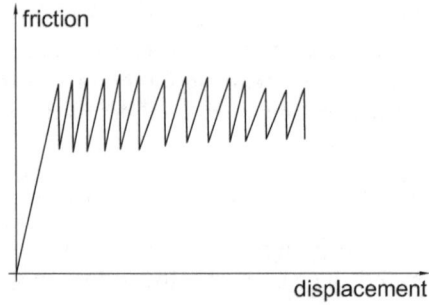

Fig. 13.13. Stick - slip

13.7 Anisotropy

Transverse isotropic elasticity is often assumed for stratified rock. If the x_3 Cartesian coordinate is perpendicular to the bedding plane (lamination), the stress strain relation reads:

$$\begin{pmatrix} \varepsilon_{11} \\ \varepsilon_{22} \\ \varepsilon_{33} \\ \varepsilon_{12} \\ \varepsilon_{23} \\ \varepsilon_{13} \end{pmatrix} = \frac{1}{E_1} \begin{pmatrix} 1 & -\nu_1 & -\nu_2 & 0 & 0 & 0 \\ -\nu_1 & 1 & -\nu_2 & 0 & 0 & 0 \\ -\nu_2 & -\nu_2 & 1 & 0 & 0 & 0 \\ 0 & 0 & 0 & 2(1+\nu_1) & 0 & 0 \\ 0 & 0 & 0 & 0 & E_1/G_2 & 0 \\ 0 & 0 & 0 & 0 & 0 & E_1/G_2 \end{pmatrix} \begin{pmatrix} \sigma_{11} \\ \sigma_{22} \\ \sigma_{33} \\ \sigma_{12} \\ \sigma_{23} \\ \sigma_{13} \end{pmatrix}$$

5 material constants, $E_1, E_2, \nu_1, \nu_2, G_2$ are now needed.

With anisotropic samples the strength depends on the direction of the applied stresses. E.g., the shear strength is smallest if the largest principal stress forms an angle of approximately 30° to the lamination plane (Fig. 13.14).

An increase of the lateral stress can increase or reduce the effect of anisotropy. The anisotropy of stratified rocks does not become apparent if the sample is rotated around axes, which are perpendicular to the lamination planes (Fig. 13.16). Other rotations however (e.g. by an angle ϑ in Fig. 13.15) affect the shear strength. A possibility to analytically express anisotropy is to consider reduced shear strength parameters c_l and φ_l at the lamination surfaces:

$$\tau_{fl} = c_l + \sigma \tan \varphi_l \quad . \tag{13.7}$$

In order to decide according to the failure criterion of MOHR-COULOMB whether a stress state $\sigma_1, \sigma_2, \sigma_3 = \sigma_2$ causes failure, one must examine all possible inclinations θ, $0° \leq \theta \leq 360°$, and calculate the shear and normal stresses acting upon these planes. For $\theta \neq \vartheta$ one sets the equation $\tau = c + \sigma \tan \varphi$ as failure criterion, and for $\theta = \vartheta$ equation 13.7. The dependence of the shear strength $(\sigma_1 - \sigma_2)_f$ on ϑ according to this theory is represented in Fig. 13.17.

13.8 Rate dependence and viscosity of soil and rock   253

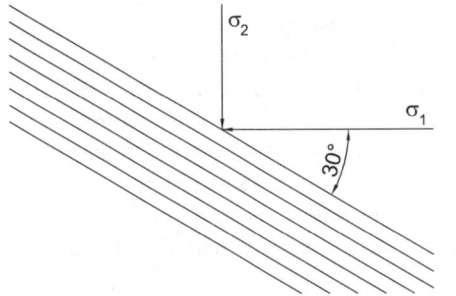

Fig. 13.14. Schistosity: with stratified rocks the shear strength is minimal, if the largest principal stress is inclined by ca 30° to the bedding.

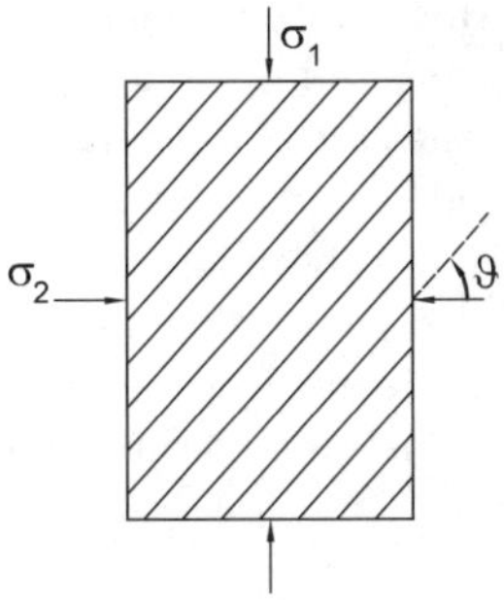

Fig. 13.15. The strength of laminated samples depends on the angle ϑ of bedding orientation.

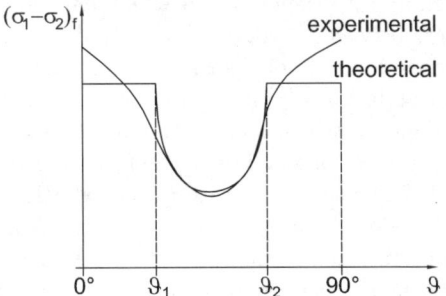

Fig. 13.16. The anisotropy of laminated samples remains undetectable, if the rotation occurs around axes, that are perpendicular to the bedding.

Fig. 13.17. Dependence of shear strength on the orientation of the lamination plane. For $\vartheta_1 \leq \vartheta \leq \vartheta_2$ the shear strength is controlled by the parameters c_l and φ_l

13.8 Rate dependence and viscosity of soil and rock

It is generally known that fluids have a property called viscosity which implies that the resistance against shearing increases with shearing rate. What about solids? An idealised concept for solids is 'rate independence', which means that the material behaviour is invariant upon changes of the time scale. E.g., doubling the rate of deformation will not have any influence on the stress to be obtained at a particular deformation. Elastic and elastoplastic materials are by definition rate independent. Thus, they do not exhibit creep or relaxation. It is recalled that creep means increase of deformation at constant stress and relaxation means decrease of stress at constant deformation (Fig. 13.18). The material constants of rate-independent materials do not bear the dimension of time.

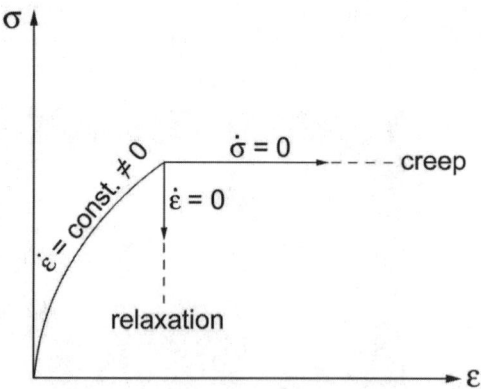

Fig. 13.18. Creep and relaxation (idealised)

In reality, solids exhibit rate dependence, which can be neglected for many processes. There are, however, processes, especially very fast or very slow ones, where rate dependence is of decisive importance. For instance, think about secondary consolidation of soil and folding of rock strata. A convenient way to detect rate dependence of soils is to conduct so-called jump tests: a deformation (e.g. a conventional triaxial test) is carried out with constant rate. At particular deformations the rate of deformation is changed from $\dot\varepsilon = \dot\varepsilon_a$ to $\dot\varepsilon = \dot\varepsilon_b$ with, say, $\dot\varepsilon_b = 10\dot\varepsilon_a$. For a rate dependent material, this will imply a corresponding stress change $\Delta\sigma$.[15,16] It has been observed that $\Delta\sigma \propto \log(\dot\varepsilon_b/\dot\varepsilon_a)$. Thus, the relation

$$\Delta\sigma = I_v \sigma \Delta(\log\dot\varepsilon)$$

---

[15] Prandtl, L.; Ein Gedankenmodell zur kinetischen Theorie der festen Körper ZAMM, 8, Heft 2, April 1928, 85-106

[16] F. Tatsuoka, et al., Time dependent deformation characteristics of stiff geomaterials in engineering practice. In: Pre-failure Deformation Characteristics of Geomaterials, Jamiolkowski et al, editors, Swets & Zeitlinger, Lisse, 2001, 1161-1262

13.8 Rate dependence and viscosity of soil and rock

has been established with the 'viscosity index' I_v being a material constant. Fig. 13.19 shows experimental results of strain controlled triaxial jump tests with dry sand.[17]

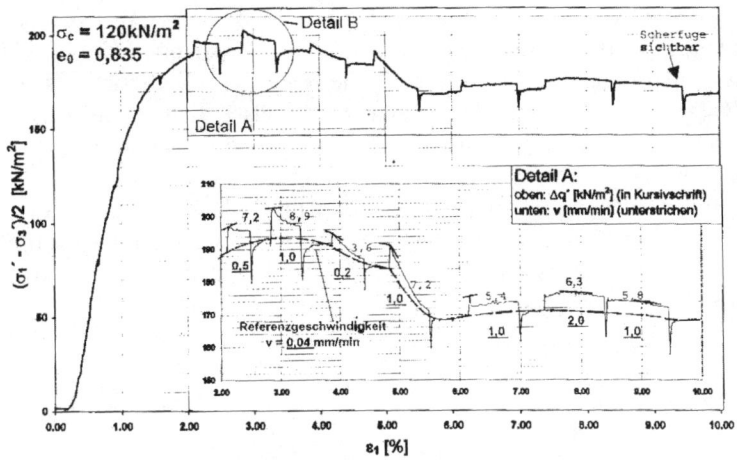

Fig. 13.19. Rate dependence of fine sand

It is interesting to note that the same relation has been found for the rate-dependent friction along rock interfaces by DIETERICH and RUINA.[18] Creep and relaxation are manifestations of rate dependence. Their interrelation is straightforward for the case of viscoelasticity (cf. Fig.14.22), it is however not so obvious for other relations, e.g. viscoplastic ones. Most of the existing experimental observations on creep and relaxation of rock are one-dimensional. I.e., only one component of stress or strain has been observed. The main findings are:

- The relaxation rate decreases with the logarithm of time, i.e. $\dot\sigma \propto \log t$.
- For deviatoric deformations, the creep rate $\dot\varepsilon$ increases with deviatoric stress σ, i.e. $\dot\varepsilon \propto \sigma^n$. This equation in often referred to as NORTON's law.
- Creep rate increases with temperature. This implies that creep is a thermally activated process. Such processes often obey the ARRHENIUS relation $\dot\varepsilon \propto \exp(-Q/RT)$, where Q and R are constants[19] and T is the absolute temperature.

---

[17] B. Eichorn, Der Einfluß der Schergeschwindigkeit beim Triaxialversuch, Diplomarbeit, Universität Innsbruck, 1999

[18] A. Ruina, Slip Instability and State Variable Friction Laws. *J. Geophys. Res.*, Vol. 88, No. B12, 10,359-10,370, Dec. 10, 1983

[19] R is the gas constant, $R = 8.314472$ J/(mol·K).

- Often, three stages of creep can be discerned: primary creep (decreasing creep rate), secondary creep (constant creep rate) and tertiary creep (increasing creep rate until, finally, collapse sets on).

The incorporation of rate-dependence into general constitutive equations is not easy. PERZYNA's approach is as follows: The plastic behaviour is modelled by means of a so-called yield function, the actual stress however may surpass the yield surface. Its distance to the yield surface determines the intensity of creep rate. A promising new concept is the theory of viscohypoplasticity[20].

An additional source of viscous behaviour in soils is due to the viscosity of pore fluids and the related dissipation of pore pressure according to the consolidation theory.

A very important issue in conventional tunnelling, the stand-up time, cannot be related to constitutive properties of the ground and, therefore, cannot be assessed on a rational basis. For safety, it should rather be considered as a fall-down time.

13.9 Size effect

The fact that the size of a sample affects its strength and other mechanical properties is called scale effect or size effect. The scale effect is more pronounced with presence of stress gradients, thus with inhomogeneous stress distribution (e.g. with the Brazilian test), and is reduced with increasing lateral pressure σ_2.

13.9.1 Size effect in rock

The strength of rock (and also of other materials like e.g. concrete) depends on the size of the tested sample in the sense that larger samples exhibit a lower strength. This scale effect cannot be modelled with a so-called simple material.[21] It is attributed to small defects which endow the continuum with an internal structure. GRIFFITH (1921) and WEIBULL (1939) explained the scale effect from the fact that the probability of a defect increases with the size of the considered sample.

---

[20]G. Gudehus, Prognose und Kontrolle von Kriechen und Relaxation in weichen Baugrund mittels Visko-Hypoplastizität, *Bauingenieur* **79**, September 2004, 400-409

[21]'Simple materials' are defined by the fact that the stress at a material point depends only on the deformation at this point and not on higher deformation gradients. The mechanical behaviour of simple materials can be completely revealed via tests with homogeneous deformation.

Fig. 13.20. Heap of sand[22]

The internal structure of rock is revealed if we consider its surface as it appears at natural or artificial rupture surfaces. Such surfaces are self-similar in the sense that a part of them is similar to the whole: If such a surface is subdivided in smaller parts, each part looks like the original surface. This is why photographs of rock surfaces and of granular soils do not reveal the size of the objects shown, unless another object of known size (e.g. a hammer or a coin) is added. Fig. 13.20 gives the impression of a high mountain although it shows a sand heap of approx. 20 cm height.

Self-similar irregular surfaces are often fractals, i.e. they have a fractal dimension. This statement has the following meaning: To cover a fractal curve or a fractal surface with squares (cubes) of the edge δ we need N squares (cubes) (Fig. 13.21). Obviously, N depends on δ: $N = N(\delta)$. The smaller δ, the larger N is. For usual (non-fractal) curves we have $N \propto \frac{1}{\delta}$ and for non-fractal surfaces we have $N \propto \frac{1}{\delta^2}$. In general $N \propto \frac{1}{\delta^D}$, where D (the so-called fractal dimension) is a fraction for fractals. E.g., the shoreline of Great Britain has the fractal dimension $D = 1.3$.

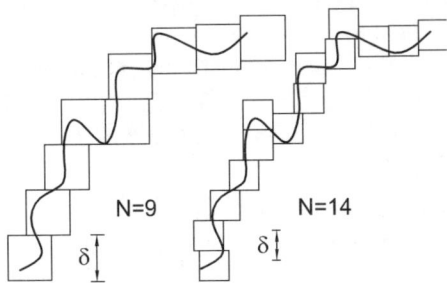

Fig. 13.21. The number N of squares needed to cover a curve depends on their edge length δ.

---

[22]photograph: Prof. G. Gudehus

The length of a fractal curve is obtained as $L \approx N\delta = \text{const} \cdot \delta^{1-D}$. Plotting L (or N) over δ in a log-log plot gives a straight line the slope of which gives D. Fractal rock rupture surfaces have a dimension $2 < D < 3$. Regarding very small cubical samples of the edge δ and the strength σ_δ [23], we obtain the compressive strength q_u of a sample with the diameter d as follows:
The compressive force at fracture is $F \propto \frac{1}{\delta^D} \sigma_d$ and the cross section of the sample is $A \propto \delta^2$. Hence, $q_u = F/A \propto \delta^{2-D}$ and with the sample diameter $d \propto \delta$ we finally have $q_u \propto d^{2-D}$. This equation explains the scale effect, i.e. the fact that q_u depends on d.

D can be obtained from experiments with samples of various sizes. The fractal dimension D can also be obtained from geometrical analysis of rock rupture surfaces. To this purpose the roughness of the surface is recorded with a mechanical pin or with a laser beam.[24] The fractal dimension of traced curves on rupture surfaces has been obtained for marble as $1.11 < D < 1.76$ and for sandstone $1.02 < D < 1.41$.

13.9.2 Size effect in soil

Also the cohesion of soil samples proves to be scale dependent: smaller samples have a larger cohesion. This can be described with fractals if we assume that the voids of the sample cross section have a fractal dimension.[25] Plotting the cohesion of London clay over the sample diameter gives the fractal dimension $D = 1.64$. There is some evidence that specific fracture processes lead to a number N of pieces which increases with decreasing size (length) r of the fragments: $N \propto r^{-D}$. If D is independent of r, then fracture is scale invariant. E.g., for morainic debris it has been found $D = 2.88$.[26]

13.9.3 Rodionov's theory

RODIONOV[27] assumes that the behaviour of rock is governed by inhomogeneities of, say, spherical form, the nature of which is not further specified. He only assumes that their sizes (e.g. diameters) l_i are distributed (in the sense of a grain size distribution) in such a way that the material does not possess any internal length. In other words, the size l_i of the largest inhomogeneity within a sample is proportional to the size of the sample itself. He further assumes

---

[23] σ_δ is called the renormalised strength

[24] C. Scavia (1996), The effect of scale on rock fracture toughness: a fractal approach. *Géotechnique* **46**, No. 4, 683-693, also: Chr.E. Krohn (1988), Sandstone Fractal and Euclidean Pore Volume Distributions. *J. of Geophys. Research*, **93**, No. B4, 3286-3296

[25] M.V.S. Bonala, L.N. Reddi (1999), Fractal representation of soil cohesion, *J. of Geotechn. and Geoenvironmental Eng.*, Oct. 1999, 901-904

[26] D.L. Turcotte (1986), Fractals and Fragmentation, *J. of Geophys. Res.* **91**, 1921-1926

[27] V.N. Rodionov, Ocherk Geomekhaniki, Nauchny Mir, Moscow, 1996.

that during loading there are being formed stress fields around the inhomogeneities. The deviatoric parts of the related stresses increase with deviatoric deformation (according to a constitutive equation such as $\mathring{\sigma}_{ij} = h_{ij}(\sigma_{kl}, \dot{\varepsilon}_{mn})$. RODIONOV assumes the linear elastic relation $\mathring{\sigma}^*_{ij} = 2G\dot{\varepsilon}^*_{ij}$ (where σ^*_{ij} is the deviatoric part of stress σ_{ij}) combined with relaxation:

$$\mathring{\sigma}^*_{mn} = 2G\dot{\varepsilon}^*_{mn} - \frac{v}{l_i}\sigma^*_{mn}.$$

Setting $\dot{\varepsilon}_{ij}$ = const and replacing tensors with scalars (e.g. components or appropriate invariants) he obtains therefrom

$$\sigma^* = 2G\varepsilon^* \frac{l_i}{v}\left(1 - \exp(-\frac{v}{l_i}t)\right).$$

Based on data from wave attenuation, RODIONOV assumes that v can be approximately considered as a 'universal' constant for rock:

$$v \approx 2 \cdot 10^{-6} \text{ cm/s}.$$

To a particular time t^* that elapsed from the begin of loading we can assign the length $l^* = vt^*$. We can then distinguish between large inhomogeneities ($l_i \gg l^*$) and small ones ($l_i \ll l^*$) noting, however, that this distinction depends on the time lapse:

Large inhomogeneities: $\exp(-l^*/l_i) \approx 1 - l^*/l_i \leadsto \sigma^* = 2G\varepsilon^*t^*$, i.e. deviatoric stress increases with time t^*, a process which eventually leads to fracture (fragile or brittle behaviour).

Small inhomogeneities: $\exp(-l^*/l_i) \approx 0 \leadsto \sigma^* = 2G\varepsilon^* l_i/v$, i.e. the material around small inhomogeneities behaves as a viscous material (cf. ductile behaviour).

We see thus that the material behaviour of the considered body, in particular the distinction between brittle and ductile behaviour, depends on the value of the parameter vt/l_i. As l_i correlates with a characteristic length (size) L of the body, we can consider the parameter vt/L as determining the material behaviour. If we consider a model test simulating a prototype (e.g. the convergence of a tunnel), where model and prototype consist of the same material, then the following similarity condition must be preserved according to RODIONOV:

$$\frac{t_{model}}{t_{prototype}} = \frac{L_{model}}{L_{prototype}}.$$

E.g., if we consider the convergence of two cylindrical cavities with radius R_1 and R_2, respectively, observed in the same squeezing rock, then the same convergence $u_1/R_1 = u_2/R_2$ occurs after the time lapses t_1 and t_2, respectively, where $t_1/t_2 = R_1/R_2$.

With increasing time, l^* increases until it obtains the value of L. Then the body necessarily behaves in a brittle way and failure sets eventually on.

13.10 Discrete models

The discontinuous nature of jointed rock can be taken into account by so-called discrete models, which separately regard each individual part ('block') and analyse the interaction between them. The equations governing this interaction are relatively simple, the large number of blocks as well as the consideration of three dimensions imply however that discrete models can be analysed only with the help of computer. Their strength is at the same time their weakness: they can be only used *ad hoc* and do not provide general statements about phenomenological-macroscopic quantities[28] (e.g. stresses). One must also consider that the precise location of the joints is in most cases not known.

The simplest discrete models are the so-called rigid body collapse mechanisms with plane or circle-cylindrical contact surfaces.[29] At the contact surfaces the shear strength is assumed to be fully mobilised, thus only limit states can be analysed.

A further step of development are the so-called *discrete element methods (DEM)*[30]. Discrete element methods (e.g. the program UDEC) are characterised by the following items:

1. They permit finite displacements and rotations (including upheavals) of the individual blocks.
2. They are equipped with algorithms, which detect the contacts of the individual blocks. Such algorithms require large computation time, which increases with the square of the number n of blocks. If the dimensions of the blocks are comparable, then one can reduce the computation time (proportional to n) by parcelling the regarded area.

The contacts of the individual blocks can be rigid *(hard contacts)* or deformable *(soft contacts)*, approximately modelled with HERTZ's law of elastic compression. Equally, the blocks can be rigid or deformable. In the latter case one computes the deformation of blocks with finite elements.

Discrete element methods with rigid contacts are applied also to so-called granular-dynamic investigations. Here granular media (e.g. soil) are considered as accumulations of balls or ellipsoids. One can simulate numerically some

---

[28] The same procedure, i.e. the consideration of individual bodies resp. particles, marks the kinetic gas theory. There, however, the temporal development of the starting situation permits — for so-called ergodic systems — the derivation of statements on phenomenological - macroscopic quantities, e.g. temperature and pressure

[29] see e.g. D. Kolymbas: Geotechnik - Bodenmechanik und Grundbau, Springer Verlag, 1997

[30] P.A. Cundall and R.D. Hart: Numerical Modeling of Discontinua; R.D. Hart: An Introduction to Distinct Element Modeling for Rock Engineering. Both in Comprehensive Rock Engineering, Volume 2, Pergamon Press, 1993, pages 231–243 and 245–261

problems of soil mechanics and chemical engineering (e.g. discharge from a silo).

Some further approaches based on rigid blocks capable to rotate, such as the key block theory of SHI and GOODMAN, are hardly tractable.

13.11 Rock mass strength

The determination of the rock mass strength is very difficult, or hardly possible. Depending upon the quantity of the individual blocks we can apply large-scale laboratory or field tests (see Section 13.13). Otherwise one has to resort to empirical values, which are obtained from back calculation of individual cases or are mere rules of thumb. Two approaches are widespread:

Approach of Protodyakonov: The unconfined strength q_d of a cubical rock sample with the edge length d is set in relationship with the uniaxial rock mass strength q_V:

$$\frac{q_d}{q_V} = \frac{d/a + m}{d/a + 1}$$

a is the joint spacing and m is an empirical reduction factor.

q_d	m
> 75 MPa	2 - 5
< 75 MPa	5 - 10

Approach of Hoek and Brown: The failure criterion is formulated as relationship between the largest principal stress σ_1 and the smallest principal stress σ_3. With soil the envelope of MOHR's circles at failure is in first approximation a straight line (Fig. 13.22), and the failure criterion reads

$$\frac{\sigma_1}{\sigma_c} = \frac{\sigma_3}{\sigma_c} \cdot \frac{1 + \sin\varphi}{1 - \sin\varphi} + 1 \quad ,$$

whereby σ_c is the unconfined strength (with $c > 0$).

For unjointed rock the failure condition (Fig. 13.22) can be represented by the equation

$$\frac{\sigma_1}{\sigma_{ci}} = \frac{\sigma_3}{\sigma_{ci}} + \sqrt{m_i \frac{\sigma_3}{\sigma_{ci}} + 1} \quad . \tag{13.8}$$

m_i is obtained by adjustment to results from triaxial tests. The index i is to signify 'intact rock'. Equ. 13.8 corresponds to a curved yield locus (i.e. envelope of MOHR circles). Note that also for soils the yield locus is curved, the linear yield locus being only a simplification.

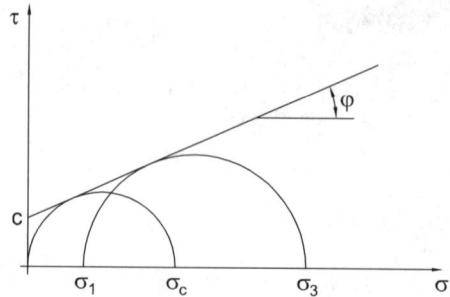

Fig. 13.22. Failure envelope for cohesive soil

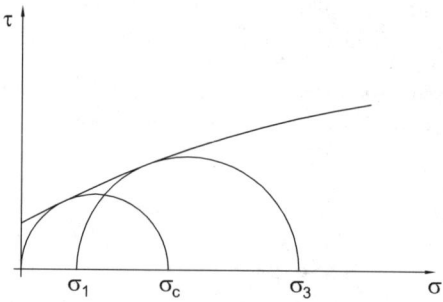

Fig. 13.23. Curved failure envelope for unjointed rock

For the strength of jointed rock HOEK and BROWN have suggested the following empirical relation:

$$\frac{\sigma_1}{\sigma_{ci}} = \frac{\sigma_3}{\sigma_{ci}} + \sqrt{m\frac{\sigma_3}{\sigma_{ci}} + s} \ ,$$

whereby m and s is to be empirically determined. Initially, m and s were given as functions of RMR, either by means of tables (e.g. Tab. 13.2) or by algebraic expressions such as

disturbed rock mass	undisturbed or interlocking rock mass
$m = m_i \ \exp\left(\dfrac{RMR-100}{14}\right)$	$m = m_i \ \exp\left(\dfrac{RMR-100}{28}\right)$
$s = \exp\left(\dfrac{RMR-100}{6}\right)$	$s = \exp\left(\dfrac{RMR-100}{9}\right)$

13.11 Rock mass strength

Later on, the values of m and s have been given in dependence of the 'Geological Strength Index' GSI and the 'Disturbance Factor' D:[31]

$$\frac{\sigma_1}{\sigma_{ci}} = \frac{\sigma_3}{\sigma_{ci}} + \left(m\frac{\sigma_3}{\sigma_{ci}} + s\right)^a$$

$$m = m \, \exp\left(\frac{GSI - 100}{28 - 14D}\right)$$

$$s = \exp\left(\frac{GSI - 100}{9 - 3D}\right)$$

$$a = \frac{1}{2} + \frac{1}{6}\left[\exp\left(-\frac{GSI}{15}\right) - \exp\left(-\frac{20}{3}\right)\right]$$

The determination of GSI and D occurs on the basis of tables and diagrams[32] and is rather vague.

According to newer results, the empirical relationship by HOEK and BROWN is well suitable for jointed *ductile* rock.[33] Thus, it is recommended rather for shallow tunnels, whereas it is not suitable for brittle rock mass, as encountered around deep tunnels.

The RMR-value is also used for further estimations for rock masses. For example YOUNG's modulus E of the ground is estimated as follows:

$$E\,[\text{GPa}] \approx 2 \cdot RMR - 100 \text{ for } RMR > 50$$
$$E\,[\text{GPa}] \approx 10^{(RMR-10)/40} \text{ for } RMR < 50$$

Note that such empirical estimations[36] are based upon specific experiences that are not generally valid. They can be consulted, if necessary, for rough estimations with the reservation of further examinations.

The popularity of the HOEK-BROWN criterion is due to the fact that it is the only available tool that gives a whatsoever answer to the main question of rock mechanics referring to the mechanical behaviour and, in particular, the strength of rock mass. This question is still unanswered, and the insistent reference to the HOEK-BROWN criterion should not hide the fact that this criterion does *not* contain any rational approach. Originally based on model

---

[31] E. Hoek, A brief history of the development of the Hoek-Brown failure criterion, www.rockscience.com

[32] E. Hoek, C. Carranza-Torres, B. Corkum, Hoek-Brown failure criterion – 2002 edition, www.rockscience.com

[33] P.K. Kaiser and others, Underground works in hard rock tunnelling and mining, GeoEng 2000, Melbourne

[35] E. Hoek, Strength of jointed rock masses, 23rd Rankine Lecture, *Géotechnique* **33**, No. 3, 187-223

[36] Virtually, every estimation is empirical. One should distinguish, however, between estimations based on comprehensible rational rules (e.g. the estimation of the friction angle with laboratory tests) and ones based on mere experience.

Rock	RMR	a	m	s
Limestone, marble, dolomite	100	∞	7	1
	85	1 … 3 m	3.5	0.1
	65	1 … 3 m	0.7	$4 \cdot 10^{-3}$
	44	0.3 … 1 m	0.14	$1 \cdot 10^{-4}$
	23	3 …		0
Slate	100	∞	10	1
	85	1 … 3 m	5	0.1
	65	1 … 3 m	1	$4 \cdot 10^{-3}$
	44	0.3 … 1 m	0.2	$1 \cdot 10^{-4}$
	23	3 … 50 cm	0.05	$1 \cdot 10^{-5}$
	3	< 5 cm	0.01	0
Sandstone, quartzite	100	∞	15	1
	85	1 … 3 m	7.5	0.1
	65	1 … 3 m	1.5	$4 \cdot 10^{-3}$
	44	0.3 … 1 m	0.3	$1 \cdot 10^{-4}$
	23	3 … 50 cm	0.08	$1 \cdot 10^{-5}$
	3	< 5 cm	0.015	0
Magmatic, fine-grained	100	∞	17	1
	85	1 … 3 m	8.5	0.1
	65	1 … 3 m	1.7	$4 \cdot 10^{-3}$
	44	0.3 … 1 m	0.34	$1 \cdot 10^{-4}$
	23	3 … 50 cm	0.09	$1 \cdot 10^{-5}$
	3	< 5 cm	0.017	0
Magmatic, coarse-grained	100	∞	25	1
	85	1 … 3 m	12.5	0.1
	65	1 … 3 m	2.5	$4 \cdot 10^{-3}$
	44	0.3 … 1 m	0.5	$1 \cdot 10^{-4}$
	23	3 … 50 cm	0.13	$1 \cdot 10^{-5}$
	3	< 5 cm	0.025	0

Table 13.2. m- and s-values for rock mass strength according to HOEK and BROWN[35]

tests with concrete bricks, thermally treated marble, and triaxial tests with jointed andesite,[37] the criterion remains a possibly useful but limited and non-tractable tool, which should not be overestimated.

A promising approach to study the mechanical behaviour of rock mass is to fracture an intact rock sample in a controlled manner (say, within a triaxial apparatus) and then to unload the sample. Reloading will now — to some extent — reveal the behaviour of the rock mass.

---

[37]E. Hoek, Strength of jointed rock masses, 23rd Rankine Lecture, *Géotechnique* 33, No. 3, 187-223

13.12 Swelling

Some minerals have the property to increase their volume when absorbing water. If the volume increase is inhibited, then a corresponding pressure is exerted upon the containment. It is distinguished between mechanical, osmotic, intracrystalline and hydration swelling[38,39]. This distinction refers to the mechanisms of water attachment but not to the phenomenology of swelling, which in all cases stated above is due to the affinity of the minerals to water. Thus, we retain only the distinction between the physicochemical swelling, which is a volume increase due to a whatsoever attachment of water, and the mechanical swelling, which is a volume increase due to unloading. With mechanical laboratory tests, swelling can be revealed in the oedometer with access to water (so-called HUDER-AMBERG test). With constant normal stress the sample expands with time. If the strain is kept constant (i.e. inhibited deformation), then the stress increases with time (Fig. 13.24).

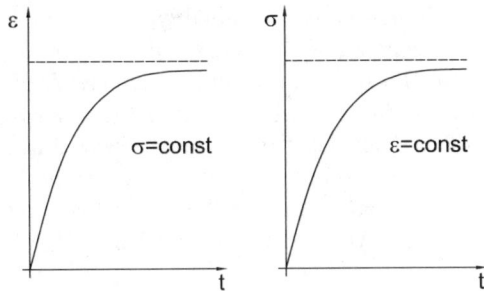

Fig. 13.24. Swelling and pressure due to inhibited swelling

In the laboratory the swelling pressure is reported to rise up to 100 MPa. Massive anhydrite ($CaSO_4$) is impermeable and, therefore, not prone to swelling. In mixtures of anhydrite and swelling clay minerals, however, clay swells first and creates thus access to water in such a way that finally also anhydrite swells. A slowdown of swelling followed by acceleration, as observed in laboratory tests, has been attributed to the transition from clay swelling to anhydrite swelling (Fig. 13.25).

A slowdown of swelling should, therefore, not be misinterpreted as termination of swelling. This would considerably underestimate the swelling potential.

---

[38] H.H. Einstein: Tunnelling in Difficult Ground - Swelling Behaviour and Identification of Swelling Rocks. *Rock Mechanics and Rock Engineering*, 1996, 29(3), 113-124

[39] An additional mechanism of swelling occurs in soils stabilised with lime. Sulfate ions may lead to formation of Etringite which can swell if brought in contact with water, see D. Dermatas: Ettringite-induced swelling in soils: State-of-the-art. *Appl. Mech. Rev.* Vol. 48, No. 10, 1995, 659-673

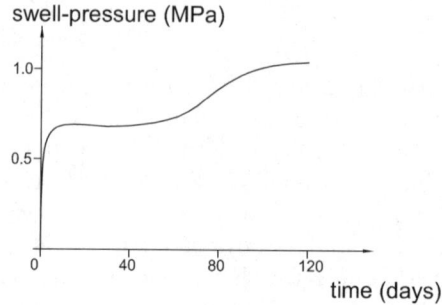

Fig. 13.25. Transition from clay swelling to anhydrite swelling

Even a test duration of two years can, in some cases, be insufficient to properly assess the swelling potential.

Within excavated tunnels, the access of water that may cause swelling is either from penetration of aquifers or due to the entrance of precipitation water from the portals. Air humidity is, presumably, not so important as swelling is observed only in the rock adjacent to the tunnel invert, up to a depth that roughly corresponds to the span of the tunnel. Upheavals of the invert due to swelling may last several decades and amount up to several meters.

During excavation of the Belchen tunnel in the Swiss Jura, swelling has destroyed the drainage. Invert reinforcement placed in the evening had to be removed next morning, because the swelling rock had taken the free space left for the concrete cover. An attempt for a rational description of swelling is given in Appendix H.

Hints for the construction of tunnels within swelling ground can be found in the related recommendations of the ISRM.[40] Upheavals due to swelling can either be resisted (so-called active- or rigid-design, cf. Engelberg-tunnel[41], Fig. 13.26) or freely allowed for (so-called passive design), in which case regular after profiling is necessary. The swelling pressure usually acts upon the lower half of the tunnel lining and is limited by the resistance of the overburden rock against upheaval (Fig. 13.26 right).

The tendency of a rock to swell can be investigated best with mineralogical analysis. There are, however, also some hints[42], e.g. shrinkage cracks may point to ability for swelling. Rocks with high contents of clay minerals may

---

[40] International Society for Rock Mechanics, Commission on Swelling Rock. Comments and recommendations on design and analysis procedures for structures in argillaceous swelling rock. *Int. J. Rock Mechanics Min. Sci. & Geomech. Abstr.* Vol. 31, No. 5, p. 535-546, 1994

[41] *Tunnels & Tunnelling International*, March 1998, 23-25

[42] International Society for Rock Mechanics, Commission on Swelling Rock. Suggested Methods for rapid field identification of swelling and slaking rocks. *Int. J. Rock Mechanics Min. Sci. & Geomechanics Abstracts* Vol. 31, No. 5, 547-550, 1994

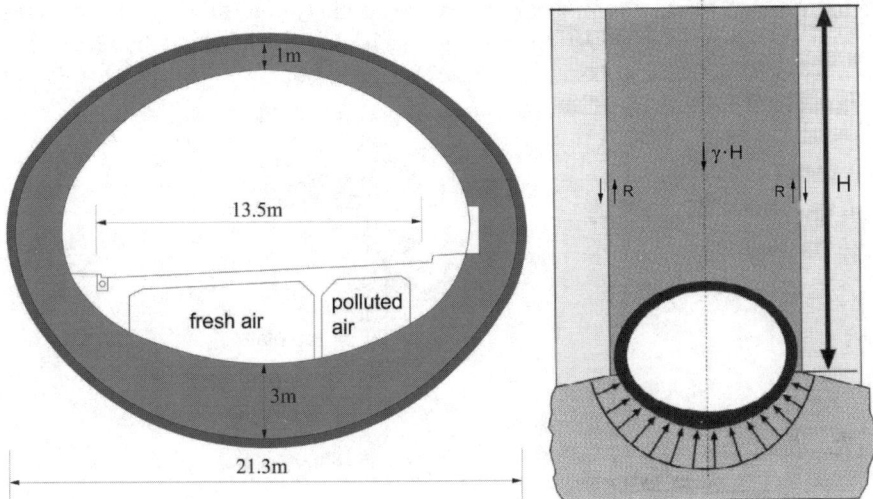

Fig. 13.26. Engelberg base tunnel: stiff lining to withstand swelling (left), limitation of pressure due to swelling (right)

also swell. Rubbed against the fingers it is felt like soap and has also a creamy taste. A 1-2 cm^3 sized piece of dry rock put into a glass of water will burst out within the first 30 seconds if it contains swelling minerals. Anhydrite can be distinguished from limestone by means of hydrochloric acid. Swelling is often confound with squeezing[43].

13.13 Field tests

By applying stresses and measurement of the resulting displacements and rotations we try to estimate the stiffness and strength of the rock in situ. The measurements are evaluated according to elastic solutions or empirical formulas. Apart from the in-situ shear and triaxial tests and the various soundings used in soil mechanics (e.g. SPT), the field tests of rock mechanics are based on the expansion of various cavities. The expansion of cavity is recorded vs. the applied pressure. Comparing the results with existing analytical solutions, the stiffness of the surrounding rock mass can be inferred. If the applied deformation is large enough to induce plastic flow of the rock, then also the strength characteristics can be — approximately — inferred.

---

[43]M. Panet: Two Case Histories of Tunnels through Squeezing Rocks. *Rock Mech. & Rock Engineering* (1996) 29 (3), 155-164; see also: G. Mesri et al.: Meaning, measurement and field application of swelling pressure of clay shales. *Géotechnique* 44, 1, 129-145 (1994), esp. Fig. 9

Shear test: Normal and shear forces are applied by means of hydraulic pistons. Their lines of action intersect eachother in the shear band, so that no tilt moments appear (Fig. 13.27). In a similar way can be carried out triaxial tests (Fig. 13.28).

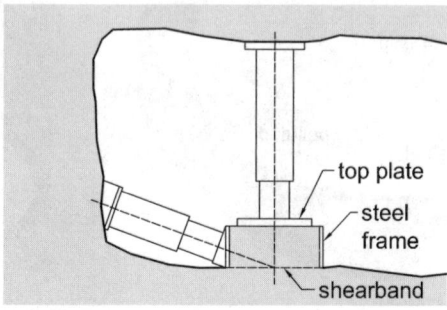

Fig. 13.27. Shear test in situ

Fig. 13.28. Exploratory gallery with field triaxial testing[44]

Flat jacks: These are placed in rock slots filled with mortar (Fig. 13.29) and subsequently expanded by application of pressure.

Pressure chamber: A closed cavity in the rock is filled with a liquid. After equalizing temperature with the surrounding rock the liquid is pressurized, and the appropriate deformations are measured.

Radial press: Between a steel ring and the wall of a tunnel or exploration adit are placed flat jacks. The expansion of the cavity in front and behind the radial press is measured[45] (Fig. 13.30, 13.31).

---

[44] *Tunel*, **9** (2000) 2, p. 19

[45] A variety of measured values can be seen in G. Seeber, Druckstollen und Druckschächte, Enke in Georg Thieme Verlag, Stuttgart-New York, 1999

13.13 Field tests   269

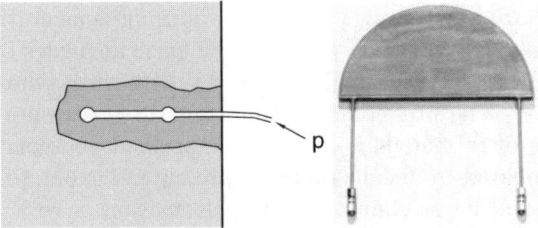

Fig. 13.29. Flat jack

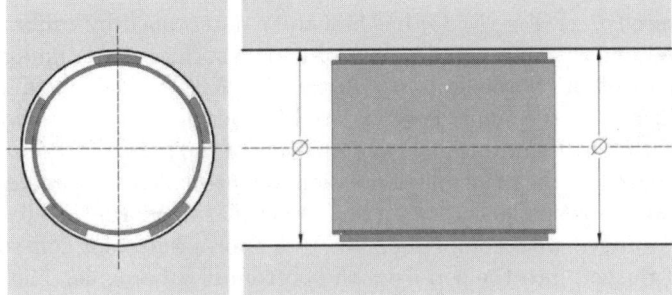

Fig. 13.30. Cross- and longitudinal sections of the radial press

Fig. 13.31. TIWAG radial press[46]

---

[46]Beiträge zur Technikgeschichte Tirols, Sonderheft 1984, Wagnersche Universitäts-Buchhandlung, Innsbruck, 1984

Borehole expansion tests: There are various implementations resulting in a rather confusing diversity of tests. The basic idea goes back to KÖGLER (1934) and has later been developed by MÉNARD, who introduced the *pressiometer*. The pressiometer comprises three pressure chambers with walls made of rubber. It is lowered to specified positions within a borehole and inflated by means of water pressure. Thanks to the inflation of the upper and lower chambers, the deformation of rock adjacent to the intermediate chamber can be considered as plane axisymmetric deformation. The pressure versus volume increase is recorded. Of course, a part of the pressure is needed to overcome the pressiometer's own stiffness. This pressure (p_i, cf. Fig. 13.33) has first to be determined by calibration tests carried out within a containment of known stiffness. It is important that the wall of the borehole is not disturbed. With the so-called *self-boring pressuremeter* the same device is used for boring (the excavated material is removed by inner flushing) and inflation of the borehole. Some other devices based on the principle of pressiometer are called *dilatometer*. The *flat dilatometer* is spade-shaped and has a circular membrane (∅ 6 cm) which is pressurized with compressed air. It is appropriated for soft soils and is being pushed into the soil from the bottom of a borehole. The *Goodman jack* consists of two jaws that are hydraulically jacked against each other and, thus, penetrate into the walls of the borehole. Force vs. displacement are recorded.

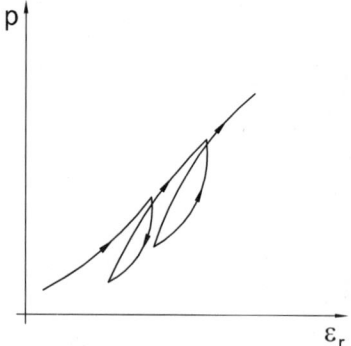
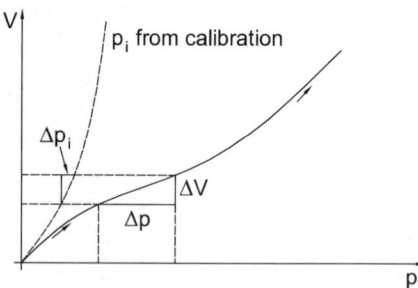

Fig. 13.32. Typical p - ε_r - curve obtained with the dilatometer test

Fig. 13.33. V vs. p curve obtained with pressiometer at loading. The deformation modulus is obtained as $(\Delta p - \Delta p_i)/\Delta V$.

14

Stress and deformation fields around a deep circular tunnel

14.1 Rationale of analytical solutions

The analytic representation of stress- and deformation fields in the ground surrounding a tunnel succeeds only in some extremely simplified special cases, which are rather academic. Nevertheless, analytical solutions offer the following benefits:

- Being exact solutions, they provide insight into the basic mechanisms (i.e. displacements, deformation and stress fields) of the considered problem.
- They provide insight of the role and the importance of the involved parameters.
- They can serve as benchmarks to check numerical solutions.

In this section, some solutions are introduced which are based on HOOKE's law, the simplest material law for solids. The underground is regarded here as linear-elastic, isotropic semi-infinite space, which is bound by a horizontal surface, the ground surface. The tunnel is idealised as a tubular cavity with circular cross section. Before its construction, the so-called primary stress state prevails. This stress state prevails also after the construction of the tunnel in a sufficiently large distance (so-called far field).

14.2 Some fundamentals

The equilibrium equation of continuum mechanics written in cylindrical coordinates reveals the mechanism of arching in terms of a differential equation. For axisymmetric problems, as they appear in tunnels with circular cross section, the use of cylindrical coordinates (Fig. 14.1) is advantageous. In axisymmetric deformation, the displacement vector has no component in θ-direction: $u_\theta \equiv 0$. The non-vanishing components of the strain tensor

14 Stress and deformation fields around a deep circular tunnel

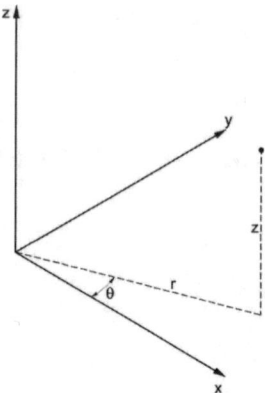

Fig. 14.1. Cylindrical coordinates r, θ, z

$$\varepsilon_{rr} = \frac{\partial u_r}{\partial r}, \quad \varepsilon_{\theta\theta} = \frac{1}{r}\frac{\partial u_\theta}{\partial \theta} + \frac{u_r}{r}, \quad \varepsilon_{zz} = \frac{\partial u_z}{\partial z}$$

$$\varepsilon_{r\theta} = \varepsilon_{\theta r} = \frac{1}{2}\left(\frac{1}{r}\frac{\partial u_r}{\partial \theta} - \frac{u_\theta}{r} + \frac{\partial u_\theta}{\partial r}\right)$$

$$\varepsilon_{rz} = \varepsilon_{zr} = \frac{1}{2}\left(\frac{\partial u_r}{\partial z} + \frac{\partial u_z}{\partial r}\right)$$

$$\varepsilon_{\theta z} = \varepsilon_{z\theta} = \frac{1}{2}\left(\frac{1}{r}\frac{\partial u_z}{\partial \theta} + \frac{\partial u_\theta}{\partial z}\right)$$

reduce, in this case, to

$$\varepsilon_r = \frac{\partial u_r}{\partial r}, \quad \varepsilon_\theta = \frac{u_r}{r}, \quad \varepsilon_z = \frac{\partial u_z}{\partial z} \quad ,$$

where u_r and u_z are the displacements in radial and axial directions, respectively.

The stress components σ_r, σ_θ, σ_z are principal stresses (Fig. 14.3). The equation of equilibrium in r-direction reads:

$$\frac{\partial \sigma_r}{\partial r} + \frac{\sigma_r - \sigma_\theta}{r} + \varrho g \cdot \boldsymbol{e}_r = 0 \tag{14.1}$$

and in z-direction:

$$\frac{\partial \sigma_z}{\partial z} + \varrho g \cdot \boldsymbol{e}_z = 0 \quad . \tag{14.2}$$

Herein, ϱ is the density, ϱg is the unit weight, $\boldsymbol{e}_r$ and $\boldsymbol{e}_z$ are unit vectors in r- and z-directions. The second term in equation 14.1 describes arching. This can be seen as follows: If r points to the vertical direction z (Fig. 14.2), then Equ. 14.1 reads:

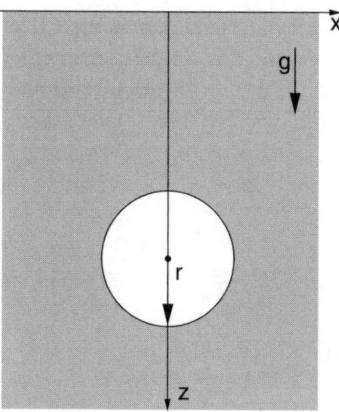

Fig. 14.2. To the explanation of Equ. 14.3. Due to gravity **g**, axisymmetric conditions prevail only for $x = 0$.

$$\frac{d\sigma_z}{dz} = \gamma - \frac{\sigma_x - \sigma_z}{r} \quad . \tag{14.3}$$

Herein, the term $(\sigma_x - \sigma_z)/r$ is responsible for that fact that σ_z does not increase linearly with depth (i.e. $\sigma_z = \gamma z$). In case of arching, i.e. for $(\sigma_x - \sigma_z)/r > 0$, σ_z increases underproportionally with z. Note that this term, and thus arching, exists only for $\sigma_r \neq \sigma_\theta$. This means that arching is due to the ability of a material to sustain deviatoric stress, i.e. shear stress. No arching is possible in fluids. This is why soil/rock often 'forgives' shortages of support, whereas (ground)water is merciless.

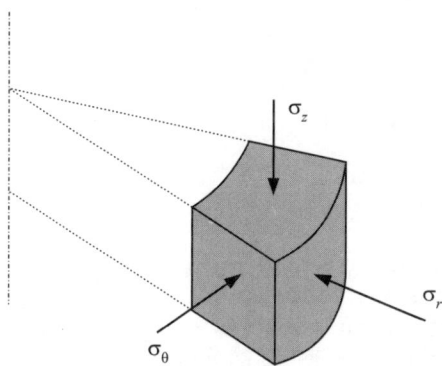

Fig. 14.3. Components of the stress tensor in cylindrical coordinates

The equilibrium equation in θ-direction,

$$\frac{1}{r} \cdot \frac{\partial \sigma_\theta}{\partial \theta} = 0, \tag{14.4}$$

is satisfied identically, as all derivatives in θ-direction vanish in axisymmetric stress fields.

The arching term can be easily explained as follows: Consider the volume element shown in Fig. 14.4.

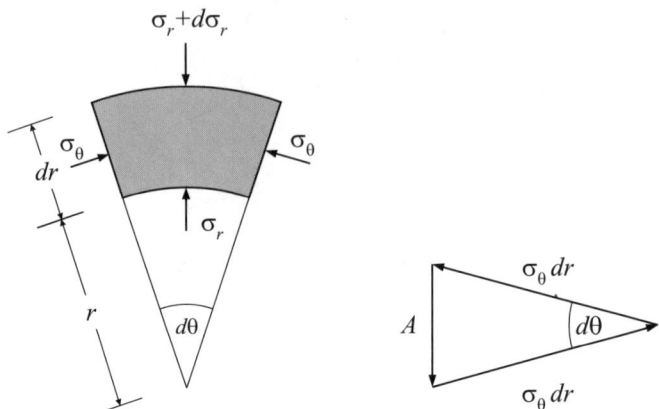

Fig. 14.4. Equilibrium of the volume element in r-direction

The resultant A of the radial stresses reads

$$A = (\sigma_r + d\sigma_r) \cdot (r + dr)d\theta - \sigma_r r d\theta \approx \sigma_r dr d\theta + d\sigma_r r d\theta$$

and should counterbalance the resultant of the tangential stresses σ_θ. The vectorial sum of forces shown in Fig. 14.4 yields

$$\frac{A}{\sigma_\theta dr} = d\theta \quad .$$

It then follows (for $\mathbf{g} \cdot \mathbf{e}_r = 0$):

$$\frac{d\sigma_r}{dr} + \frac{\sigma_r - \sigma_\theta}{r} = 0 \quad .$$

With reference to the arching term $\frac{\sigma_\theta - \sigma_r}{r}$ attention should be paid to r. At the tunnel crown, r is often set equal to the curvature radius of the crown. However, this is not always true. If we consider the distribution of σ_z and σ_x above the crown at decreasing support pressures p, we notice that for $K = \sigma_x/\sigma_z < 1$ the horizontal stress trajectory has the opposite curvature than the tunnel crown (Fig. 14.5)

Equations 14.1, 14.2 and 14.4 are the special case of the equilibrium equations in cylindrical coordinates, which in the general case read:

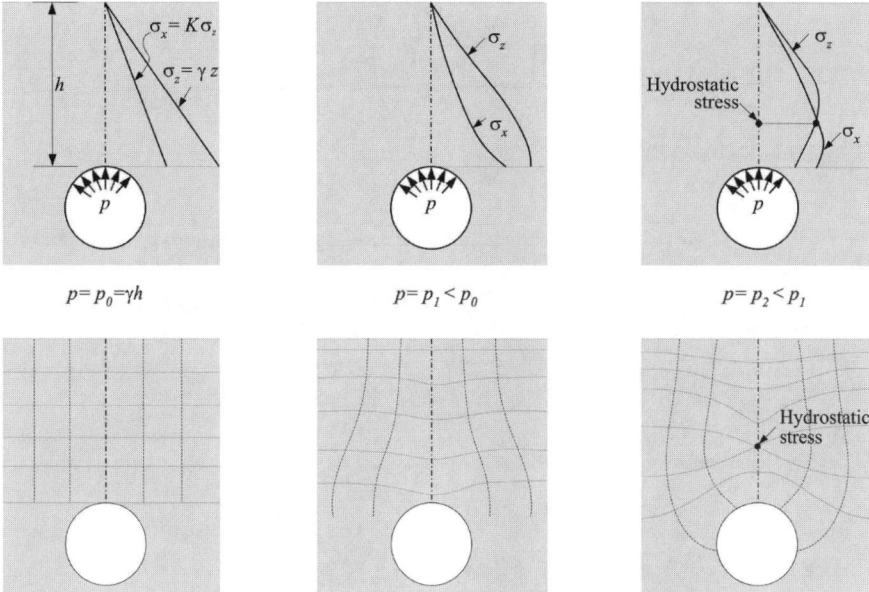

Fig. 14.5. Stress distributions above the crown for various values of support pressure p, and corresponding stress trajectories. At the hydrostatic point (right), the curvature radius of the stress trajectory vanishes, $r = 0$.

$$\frac{\partial \sigma_{rr}}{\partial r} + \frac{1}{r} \cdot \frac{\partial \sigma_{\theta r}}{\partial \theta} + \frac{\partial \sigma_{zr}}{\partial z} + \frac{1}{r} \cdot (\sigma_{rr} - \sigma_{\theta\theta}) + \varrho b_r = \varrho a_r$$

$$\frac{\partial \sigma_{r\theta}}{\partial r} + \frac{1}{r} \cdot \frac{\partial \sigma_{\theta\theta}}{\partial \theta} + \frac{\partial \sigma_{z\theta}}{\partial z} + \frac{1}{r} \cdot (\sigma_{r\theta} + \sigma_{\theta r}) + \varrho b_\theta = \varrho a_\theta$$

$$\frac{\partial \sigma_{rz}}{\partial r} + \frac{1}{r} \cdot \frac{\partial \sigma_{\theta z}}{\partial \theta} + \frac{\partial \sigma_{zz}}{\partial z} + \frac{1}{r} \cdot \sigma_{rz} + \varrho b_z = \varrho a_z$$

$\boldsymbol{b} = \{b_r, b_\theta, b_z\}$ and $\boldsymbol{a} = \{a_r, a_\theta, a_z\}$ are the mass forces (i.e. force per unit mass) and acceleration, respectively.

For problems with plane deformation (plane strain), the stress is usually represented in Cartesian coordinates as:

$$\begin{pmatrix} \sigma_{xx} & \sigma_{xy} & 0 \\ \sigma_{xy} & \sigma_{yy} & 0 \\ 0 & 0 & \sigma_{zz} \end{pmatrix}$$

or in cylindrical coordinates as

14 Stress and deformation fields around a deep circular tunnel

$$\begin{pmatrix} \sigma_{rr} & \sigma_{r\theta} & 0 \\ \sigma_{r\theta} & \sigma_{\theta\theta} & 0 \\ 0 & 0 & \sigma_{zz} \end{pmatrix} .$$

The transformation rules are:

$$\sigma_{xx} = \frac{\sigma_{rr}+\sigma_{\theta\theta}}{2} + \frac{\sigma_{rr}-\sigma_{\theta\theta}}{2}\cos 2\theta - \sigma_{r\theta}\sin 2\theta$$

$$\sigma_{yy} = \frac{\sigma_{rr}+\sigma_{\theta\theta}}{2} - \frac{\sigma_{rr}-\sigma_{\theta\theta}}{2}\cos 2\theta + \sigma_{r\theta}\sin 2\theta$$

$$\sigma_{xy} = \frac{\sigma_{rr}-\sigma_{\theta\theta}}{2}\sin 2\theta + \sigma_{r\theta}\cos 2\theta$$

and

$$\sigma_{rr} = \frac{\sigma_{xx}+\sigma_{yy}}{2} + \frac{\sigma_{xx}-\sigma_{yy}}{2}\cos 2\theta + \sigma_{xy}\sin 2\theta$$

$$\sigma_{\theta\theta} = \frac{\sigma_{xx}+\sigma_{yy}}{2} - \frac{\sigma_{xx}-\sigma_{yy}}{2}\cos 2\theta - \sigma_{xy}\sin 2\theta$$

$$\sigma_{r\theta} = -\frac{\sigma_{xx}-\sigma_{yy}}{2}\sin 2\theta + \sigma_{xy}\cos 2\theta \quad .$$

14.3 Geostatic primary stress

An often encountered primary stress (for horizontal ground surface) is $\sigma_{zz} = \gamma z$, $\sigma_{xx} = \sigma_{yy} = K\sigma_{zz}$, whereby z is the Cartesian coordinate pointing downward, γ is the specific weight of the rock and K is the so-called lateral stress coefficient. For non-cohesive materials K has a value between the active and the passive earth pressure coefficient $K_a \leq K \leq K_p$, and we can often set $K = K_0 = 1 - \sin\varphi$. The stress field around a tunnel has to fulfil the equations of equilibrium

$$\frac{\partial \sigma_{zz}}{\partial z} + \frac{\partial \sigma_{zx}}{\partial x} = \gamma \quad , \quad \frac{\partial \sigma_{zx}}{\partial z} + \frac{\partial \sigma_{xx}}{\partial x} = 0 \quad ,$$

as well as the boundary conditions at the ground surface ($z = 0$) and at the tunnel wall. We assume that the tunnel is unsupported, so the normal and the shear stress at the tunnel wall must disappear. The analytical solution of this problem is extremely complicated[1] and consequently offers no advantages compared to numerical solutions (e.g. according to the method of finite elements, Fig. 14.6).

---

[1] R.D. Mindlin: Stress distribution around a tunnel, *ASCE Proceedings*, April 1939, 619-649

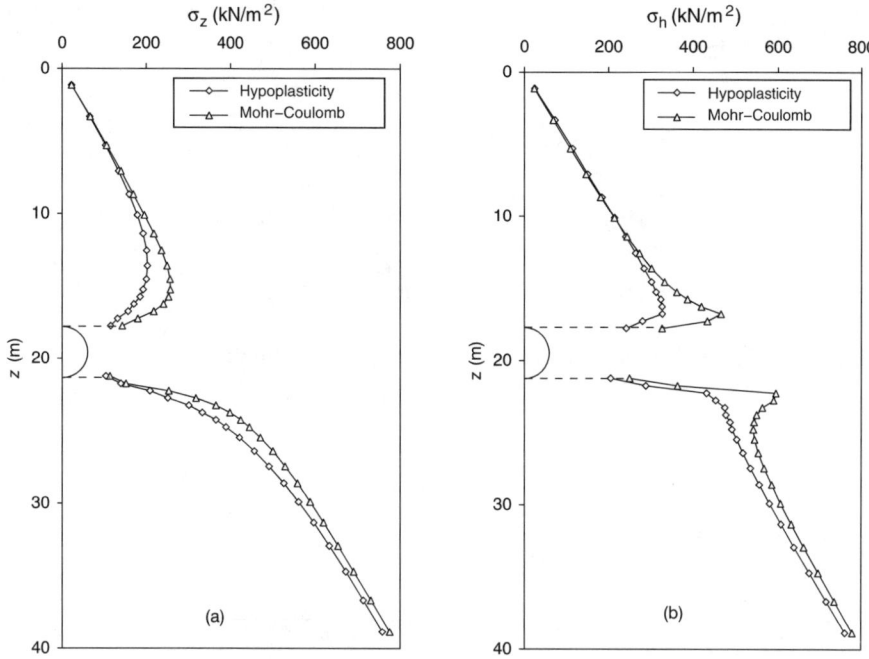

Fig. 14.6. Distributions of vertical and horizontal stress along the vertical symmetry axis. Circular tunnel cross section ($r = 1.0$m). Numerically obtained with hypoplasticity and Mohr-Coulomb elastoplasticity[3]

The analytical solution can be simplified, if one assumes the primary stress in the neighbourhood of the tunnel as constant (Fig. 14.7) and not as linear increasing: $\sigma_{zz} \approx \gamma H$, $\sigma_{xx} \approx K\gamma H$. This approximation is meaningful for deep tunnels ($H \gg r$). The stress field around the tunnel can then be represented in polar coordinates as follows:

$$\sigma_{rr} = \gamma H \left[\frac{1+K}{2}\left(1 - \frac{r_0^2}{r^2}\right)\right] + \gamma H \left[\frac{1-K}{2}\left(1 + 3\frac{r_0^4}{r^4} - 4\frac{r_0^2}{r^2}\right)\cos 2\vartheta\right]$$

$$\sigma_{\vartheta\vartheta} = \gamma H \left[\frac{1+K}{2}\left(1 + \frac{r_0^2}{r^2}\right)\right] - \gamma H \left[\frac{1-K}{2}\left(1 + 3\frac{r_0^4}{r^4}\right)\cos 2\vartheta\right] \quad (14.5)$$

$$\sigma_{r\vartheta} = -\gamma H \frac{1-K}{2}\left(1 - 3\frac{r_0^4}{r^4} + 2\frac{r_0^2}{r^2}\right)\sin 2\vartheta$$

---

[3]Tanseng, P., Implementations of Hypoplasticity and Simulations of Geotechnical Problems: Including Shield Tunnelling in Bangkok Clay, PhD Thesis at University of Innsbruck, 2004

14 Stress and deformation fields around a deep circular tunnel

One can easily verify that this solution fulfils the boundary conditions: For $r = r_0$ obviously $\sigma_{rr} = \sigma_{r\vartheta} = 0$, and for $r \to \infty$ is

$$\sigma_{rr} = \gamma H \left(\frac{1+K}{2} + \frac{1-K}{2} \cos 2\vartheta \right)$$
$$\sigma_{\vartheta\vartheta} = \gamma H \left(\frac{1+K}{2} - \frac{1-K}{2} \cos 2\vartheta \right) \quad (14.6)$$
$$\sigma_{r\vartheta} = -\gamma H \frac{1-K}{2} \sin 2\vartheta$$

Thus, the far field is identical to the primary stress field:

$$\sigma_{zz} = \gamma H$$
$$\sigma_{xx} = K\gamma H$$
$$\sigma_{xz} = 0 \quad .$$

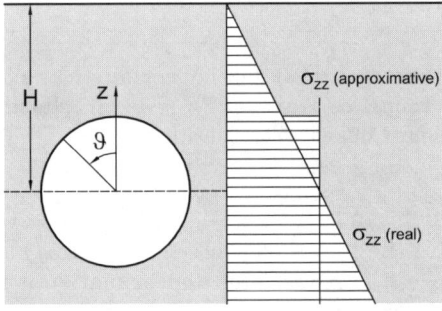

Fig. 14.7. Distribution of vertical stress in the environment of a tunnel.

The problem here is known as the 'generalised KIRSCH-problem'.[4] From equation 14.5 (2) follows that at the invert and at the crown (for $r = r_0$ and $\vartheta = 0$ or π) the tangential stress reads $\sigma_{\vartheta\vartheta} = \gamma H(3K - 1)$. Thus, for $K < \frac{1}{3}$ one obtains tension, i.e. $\sigma_{\vartheta\vartheta} < 0$.

---

[4] For the general case that the tunnel axis coincides with none of the principal directions of the primary stress, see the elastic solution of F.H. Cornet, Stress in Rock and Rock Masses. *In:* Comprehensive Rock Engineering, edited by J.A. Hudson, Pergamon Press, 1993, Volume **3**, p. 309

14.4 Hydrostatic primary stress

The special case of equation 14.5 for $K = 1$ (i.e. for $\sigma_{xx} = \sigma_{yy} = \gamma H$) reads with $\sigma_\infty := \gamma H$:[5]

$$\sigma_r = \sigma_\infty \left(1 - \frac{r_0^2}{r^2}\right)$$

$$\sigma_\vartheta = \sigma_\infty \left(1 + \frac{r_0^2}{r^2}\right) \quad (14.7)$$

$$\sigma_{r\vartheta} = 0$$

This solution exhibits axial symmetry, as the radius r is the only independent variable and ϑ does not appear.

We generalise now equation 14.7 regarding the case that the tunnel wall is subjected to a constant pressure p (so-called support pressure). One obtains the stress field as a special case of the solution which LAMÉ found in 1852 for a thick tube made of linear-elastic material.[6] r_i and r_a are the internal- and external radius of the tube, furthermore p_i and p_a are the internal- and external pressures, respectively. LAMÉ's solution reads:

$$\sigma_r = \frac{p_a r_a^2 - p_i r_i^2}{r_a^2 - r_i^2} - \frac{p_a - p_i}{r_a^2 - r_i^2} \frac{r_i^2 r_a^2}{r^2}$$

$$\sigma_\vartheta = \frac{p_a r_a^2 - p_i r_i^2}{r_a^2 - r_i^2} + \frac{p_a - p_i}{r_a^2 - r_i^2} \frac{r_i^2 r_a^2}{r^2} \quad (14.8)$$

$$\sigma_z = 2\nu \frac{p_a r_a^2 - p_i r_i^2}{r_a^2 - r_i^2} \quad , \quad \sigma_{r\vartheta} = 0 \quad .$$

For $r_a \to \infty$, $p_a \to \sigma_\infty$ one obtains the searched stress field (based on the assumption of elastic behaviour):

$$\sigma_r = \sigma_\infty \left(1 - \frac{r_0^2}{r^2}\right) + p\frac{r_0^2}{r^2} = \sigma_\infty - (\sigma_\infty - p)\frac{r_0^2}{r^2}$$

$$\sigma_\vartheta = \sigma_\infty \left(1 + \frac{r_0^2}{r^2}\right) - p\frac{r_0^2}{r^2} = \sigma_\infty + (\sigma_\infty - p)\frac{r_0^2}{r^2} \quad (14.9)$$

$$\sigma_{r\vartheta} = 0$$

The solution (14.8) is obtained from the equation of radial equilibrium, which can be written for an elastic material as:

$$\frac{d}{dr}\left[\frac{1}{r}\frac{d}{dr}(ru)\right] = 0 \quad ,$$

---

[5] In denoting principal stresses, double indices can be replaced by single ones, i.e. $\sigma_{rr} \equiv \sigma_r$ etc.

[6] L. Malvern, Introduction to the Mechanics of a Continuous Medium, Prentice-Hall, 1969, p. 532

where u is the radial displacement. Integration of this equation yields

$$u = Ar + \frac{B}{r} \quad , \tag{14.10}$$

where the integration constants A and B can be determined from the boundary conditions. From 14.10 we obtain $\varepsilon_r = du/dr = A - B/r^2$ and $\varepsilon_\vartheta = u_r/r = A + B/r^2$. Introducing ε_r and ε_ϑ into HOOKE's law we obtain

$$\sigma_r = 2A\lambda + 2GA - 2GB/r^2 \tag{14.11}$$
$$\sigma_\vartheta = 2A\lambda + 2GA + 2GB/r^2 \tag{14.12}$$

From the boundary conditions

$$r = a: \quad -p_i = 2A(\lambda + G) - 2GB/a^2 \tag{14.13}$$
$$r = b: \quad -p_a = 2A(\lambda + G) - 2GB/b^2 \tag{14.14}$$

we obtain Equ. 14.8. Requiring $p_a \to -\sigma_\infty$ for $b \to \infty$ we obtain $A = \sigma_\infty \frac{1}{2(\lambda + G)}$. With $p_i = -p$ and $a = r_0$ we finally obtain $B = \frac{\sigma_\infty - p}{2G} r_0^2$, hence

$$u = \frac{\sigma_\infty}{2(\lambda + G)} r + \frac{\sigma_\infty - p}{2G} \cdot \frac{r_0^2}{r} \quad . \tag{14.15}$$

The first part of (14.15) is independent of p and represents the displacement which results from application of the hydrostatic pressure σ_∞. Only the second part is due to the excavation of the tunnel (i.e. due to the reduction of the stress at the cavity wall from σ_∞ to p).[7]

With the application of the pressure p the wall of the tunnel yields by the amount $u|_{r_0}$. The displacement $u|_{r_0}$ can be computed as a function of p with the help of the solution by LAMÉ.[8] One obtains the following linear relationship between $u|_{r_0}$ and p:

$$u|_{r_0} = r_0 \frac{\sigma_\infty}{2G} \left(1 - \frac{p}{\sigma_\infty}\right) \quad . \tag{14.16}$$

Fig. 14.8 shows the plot of equation 14.16. If the support pressure p is smaller than σ_∞, then one obtains from Equ. (14.9) a radial stress σ_r which increases with r and a tangential stress σ_ϑ which decreases with r, Fig. 14.9.

---

[7]For the case of the *spherical* symmetry the corresponding equations read: $\frac{d}{dr}\left[\frac{1}{r^2}\frac{d}{dr}(r^2 u)\right] = 0$, $u = Ar + \frac{B}{r^2}$, $\sigma_r = \sigma_\infty - (\sigma_\infty - p)\left(\frac{r_0}{r}\right)^3$ and $\sigma_\phi = \sigma_\theta = \sigma_\infty + \frac{1}{2}(\sigma_\infty - p)\left(\frac{r_0}{r}\right)^3$.

[8]thereby u is presupposed as small, so that the tunnel radius r_0 may be regarded as constant

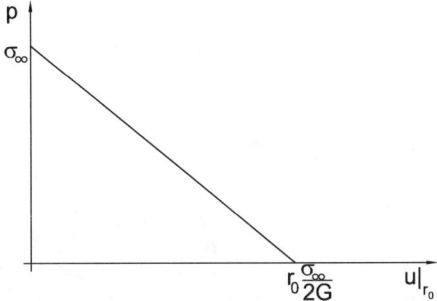

Fig. 14.8. Relationship between p and $u|_{r_0}$ for linear elastic ground

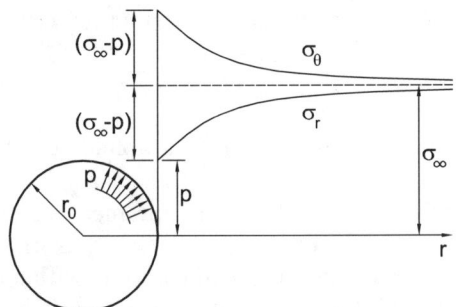

Fig. 14.9. Stress field in linear elastic ground

14.5 Plastification

According to equation 14.9, the principal stress difference

$$\sigma_\vartheta - \sigma_r = 2(\sigma_\infty - p)\frac{r_0^2}{r^2}$$

increases when p decreases. Now we can take into account that rock is not elastic, resp. it may be regarded as elastic only as long, as the principal stress difference $\sigma_\vartheta - \sigma_r$ does not exceed a threshold which is given by the so-called limit condition. For a frictional material the limit condition (Fig. 14.10) reads

$$\sigma_\vartheta - \sigma_r = (\sigma_\vartheta + \sigma_r)\sin\varphi \;, \tag{14.17}$$

where φ is the so-called friction angle (or the angle of internal friction). Thus, only such stress states are feasible, to which applies:

$$\sigma_\vartheta - \sigma_r \leq (\sigma_\vartheta + \sigma_r)\sin\varphi \;. \tag{14.18}$$

For a material with friction and cohesion c the limit condition reads

$$\sigma_\vartheta - \sigma_r = (\sigma_\vartheta + \sigma_r)\sin\varphi + 2c\cos\varphi \;. \tag{14.19}$$

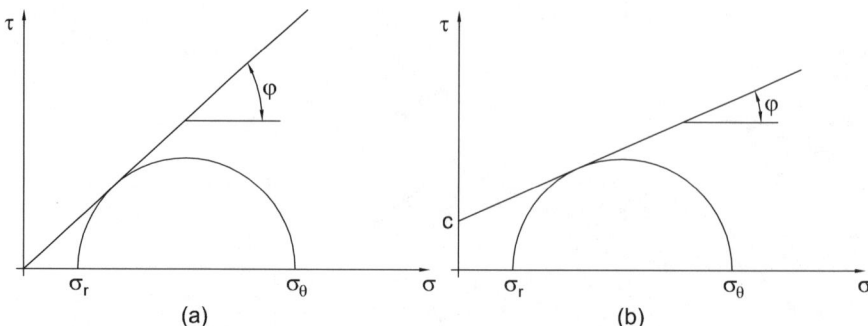

Fig. 14.10. Limit condition in the MOHR diagram for non-cohesive (a) and for cohesive (b) frictional material

The limit condition can also be represented graphically in the MOHR diagram as shown in Fig. 14.10.

If, therefore, p is sufficiently small, the requirement 14.18 expressed by the equation 14.9 is violated in the range $r_0 < r < r_e$ (r_e is still to be determined). Consequently, the elastic solution 14.9 cannot apply within this range. Rather, another relation applies which is to be deduced as follows: We regard the equation of equilibrium in radial direction

$$\frac{d\sigma_r}{dr} + \frac{\sigma_r - \sigma_\vartheta}{r} = 0 \qquad (14.20)$$

and set into it the equation derived from relationship 14.17

$$\sigma_\vartheta = K_p \sigma_r \qquad \text{with} \qquad K_p := \frac{1 + \sin\varphi}{1 - \sin\varphi}$$

In soil mechanics K_p is called the coefficient of passive earth pressure. One obtains

$$\frac{d\sigma_r}{\sigma_r} = (K_p - 1)\frac{dr}{r}$$

or

$$\ln \sigma_r = (K_p - 1)\ln r + \ln C_1$$

i.e.

$$\sigma_r = C_1 r^{K_p - 1}.$$

The integration constant C_1 follows from the requirement $\sigma_r = p$ for $r = r_0$, so that we finally obtain:

14.5 Plastification

$$\sigma_r = p\left(\frac{r}{r_0}\right)^{K_p-1}$$

$$\sigma_\vartheta = K_p p\left(\frac{r}{r_0}\right)^{K_p-1} \quad .$$

The range $r_0 < r < r_e$ where solution 14.21 applies, is called the plastified zone. Instead, the requirement 14.17 (limit condition) applies, also known as the condition for plastic flow. The word 'plastic' highlights that with stress states that fulfil conditions 14.17 and 14.19, deformations occur without stress change (so-called plastic deformations or 'flow').

If the range $r_0 \leq r \leq r_e$ is plastified, then the elastic solution 14.9 must be slightly modified: σ_e is the value of σ_r for $r = r_e$. In place of equation 14.9 now applies for $r_e \leq r < \infty$:

$$\sigma_r = \sigma_\infty - (\sigma_\infty - \sigma_e)\frac{r_e^2}{r^2}$$

$$\sigma_\vartheta = \sigma_\infty + (\sigma_\infty - \sigma_e)\frac{r_e^2}{r^2}$$

$$\sigma_{r\vartheta} = 0$$

For $r = r_e$ the stresses read: $\sigma_r = \sigma_e$, $\sigma_\vartheta = 2\sigma_\infty - \sigma_e$. They must fulfil the limit condition, i.e. $\sigma_\vartheta = K_p \sigma_r$ or $K_p \sigma_e = 2\sigma_\infty - \sigma_e$. It then follows

$$\sigma_e = \frac{2}{K_p+1}\sigma_\infty \quad . \tag{14.21}$$

At the boundary $r = r_e$ the radial stresses of the elastic and the plastic ranges must coincide:

$$p\left(\frac{r_e}{r_0}\right)^{K_p-1} = \sigma_e \tag{14.22}$$

From (14.21) and (14.22), finally, follows the radius r_e of the plastic range

$$r_e = r_0\left(\frac{2}{K_p+1}\frac{\sigma_\infty}{p}\right)^{\frac{1}{K_p-1}} \quad . \tag{14.23}$$

If we evaluate equation 14.23 for the case $r_e = r_0$ we obtain the support pressure p^* at which plastification sets on:

$$r_0 = r_0\left(\frac{2}{K_p+1}\frac{\sigma_\infty}{p}\right)^{\frac{1}{K_p-1}}$$

$$\leadsto p = p^* = \frac{2}{K_p+1}\sigma_\infty = (1-\sin\varphi)\sigma_\infty$$

The distributions of σ_r and σ_ϑ in the case of plastification are represented in Fig. 14.11.

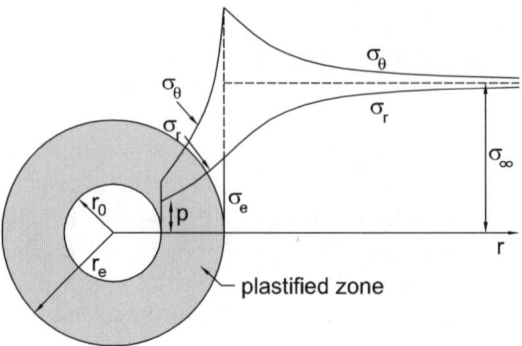

Fig. 14.11. Distributions of σ_r and σ_ϑ within the plastic and elastic ranges. Note that σ_ϑ is continuous at $r = r_e$, and therefore $\sigma_r(r)$ is smooth at $r = r_e$ (by virtue of Equ. 14.20)

14.5.1 Consideration of cohesion

If the rock exhibits friction and cohesion, one obtains from equation 14.19:

$$\sigma_\vartheta = K_p \cdot \sigma_r + 2c \frac{\cos\varphi}{1 - \sin\varphi} = K_p \cdot \sigma_r + C \quad.$$

Introducing in Equ. 14.20 yields:

$$\frac{d\sigma_r}{dr} + \frac{\sigma_r(1 - K_p) - C}{r} = 0 \quad.$$

With the substitution $s := \sigma_r(1 - K_p) - C$ we obtain:

$$\frac{ds}{dr} + (1 - K_p)\frac{s}{r} = 0 \quad.$$

Separation of variables yields:

$$\ln s + \ln r^{1-K_p} = \text{const}_1$$

or

$$s = \text{const}_2 \, r^{K_p - 1} \quad.$$

Introducing the boundary condition $\sigma_r(r = r_0) = p$ yields

$$s = s_0 \left(\frac{r}{r_0}\right)^{K_p - 1}$$

or

$$\sigma_r(1 - K_p) - C = [\, p(1 - K_p) - C\,] \left(\frac{r}{r_0}\right)^{K_p - 1}$$

$$\sigma_r = (p + c\cot\varphi)\left(\frac{r}{r_0}\right)^{K_p-1} - c\cot\varphi$$

$$\sigma_\vartheta = K_p(p + c\cot\varphi)\left(\frac{r}{r_0}\right)^{K_p-1} - c\cot\varphi \qquad (14.24)$$

For $r = r_e$ the elastic stresses $\sigma_r = \sigma_e$ and $\sigma_\vartheta = 2\sigma_\infty - \sigma_e$ must also fulfil the limit condition. Thus

$$\sigma_e = \sigma_\infty(1 - \sin\varphi) - c\cos\varphi \quad . \qquad (14.25)$$

At the boundary $r = r_e$ the radial stresses of the elastic and the plastic zones must coincide

$$(p + c\cot\varphi)\left(\frac{r_e}{r_0}\right)^{K_p-1} - c\cot\varphi = \sigma_\infty(1 - \sin\varphi) - c\cos\varphi \quad .$$

Thus, the radius r_e of the plastic zone is obtained as:

$$r_e = r_0\left(\frac{\sigma_\infty(1-\sin\varphi) - c(\cos\varphi - \cot\varphi)}{p + c\cot\varphi}\right)^{\frac{1}{K_p-1}} \quad . \qquad (14.26)$$

Again, $\sigma_\vartheta(r)$ is continuous at $r = r_e$ and, consequently, $\sigma_r(r)$ is smooth at $r = r_e$.

14.6 Ground reaction line

The linear relationship 14.16 shows how the tunnel wall moves into the cavity, if the support pressure is reduced from σ_∞ to p. This relationship applies to linear-elastic ground. Now we want to see how the relationship between p and $u|_{r_0}$ reads, if the ground is plastified in the range $r_0 \leq r < r_e$. Within this range plastic flow takes place, i.e. the deformations increase, without stress change (Fig. 13.2 b).

In order to specify plastic flow, one needs an additional constitutive relationship, the so-called flow rule. This is a relationship between the strains ε_r and ε_ϑ (ε_z vanishes per definition for the plane deformation we are considering). With the volumetric strain $\varepsilon_v := \varepsilon_r + \varepsilon_\vartheta$ the flow rule reads in a simplified and idealised form:

$$\varepsilon_v = b\varepsilon_r \quad .$$

b is a material constant, which describes the dilatancy (loosening) of the material.[9] Note, however, that several definitions of dilatancy exist. For $b = 0$

---

[9] The angle $\psi := \arctan b$ can be called angle of dilatancy

isochoric (i.e. volume preserving, $\varepsilon_v = 0$) flow occurs. We express the strains with the help of the radial displacement u:

$$\varepsilon_r = \frac{du}{dr} \quad , \quad \varepsilon_\vartheta = \frac{u}{r}$$

and obtain thus

$$\frac{du}{dr} + \frac{u}{r} = b\frac{du}{dr} \quad ,$$

from which follows

$$u = \frac{C}{r^{\frac{1}{1-b}}} \quad .$$

The integration constant C follows from the displacement $u = u_e$ at $r = r_e$ according to the elastic solution (equation 14.16):

$$u_e = r_e \frac{\sigma_\infty}{2g}\left(1 - \frac{\sigma_e}{\sigma_\infty}\right) = \frac{C}{r_e^{\frac{1}{1-b}}} \quad . \tag{14.27}$$

From the two last equations we obtain

$$u = r_e \frac{\sigma_\infty}{2G}\left(1 - \frac{\sigma_e}{\sigma_\infty}\right)\left(\frac{r_e}{r}\right)^{\frac{1}{1-b}} \quad . \tag{14.28}$$

If we introduce here the relations 14.21 and 14.23 for σ_e and r_e, we obtain finally for $r = r_0$, $c = 0$ and $p < p^*$:

$$u|_{r_0} = r_0 \sin\varphi \frac{\sigma_\infty}{2G}\left(\frac{2}{K_p+1}\frac{\sigma_\infty}{p}\right)^{\frac{2-b}{(K_p-1)(1-b)}} \quad . \tag{14.29}$$

This relationship applies to $p < p^*$, while for $p \geq p^*$ applies the elastic relationship 14.16, Fig. 13.2 b. In Fig. 14.12 is shown the relationship between p and $u|_{r_0}$, which is called the characteristic of the ground (also called 'FENNER-PACHER-curve', or 'ground reaction curve' or 'ground line').
For the case $\varphi > 0$, $c > 0$ we can obtain the relationship between the cavity wall displacement $u|_{r_0}$ and p, if we use equation 14.28 with equations 14.25 and 14.26:

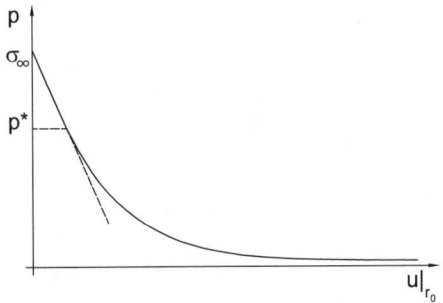

Fig. 14.12. Ground reaction line with plastification (non-cohesive ground)

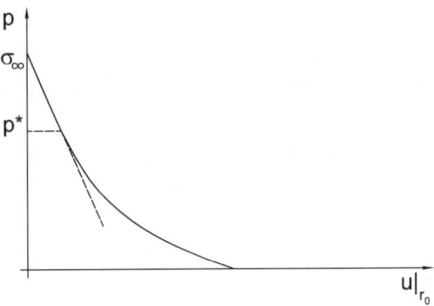

Fig. 14.13. Ground reaction line with plastification (cohesive ground)

$$u|_{r_0} = r_0 \left[\frac{\sigma_\infty(1-\sin\varphi) - c(\cos\varphi - \cot\varphi)}{p + c\cot\varphi} \right]^{\frac{2-b}{(K_p-1)(1-b)}}$$

$$\times \frac{\sigma_\infty}{2G}\left(\sin\varphi + \frac{c}{\sigma_\infty}\cos\varphi\right) \quad (14.30)$$

Contrary to equation 14.29, equation 14.30 supplies a finite displacement $u|_{r_0}$ for $p = 0$ (Fig. 14.13). I.e., in a cohesive material the cavity can persist also without support, whereas it closes up in non-cohesive material (if unsupported). The tunnel wall displacement $u|_{r_0}$ is called in tunnelling 'convergence'.

The equations deduced in this section cannot easily be evaluated for the case $\varphi = 0$ and $c > 0$. In this case it can be obtained from (14.19) and (14.20):
In the plastic range $r_0 < r \leq r_e$:

$$\sigma_r = 2c\ln\frac{r}{r_0} + p$$
$$\sigma_\vartheta = \sigma_r + 2c$$

furthermore

$$r_e = r_0 \exp\frac{\sigma_\infty - c - p}{2c}$$
$$\sigma_e = \sigma_\infty - c \quad .$$

With (14.28) we obtain for $p < p^* = \sigma_\infty$:

$$u|_{r_0} = r_0 \frac{c}{2G} \left[\exp\left(\frac{\sigma_\infty - c - p}{2c}\right) \right]^{\frac{2-b}{1-b}} . \qquad (14.31)$$

The ground reaction line, i.e. the dependence of p on $u|_{r_0}$, clearly shows that the pressure exerted by the rock upon the lining is not a fixed quantity but depends on the rock deformation and, thus, on the rigidity of the lining. This is a completely different perception of load and causes difficulties to many civil engineers, who are used to consider the loads acting upon, say, a bridge as given quantities.

14.7 Pressuremeter, theoretical background

There is an abundance of theoretical solutions of cavity expansion problems. In this section it is shown how the equations derived above can be applied to the problem of the expansion of a cylindrical cavity within a hydrostatically stressed elastoplastic medium.

If we increase p above the value σ_∞, then we obtain from the equations presented in Section 14.6 a solution for the problem of the pressuremeter. The elastic solution (14.16) still applies, whereby $u|_{r_0}$ becomes negative because of $p > \sigma_\infty$. With the plastic solution ($\varphi > 0$, $c = 0$, Equ. 14.21) one needs only to replace K_p by $K_a = (1 - \sin\varphi)/(1 + \sin\varphi)$ to obtain the following equation instead of (14.29):

$$u|_{r_0} = -r_0 \sin\varphi \frac{\sigma_\infty}{2G} \left(\frac{2}{K_a + 1} \frac{\sigma_\infty}{p} \right)^{\frac{2-b}{(K_a - 1)(1-b)}} \qquad (14.32)$$

The rock plastifies for $p > p^* = (1 + \sin\varphi)\sigma_\infty$. Equ. 14.32 applies for $p > p^*$. The relation (14.32) is shown in Fig. 14.14.
The plot of $u|_{r_0}$ over the logarithm of $\frac{2}{K_a+1}\frac{\sigma_\infty}{p}$ is a straight line. If one assumes a value for b, then one obtains from the slope of this straight line the friction angle φ.[10] Thus, the evaluation of pressuremeter-tests depends on assumptions. Often one assumes $b = 0$ and regards thus the problem of the undrained cavity expansion. There is also a solution if the stress-strain curve is given by a power law before reaching ideal-plastic flow.[11]

---

[10] J.M.O. Hughes, C.P. Wroth, D. Windle: Pressuremeter test in sands. *Géotechnique* 27, 455-477 (1977)
[11] R.W. Whittle, Using non-linear elasticity to obtain the engineering properties of clay – a new solution for the self boring pressuremeter. *Ground Engineering*, May 1999, 30-34.

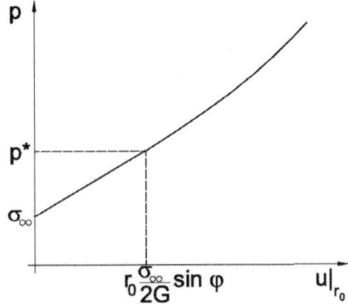

Fig. 14.14. Pressuremeter problem ($c = 0$): Cavity wall pressure p as a function of the cavity wall displacement $u|_{r_0}$

14.8 Support reaction line

Now we want to see, how the resistance p of the support changes with increasing displacement u. Considering equilibrium (Fig. 14.15) the compressive stress in the support is easily obtained as $\sigma_a = pr_0/d$ (Fig. 14.15). This stress causes the support to compress by $\varepsilon = \sigma_a/E$ (E is the YOUNG's modulus of the support).

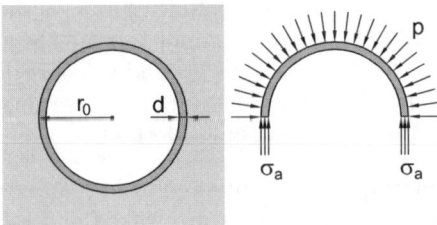

Fig. 14.15. Forces acting upon and within the support

The circumference of the support shortens by the amount $\varepsilon 2\pi r_0$, i.e. the radius shortens by the amount $u = \varepsilon r_0$. From here follows a linear relationship between u and p ('characteristic of the support', 'support reaction line' or 'support line'):

$$p = \frac{Ed}{r_0^2} u \quad \text{or} \quad u = \frac{r_0^2}{Ed} p .$$

We assume, for simplicity, that this linear relationship applies up to the collapse of the support, where $p = p_l$ (Fig. 14.16).

The reaction lines of the ground and the support serve to analyse the interaction between ground and support. To this purpose they are plotted together in a p-u-diagram. The characteristic of the support is represented by

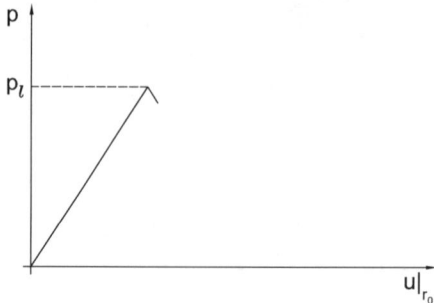

Fig. 14.16. Support reaction line

$$u(p) = u_0 + \frac{r_0^2}{Ed}p \quad .$$

Here, the part u_0 takes into account that the installation of the support cannot take place immediately after excavation. Tunnelling proceeds with a finite speed, and before the support can be installed, the tunnel remains temporarily unsupported in the proximity of the excavation face, Fig. 14.17. Therefore, at the time of the construction of the support the cavity wall has already moved by the amount u_0 into the cavity. The influence of u_0 is shown in Fig. 14.18: If u_0 is small (case 1), then the support cannot take up the ground pressure and collapses. If u_0 is large (case 2) then the ground pressure decreases with deformation so that it can be carried by the support. According to the terminology of the NATM (New Austrian Tunnelling Method)

> ... ground deformations must be allowed to such an extent that around the tunnel deformation resistances are waked and a carrying ring in the ground is created, which protects the cavity ...

The influence of the stiffness (resp. Young's modulus E) of the support can be similarly illustrated in a diagram (Fig. 14.19). A stiff support (case 1) cannot carry the ground pressure and collapses, while a flexible support (case 2) possesses sufficient carrying reserves.

14.9 Rigid block deformation mechanism for tunnels and shafts

Of particular simplicity are deformation mechanisms consisting of rigid blocks that slip relative to each other. They are realistic, in view of the always observed localisation of deformation of soil and rock into thin shear bands. The first rigid block mechanism has been introduced into soil mechanics by

14.9 Rigid block deformation mechanism for tunnels and shafts

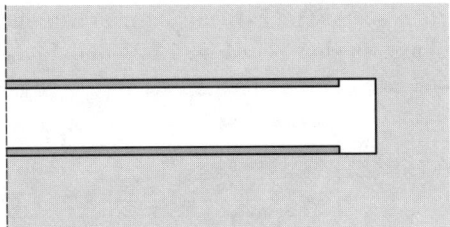

Fig. 14.17. Lack of support at the excavation face

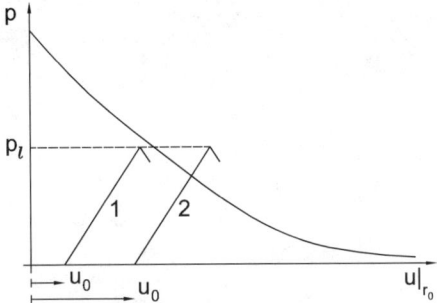

Fig. 14.18. To the influence of u_0

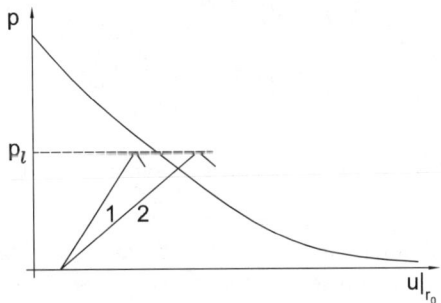

Fig. 14.19. Influence of the support stiffness

COULOMB in 1776, when he analysed the active earth pressure on the basis of a sliding wedge. With respect to tunnels, the failure mechanism shown in Fig. 14.20 proves to be useful. It was originally introduced by LAVRIKOV and REVUZHENKO[12] and, therefore, will be referred as L-R-mechanism. Its geometry is determined by r_0, r_1 and n, where the variable n denotes the number of rigid blocks. The following geometric relations can be established:
$\delta = 2\pi/n$, $\kappa = \pi \cdot (n-2)/(2n)$, $a_1 = 2r_1 \sin(\delta/2)$, $r_1/r_0 = \sin\kappa / \sin(\kappa - \alpha)$,

---

[12]S.V. Lavrikov, A.F. Revuzhenko, O deformirovanii blochnoy sredy vokrug vyrabotki, *Fizikotekhnicheskie Problemy Razrabotki Poleznykh Iskopaemykh*, Novosibirsk, 1990, 7-15

$\alpha = \kappa - \arcsin(\sin\delta/(2\cos\kappa))$. The L-R-mechanism can undergo a multiplicity of various displacements, as shown in Fig. 14.21 and 14.22.

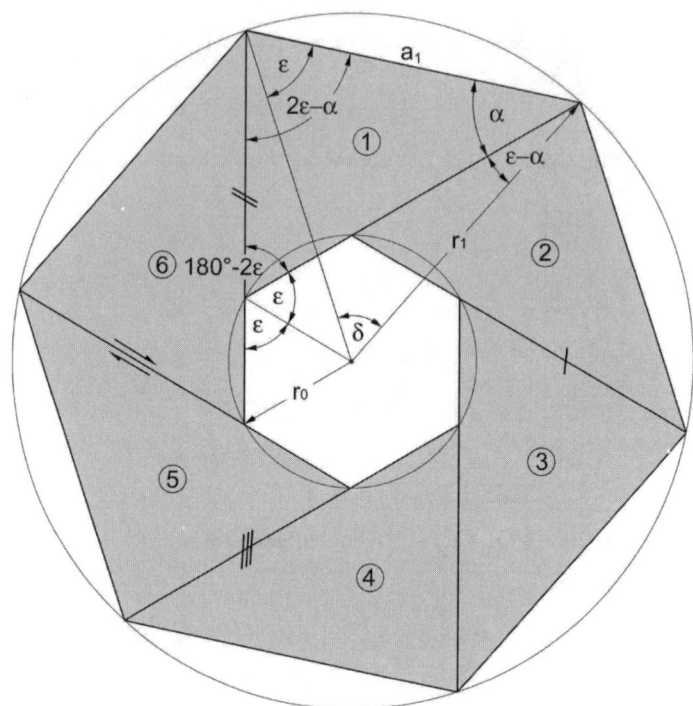

Fig. 14.20. L-R-mechanism consisting of $n=6$ blocks

The hodograph shows the displacements of the individual blocks. Denoting the reference ground with 0 we use the notation u_{ij} where i denotes the observer and j denotes the considered body. Thus, u_{43} denotes the displacement of block 3 as viewed by an observer situated on block 4. $u_{03} \equiv u_3$ is the absolute displacement of block 3 with reference to the fixed ground. Obviously, $u_{ij} = -u_{ji}$. Relative motion implies the vectorial equation

$$u_{0i} = u_{ki} + u_{ji},$$

where k and j are the numbers of any two adjacent blocks.

The oriented relative displacements are shown in Fig. 14.21. The cohesion force C_{ij} is the force exerted by the block i upon the block j. Obviously, this force is oriented in such a way that it counteracts the displacement u_{ij}.

The L-R-mechanism does not yield closed-form solutions. However, it is useful for ad hoc analyses introducing appropriate geometries and shear strength parameters. Consideration of equilibrium of the individual blocks yields estimation of loads at collapse.

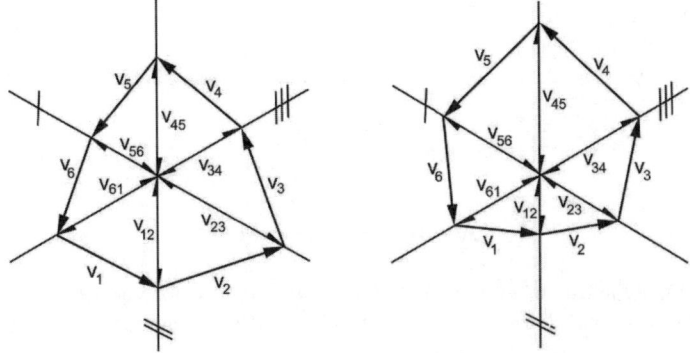

Fig. 14.21. Possible hodographs for the L-R-mechanism

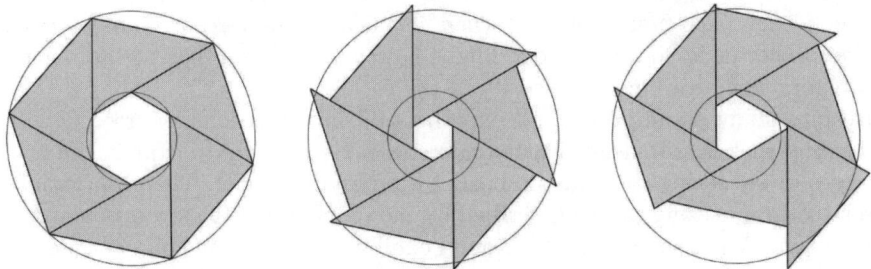

Fig. 14.22. Possible shifts of blocks

14.10 Squeezing

In so-called squeezing rock, the cavity wall displacement $u|_{r_0}$ resp. the support pressure p increases with time. With constant support pressure p (e.g. $p = 0$), $u|_{r_0}$ increases with time. Convergences up to amounts of meters can be observed and make the excavation substantially more difficult. If the convergence is prevented, e.g. by a shield, then p can increase so that the shield cannot be advanced any longer.

14.10.1 Squeezing as a time-dependent phenomenon

In continuum mechanics we refer to creep if the strain ε increases with constant stress and to relaxation if the stress decreases with constant strain. The consideration of time-dependent problems is difficult. So-called viscoelastic equations (Fig. 14.23) are, often, unrealistic for rock.
The various suggested one-dimensional creep laws, i.e. functions $\varepsilon = \varepsilon(t)$, are similarly unsatisfactory. They can hardly be represented in tensorial form, resp. they can hardly be experimentally justified for three-dimensional deformations. In addition, one must consider that a function $\varepsilon = \varepsilon(t)$ depends on

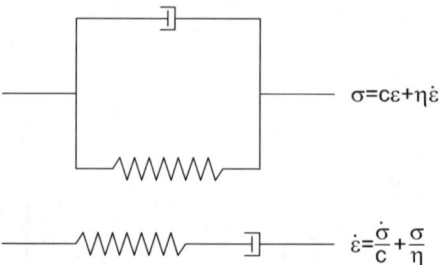

Fig. 14.23. One-dimensional models (rheologic models) for viscoelastic behaviour. Upper: KELVIN-solid, lower: MAXWELL-fluid. c and η are material constants.

the arbitrary definition of the time zero ($t = 0$) and is therefore not objective. Strictly speaking, one should differentiate between time dependence and rate dependence. The first is present when chemical changes resp. ageing occurs. The processes in squeezing rock might however belong rather to the second category and be a kind of viscosity of the rock.

Despite many assumptions and experimental investigations the rate dependence of rock is still insufficiently understood. Here we try to explain the processes in squeezing rock on the basis a simplifying model. The observations related to squeezing can be described by the reduction of the stress deviator[13] with time. We assume that the ground exhibits a short term cohesion. I.e., in the course of the excavation the cohesion c_0 is mobilised. Here we denote by the term 'cohesion' the maximum of $(\sigma_1 - \sigma_2)/2$. For the reduction of cohesion with time we meet the assumption

$$\dot{c} = -\alpha c \quad , \quad \alpha > 0 \; , \tag{14.33}$$

or

$$c = c_0 e^{-\alpha t} \; .$$

The value of α can be obtained from relaxation tests. The assumption (14.33) is meaningful, however arbitrary.[14] For simplicity, we assume $\varphi = 0$ and $\beta = 0$. From (14.31) then follows:

$$u|_{r_0} = r_0 \frac{c}{2G} \left(\exp \frac{\sigma_\infty - c - p}{2c} \right)^2 = r_0 \frac{c}{2G} \exp \frac{\sigma_\infty - c - p}{c} \; . \tag{14.34}$$

Deriving with respect to t and using (14.33) supplies (with $\dot{u}|_{r_0} = -\dot{r}_0$) :

$$\dot{u}|_{r_0} \left(1 + \frac{c}{2Ge} e^{\frac{\sigma_\infty - p}{c}} \right) = \frac{r_0 c}{2Ge} e^{\frac{\sigma_\infty - p}{c}} \left(-\frac{\dot{p}}{c} + \alpha \frac{\sigma_\infty - p}{c} - \alpha \right) \tag{14.35}$$

---

[13] i.e. the principal stress difference

[14] With given results of measurements it can be improved, possibly as $\dot{c} = -\alpha(c - c_{min})^\beta$. Appropriate data are yet missing.

For $\dot{u}|_{r_0} = 0$ one obtains a differential equation, which describes the increase of p with time if convergence is inhibited:

$$\dot{p} + \alpha p = \alpha\left(-c_0 e^{-\alpha t} + \sigma_\infty\right) \quad .$$

Its solution reads:

$$p(t) = \sigma_\infty \left(1 - e^{-\alpha t}\right) - \alpha c_0 t e^{-\alpha t} \qquad (14.36)$$

Hence, $p = \sigma_\infty$ for $t \to \infty$.
For an unsupported tunnel ($p = 0$) we obtain from (14.35) the law for the increase of convergence with time:[15]

$$\dot{u}|_{r_0} = \frac{r_0 \alpha}{2G e^{1-\frac{\sigma_\infty}{c}} + c} \cdot (\sigma_\infty - c) \quad .$$

Taking into account that for squeezing rock $\sigma_\infty \gg c$, we finally obtain

$$\dot{u}|_{r_0} \approx r_0 \alpha \frac{\sigma_\infty}{c} = r_0 \alpha \frac{\sigma_\infty}{c_0} e^{\alpha t} \quad . \qquad (14.37)$$

This equation agrees with the experience in as much as one observes squeezing behaviour at high stress (σ_∞ large), low strength (c_0 small, e.g. in distorted zones) and creeping minerals (usually clay minerals, large α). It is also reported that a high pore water pressure favours squeezing.
Equ. 14.37 predicts that the rate of convergence due to squeezing, i.e. $\dot{u}|_{r_0}$, increases with time. This is, admittedly, unrealistic and should be attributed to the highly simplified relation (14.33). However, with equations 14.36 and 14.37 we can relate the initial convergence rate $\dot{w}_0 := \dot{u}|_{r_0}(t = 0)$ with the initial rate of pressure increase $\dot{p}_0 := \dot{p}(t = 0)$:

$$\dot{p}_0 = \left(1 - \frac{c_0}{\sigma_\infty}\right) \frac{c_0 \dot{w}_0}{r_0} \quad .$$

Thus, from the knowledge of $\dot{w}_0$ we can predict the initial rate of pressure that will act upon a shield operating in the same rock.

14.10.2 Neglecting time-dependence

Many authors refer to squeezing whenever large convergences appear, no matter whether they appear simultaneously with excavation or with time delay. Let us introduce the symbol ζ to denote the ratio of convergence to tunnel radius, $\zeta := u|_{r_0}/r_0$. Using equation 14.34, ζ can be expressed in dependence of the ratio p/σ_∞ (i.e. support pressure p / in situ stress σ_∞) and of the ratio $\frac{c}{\sigma_\infty}$ (i.e. rock mass strength $q_u = 2c$ / in situ stress σ_∞) as follows:

---
[15] Note that the equations, presented here, are based on the assumption of small deformations. With large deformations they have to be modified.

$$\zeta = \frac{c}{2G} \exp\left[\frac{\sigma_\infty}{c}\left(1 - \frac{p}{\sigma_\infty}\right) - 1\right] \tag{14.38}$$

This equation, which is graphically shown in Fig. 14.24, presupposes $\varphi = 0$. The corresponding expression for $\varphi > 0$ can be derived from Equ. 14.30.[16]

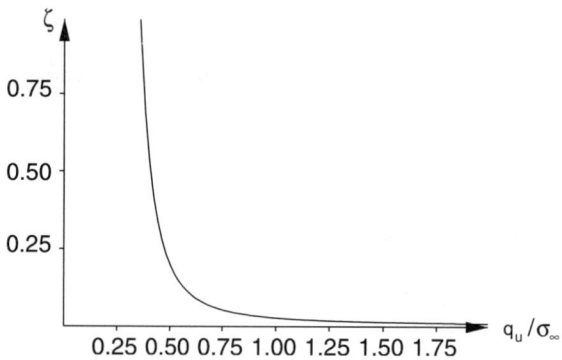

Fig. 14.24. Graphical representation of Equ. 14.38 for the case $p = 0$.

Considering different levels of ζ, HOEK assigns to them various tunnelling problems given in Table 14.1. In this way, squeezing is seen as large convergence due to low rock strength.

ζ	q_u/σ_∞	Tunnelling problems
0 ... 1%	> 0.36	Few support problems
1 ... 2.5%	0.22 ... 0.36	Minor squeezing problems
2.5 ... 5%	0.15 ... 0.22	Severe squeezing problems
5 ... 10%	0.10 ... 0.15	Very severe squeezing problems
> 10%	< 0.10	Extreme squeezing problems

Table 14.1. Squeezing problems related to convergence and rock strength according to HOEK.

In practise, squeezing is waited to fade away. The permanent lining is built only as soon as the rate of convergence falls below 2 mm/month in case of concrete lining (C20/25), or below 6-10 mm/month in case of steel reinforced concrete (≥ 50 kg/m^3) or concrete C30/35.

---

[16] Similar curves can be numerically obtained if the yield criterion of MOHR-COULOMB is replaced by that of HOEK and BROWN, see E. Hoek, Big Tunnels in Bad Rock, 36th Terzaghi Lecture, *Journal of Geotechn. and Geoenvir. Eng.*, Sept. 2001, 726-740

14.10.3 Interaction with support

Let us now consider the interaction of a squeezing rock with the support, which is here assumed as a lining (shell) of shotcrete. We neglect hardening and assume that shotcrete attains immediately its final stiffness and strength. Usually, the rock reaction line (i.e. the curve p vs. convergence w)[17] and the support reaction lines are plotted in the same diagram. Their intersection determines the convergence and the pressure acting upon the support (Fig. 14.25).

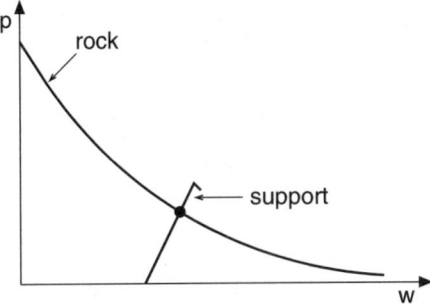

Fig. 14.25. The intersection of the reaction lines determines the pressure p and the convergence w of the support (non-squeezing case)

For squeezing rock, the rock reaction line is not unique. There are infinitely many rock reaction lines. Each of them corresponds to a particular c-value (Fig. 14.26). In this case the intersection point of rock and support lines moves along the support reaction line and crosses the various rock reaction lines. The evolution in time of this process can be obtained if we introduce into Equ. 14.34 the support reaction line $p = p_s(w)$, e.g. $p = k \cdot (w - w_0), k =$ const, for $w > w_0$:

$$w = \frac{c}{2G} r_0 \exp\left(\frac{\sigma_\infty - c - p_s(w)}{c}\right) \qquad (14.39)$$

With $c = c(t)$, equation 14.39 determines the relation $w(t)$. No matter how the relation $w(t)$ looks like, after a sufficiently long time lapse the ultimate rock reaction line, i.e. the one corresponding to c_∞, will be reached. It may happen, however, that the lining cannot support the corresponding load. This can be the case with a brittle lining, whereas a ductile lining will yield until an intersection with the ultimate rock reaction line is obtained (Fig. 14.27). Examples of yielding support show figures 14.28, 14.29 and 14.30 that refer to the Strengen tunnel (Austria). A yielding ('elastic-ideal plastic') support can be achieved by interrupting the lining with arrays of steel tubes, which

---

[17] For brevity, the symbol w is used here instead of $u|_{r_0}$.

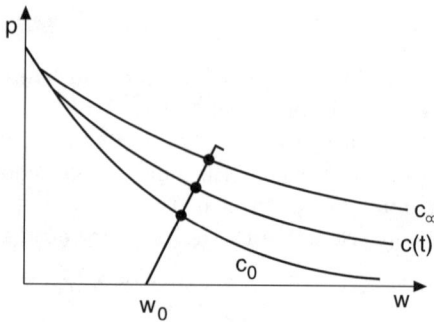

Fig. 14.26. Fan of rock reaction lines for squeezing rock

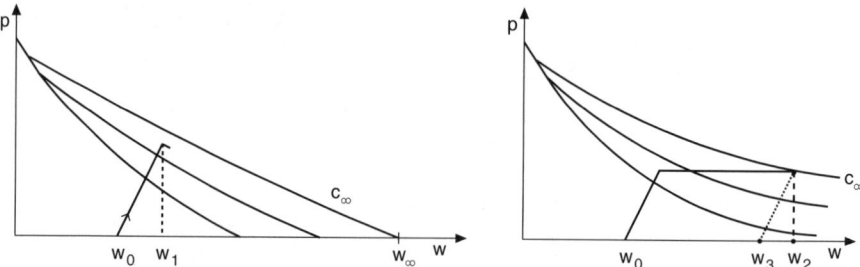

Fig. 14.27. A brittle support collapses at $w = w_1$ (left) and the convergence continues until the ultimate value $w = w_3$. A ductile (or yielding) support remains active until the ultimate convergence $w = w_2$ is obtained.

are designed to buckle under a specific load (Fig. 14.31). The positions of the tube arrays within an idealised circular lining are shown in Fig. 14.32. The circumferential length of the lining is $L = 2\pi r$. The total length of the embedded steel tubes is ml_s (m = number of tube arrays). With
E_s: YOUNG's modulus of shotcrete
d: shotcrete thickness
n: number of steel tubes per unit length of tunnel
we obtain the following relation between the thrust N and the shortening ΔL of the lining

$$\Delta L = \left[(2\pi r - ml_s) \cdot \frac{1}{d \cdot E_s} + \frac{1}{n} \frac{s_e}{F_e} \right] N$$

for $N/n < F_e$. With $w = \Delta L/(2\pi)$ and $p = N/r$ we obtain the support reaction line $p = p(w)$ (Fig. 14.33) as

$$p = w \frac{2\pi}{r[(2\pi r - ml_s)/(d \cdot E_s) + s_e/(n \cdot F_e)]}$$

for $p < nF_e/r$.

Fig. 14.28. In the upper left part is visible the array of pipes

Fig. 14.29. Anchor headplates and pipe array

The question arises whether a yielding support should be installed, which is costly. The alternative would be to install a usual support at a *later* time, when the rock has converged by an amount w_3 (see dotted line in Fig. 14.27). Of course, one has to determine the value of w_3. This can be achieved with borehole tests, i.e. the borehole is a model of the tunnel and the convergence is measured within the borehole. Equation 14.39 gives a relation between w and c for an uncased borehole. Measuring w-values at various times yields (with

Fig. 14.30. Squeezed anchor headplate. The two steel tubes had initially circular cross sections. Their squeezing indicates that this anchor has been overloaded.

numerical elimination of c from this equation) the corresponding c-values. Thus, the function $c(t)$ can be determined (for some discrete values of time t). With knowledge of $c(t)$, the complete rock reaction line and its interaction with the support can be determined as shown before.

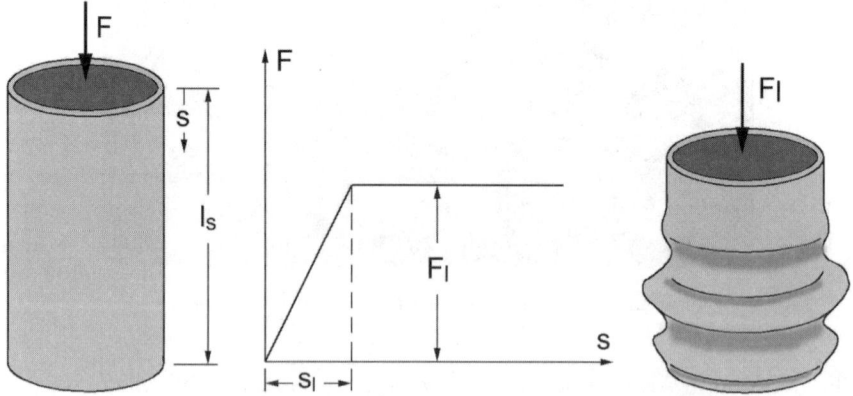

Fig. 14.31. Force-displacement characteristic of embedded tube

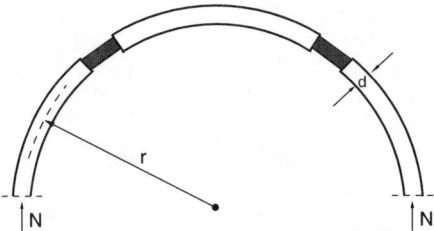

Fig. 14.32. Steel tubes embedded in shotcrete lining

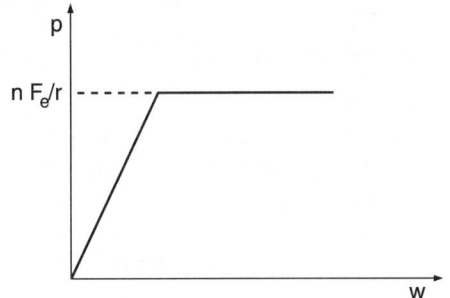

Fig. 14.33. Reaction line of yielding support

14.10.4 Squeezing in anisotropic rock

As already mentioned, squeezing rocks have low strength and are characterised by layered silicates that predominate in schists. The orientation of schistosity (or foliation) imposes a mechanical anisotropy to such rocks. It appears reasonable to assume that stress relaxation affects only shear stresses acting upon planes of schistosity. As a consequence, tunnels that perpendicular cross the planes of schistosity (Fig. 14.34 a) are not affected by squeezing, even at high depths. In contrast, tunnels whose axes have the same strike as the schistosity planes (Fig. 14.34 b) can be considerably affected by squeezing.

Typical examples for the two cases are the Landeck and Strengen tunnels in western Austria. Both tunnels have been headed within the same type of phyllitic rock. The Landeck tunnel was oriented as in Fig. 14.34 a and encountered, therefore, no squeezing problems despite a considerable overburden of up to 1,300 m. In contrast, the Strengen tunnel was oriented as in Fig. 14.34 b and had considerable problems with squeezing. Its overburden was up to 600 m.

14.11 Softening of the ground

The requirement that the lining should be flexible is limited by the fact that deformations that are too large imply reduction of the strength of the ground (so-called softening) so that the load upon the lining increases again. Softening

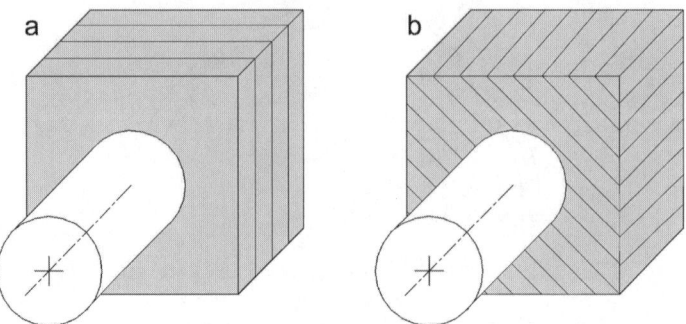

Fig. 14.34. Two different orientations of schistosity relative to tunnel axis.

is related with loosening (dilatancy, i.e. increase of volume and porosity with shear), as shown in Fig. 13.4. Softening (i.e. the reduction of stress beyond the peak, shown in Fig. 13.4) does not comply with the concept of plastic flow, which states that the deformation increases at *constant stress* (cf. Fig. 13.2 b), and is responsible for the increase of the ground pressure related to the increase of convergence $u|_{r_0}$ (Fig. 14.35).

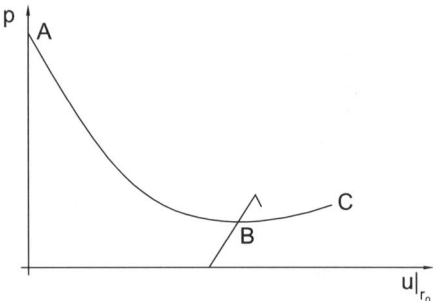

Fig. 14.35. Assumed ground reaction line for softening rock (schematically). The rising branch BC is due to the reduction of the rock strength

In order to keep the load upon the lining as low as possible, the support reaction line should intersect the ground reaction line at point B (minimum of ground reaction). This is one of the main requirements of NATM. Though this is in principle correct, it is hardly applicable in practise, as this minimum cannot be determined, neither by field measurements nor by numerical simulation.

It is worth mentioning that for a weightless rock and axisymmetric case (circular cross section, hydrostatic primary stress, i.e. $K = 1$) a rising branch of the ground reaction line is *not* obtained, even if the strength of the rock com-

14.11 Softening of the ground

pletely vanishes after the peak, as shown in Fig. 14.36.[18] This drastic softening implies $\sigma_\vartheta - \sigma_r = 0$ resp. $\sigma_\vartheta = \sigma_r$. From the equation of equilibrium (14.20) it follows that in the plastified zone $d\sigma_r/dr = 0$ resp. $\sigma_r = \sigma_\vartheta =$ const. The resulting stress distribution is shown in Fig. 14.37. Since for $r = r_e + 0$ the elastic solution (equation 14.21) must fulfil the strength condition $\sigma_\vartheta - \sigma_r = 2c$, it follows

$$\sigma_e = p = \sigma_\infty - c$$

Thus, the load on the lining cannot be less than $\sigma_\infty - c$. This result does not depend on $u|_{r_0}$: With increasing $u|_{r_0}$ the radius r_e of the plastified zone increases (according to equation 14.28), however p remains constant for $r \geq r_0$. Thus, we obtain the ground reaction line shown in Fig. 14.38, which does not exhibit any rising branch.

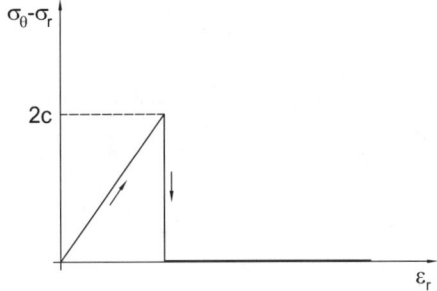

Fig. 14.36. Total loss of strength of a cohesive material

---

[18]Bliem, C. and Fellin, W. (2001): Die ansteigende Gebirgskennlinie (On the increasing ground reaction line) *Bautechnik* **78**(4): 296-305.

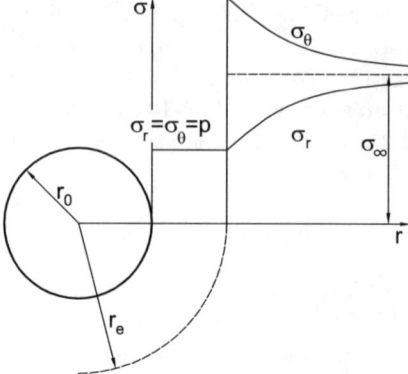

Fig. 14.37. Stress distribution in softening rock

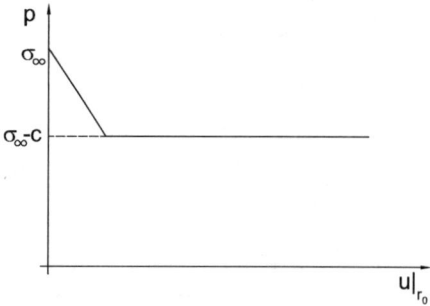

Fig. 14.38. Ground reaction line for a weightless rock with total loss of strength (axisymmetric case)

15

Supporting action of anchors/bolts

Anchors or rockbolts are reinforcements (usually made of steel) which are inserted into the ground to increase its stiffness and strength. There are various sorts of reinforcement actions and the corresponding terminology is not uniform.[1] The following terminology is used in soil mechanics:

1. If the reinforcement bar is fixed only at its both ends, then it is called an 'anchor'. Anchors can be pre-stressed or not, in the latter case they assume force only after some extension (e.g. due to convergence of the tunnel).
2. If the reinforcement bar is connected to the surrounding ground over its entire length, then it is called a 'nail' or 'bolt'. The connection can be achieved with cement mortar (Fig. 15.1).

Fig. 15.1. Nail

In jointed rock, reinforcement bars are placed ad hoc to prevent collapse of individual blocks (Fig. 15.2). Anchoring or bolting in a regular array is called 'pattern bolting'.

---

[1] see also C.R. Windsor, A.G. Thompson: Rock Reinforcement - Technology, Testing, Design and Evaluation. In: Comprehensive Rock Engineering, Vol. **4**, Pergamon Press 1993, 451-484

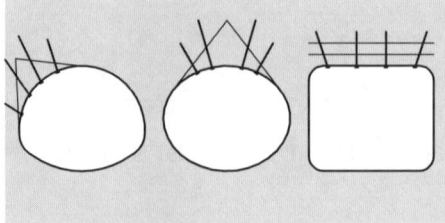

Fig. 15.2. Individual application of anchors to prevent downfall of blocks

15.1 Impact of pattern bolting

It is generally believed that reinforcing improves the mechanical behaviour of ground. Despite several attempts however, the reinforcing action of stiff inlets is not yet satisfactorily understood and their application is still empirical. In some approaches, reinforced ground is considered as a two-phase continuum in the sense that both constituents are assumed to be smeared and present everywhere in the considered body. Thus, their mechanical properties prevail everywhere, provided that they are appropriately weighed (Appendix F). The stiffening action of inlets can be demonstrated if we consider a conventional triaxial test on a soil sample containing a thin pin of, say, steel (Fig. 15.3).

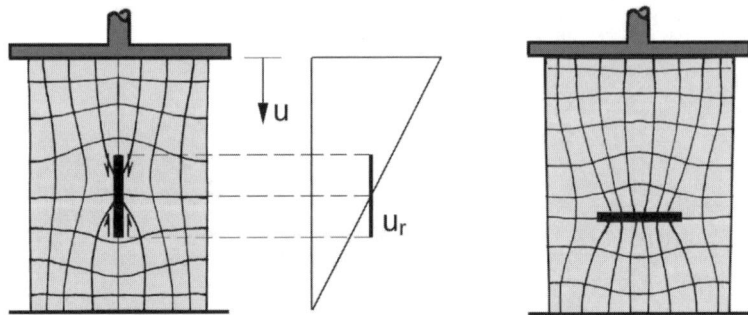

Fig. 15.3. Steel inlet in triaxial sample, distribution of vertical displacements and stress trajectories for two different orientations of the inlet.

The stiff inlet is here assumed as non-extendable (i.e. rigid). Therefore, its vertical displacement is constant as shown in Fig. 15.3. This implies a relative slip of the adjacent soil, which is oriented downwards in the upper half and upwards in the lower half. Being stiffer, the pin 'attracts' force and, thus, the adjacent soil is partly relieved from compressive stresses. As a result, the triaxial sample, viewed as a whole, is now stiffer. This effect is closely related to 'tension stiffening' known in concrete engineering.

Another way to increase the stiffness of reinforced soil is given by increasing the pressure level. As known, the stiffness of granular materials increases almost linearly with stress level. The latter can be increased by pre-stressing an array of anchors, i.e. of reinforcing inlets that transmit the force to the surrounding ground only at their ends ad not over their entire length. An analysis of this mechanism is presented in the next section.

15.1.1 Ground stiffening by pre-stressed anchors

The strengthening effect of pre-stressed pattern bolting will be considered for the case of a tunnel with circular cross section within a hydrostatically stressed elastoplastic ground. The primary hydrostatic stress is σ_∞. If the spacing of the anchors is sufficiently small, their action upon the ground can be approximated with a uniform radial stress σ_A (Fig. 15.4).

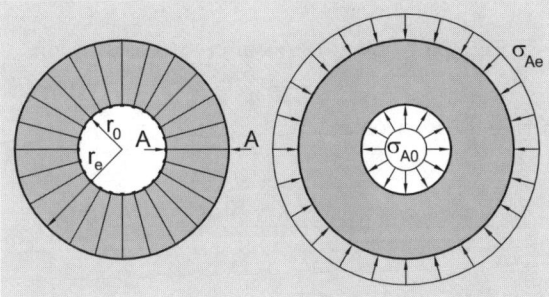

Fig. 15.4. Idealised pattern bolting

The radial stress σ_A is obtained by dividing the anchor force with the pertaining surface. Let n be the number of anchors per one meter of tunnel length. We then obtain

$$\sigma_{A0} = \frac{nA}{2\pi r_0} \quad , \quad \sigma_{Ae} = \frac{nA}{2\pi r_e}$$

or

$$\sigma_{Ae} = \sigma_{A0} \cdot \frac{r_0}{r_e} \quad .$$

It is, thus, reasonable to assume the following distribution of σ_A within the range $r_0 < r < r_e$

$$\sigma_A = \sigma_{A0} \cdot \frac{r_0}{r} \quad . \tag{15.1}$$

We consider the entire stress in the range $r_0 < r < r_e$. Pre-stressing of the anchors increases the radial stress from σ_r to $\sigma_r + \sigma_A$ (Fig. 15.5).

15 Supporting action of anchors/bolts

We now assume that in the range $r_0 < r < r_e$ the shear strength of the ground is fully mobilised. For this case we will determine the support pressure p. For simplicity, we consider a cohesionless ground ($c = 0$) and obtain

$$\sigma_\theta = K_p(\sigma_r + \sigma_A) \tag{15.2}$$

with $K_p = \dfrac{1 + \sin\varphi}{1 - \sin\varphi}$.

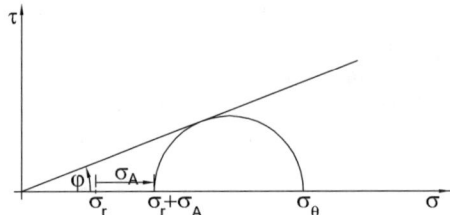

Fig. 15.5. Limit stress in pre-stressed region

Equilibrium in radial direction reads

$$\frac{d(\sigma_r + \sigma_A)}{dr} + \frac{\sigma_r + \sigma_A - K_p(\sigma_r + \sigma_A)}{r} = 0 \quad . \tag{15.3}$$

Introducing (15.1) into (15.3) yields

$$\frac{d\sigma_r}{dr} + \frac{1}{r}\left[\sigma_r(1 - K_p) - K_p \sigma_{A0}\frac{r_0}{r}\right] = 0 \quad . \tag{15.4}$$

The solution of the differential equation (15.4) is obtained as

$$\sigma_r = \text{const} \cdot r^{K_p - 1} - \sigma_{A0}\frac{r_0}{r} \quad .$$

The integration constant is obtained from the boundary condition $\sigma_r(r_0) \stackrel{!}{=} p$ where p is the pressure exerted by the ground upon the lining. We finally obtain

$$\sigma_r = (p + \sigma_{A0}) \cdot \left(\frac{r}{r_0}\right)^{K_p - 1} - \sigma_{A0}\frac{r_0}{r} \quad . \tag{15.5}$$

At the boundary of the elastic region (at $r = r_e$) it must be $\sigma_r = \sigma_e$, where σ_e is obtained from equation 14.21:

15.1 Impact of pattern bolting

$$(p+\sigma_{A0})\left(\frac{r_e}{r_0}\right)^{K_p-1} - \sigma_{A0} \cdot \frac{r_0}{r_e} = \frac{2}{K_p+1} \cdot \sigma_\infty \qquad (15.6)$$

We meet the simplifying assumption that the plastified zone coincides with the anchored ring, i.e. we introduce $r_e = r_0 + l$, where l is the theoretical anchor length, into Equ. 15.6 and eliminate p. We thus obtain the support pressure in dependence of the pre-stressing force A of the anchors, their number n per tunnel meter, the theoretical anchor length l, the tunnel radius r_0, the primary stress σ_∞ and the friction angle φ:

$$p = \left(\frac{2\sigma_\infty}{K_p+1} + \frac{nA}{2\pi r_0} \cdot \frac{r_0}{r_0+l}\right)\left(\frac{r_0}{r_0+l}\right)^{K_p-1} - \frac{nA}{2\pi r_0} \qquad (15.7)$$

The real anchor length L should be greater than the theoretical one, in such a way that the anchor force can be distributed along the boundary $r = r_e$ (Fig. 15.6). In practice, the anchor lengths are taken as 1.5 to 2 times the thickness of the plastified zone.

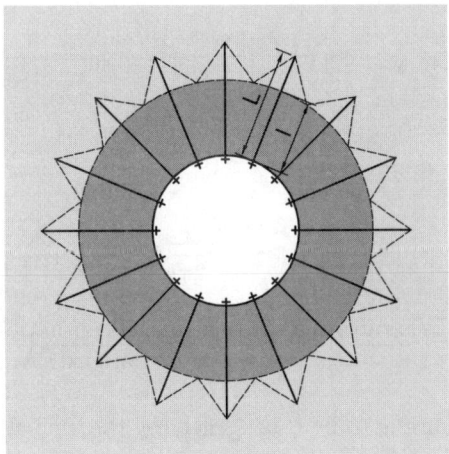

Fig. 15.6. Theoretical (l) and real (L) anchor lengths

15.1.2 Pre-stressed anchors in cohesive soils

To consider cohesion, equation 15.2 is replaced by

$$\sigma_\theta = K_p(\sigma_r + \sigma_A) + 2c\frac{\cos\varphi}{1-\sin\varphi} \quad .$$

Thus, equilibrium in radial direction reads

$$\frac{d\sigma_r}{dr} + \frac{1}{r} \cdot \left[\sigma_r(1 - K_p) - 2c \frac{\cos\varphi}{1 - \sin\varphi} - K_p \sigma_{A0} \frac{r_0}{r} \right] = 0 \quad .$$

The solution of this differential equation reads:

$$\sigma_r = \text{const} \cdot r^{K_p - 1} - \sigma_{A0} \left(\frac{r_0}{r}\right) - c \frac{2\cos\varphi}{(K_p - 1)(1 - \sin\varphi)} \quad .$$

With the boundary condition $\sigma_r(r_0) \stackrel{!}{=} p$ and with $\frac{2\cos\varphi}{(K_p-1)(1-\sin\varphi)} = \cot\varphi$ one finally obtains

$$\sigma_r = (p + \sigma_{A0} + c \cdot \cot\varphi) \left(\frac{r}{r_0}\right)^{K_p - 1} + \sigma_{A0} \frac{r_0}{r} - c \cdot \cot\varphi \quad .$$

From the requirement $\sigma_r(r_e) = \sigma_e$ with σ_e according to equation 14.25 it is obtained:

$$(p + \sigma_{A0} + c \cdot \cot\varphi) \left(\frac{r_e}{r_0}\right)^{K_p - 1} - \sigma_{A0} \frac{r_0}{r_e} - c \cdot \cot\varphi$$
$$= \sigma_\infty (1 - \sin\varphi) - c \cdot \cos\varphi \quad ,$$

With $r_e = r_0 + l$ it finally follows:

$$p = \sigma_\infty(1 - \sin\varphi) \left(\frac{r_0}{r_0 + l}\right)^{K_p - 1} - \frac{nA}{2\pi r_0} \left[1 - \left(\frac{r_0}{r_0 + l}\right)^{K_p} \right]$$
$$- c \cdot \cot\varphi \left[1 - \left(\frac{r_0}{r_0 + l}\right)^{K_p - 1} \right] - c \cdot \cos\varphi \left(\frac{r_0}{r_0 + l}\right)^{K_p - 1} \quad . \tag{15.8}$$

If the ground pressure is to be taken solely by the anchors (i.e. $p = 0$), then:

$$nA \geq \frac{2\pi r_0}{1 - \left(\frac{r_0}{r_0 + l}\right)^{K_p}} \cdot \left\{ \sigma_\infty(1 - \sin\varphi) \left(\frac{r_0}{r_0 + l}\right)^{K_p - 1} \right.$$
$$\left. - c \cdot \cot\varphi \left[1 - \left(\frac{r_0}{r_0 + l}\right)^{K_p - 1} \right] - c \cdot \cos\varphi \left(\frac{r_0}{r_0 + l}\right)^{K_p - 1} \right\} \quad .$$

In case of large convergences, support by anchors is preferable to shotcrete which is not sufficiently ductile and may fracture. However, adjustable anchors should be used.

15.1.3 Stiffening effect of pattern bolting

In this section we consider the stiffening effect of arrays of bolts, i.e. reinforcing elements that are not pre-stressed and transmit shear forces to the surrounding ground over their entire length. Considering equilibrium of the normal force N and the shear stress τ applying upon the periphery of a bolt element of the length dx (Fig. 15.7) we obtain $dN = \tau \pi d\, dx$. With $N = \sigma \pi d^2/4$, $\sigma = E\varepsilon$ and $\varepsilon = du_s/dx$ we obtain

$$\frac{d^2 u_s}{dx^2} = \frac{4\tau}{Ed} \quad ,$$

with u_s being the displacement of the bolt. Obviously, the shear stress τ acting between bolt and surrounding ground is mobilised with the relative displacement, $\tau = \tau(s)$, $s = u_s - u$, where u is the displacement of the ground.[2]

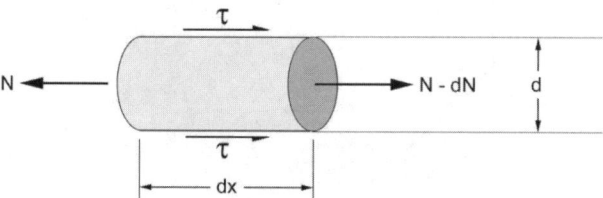

Fig. 15.7. Forces upon a bolt element

Of course, u depends on τ: In a first step of simplified (uncoupled) analysis we assume that u does not depend on τ and is given by the elastic solution (cf. Equ. 14.16):

$$u = \frac{\sigma_\infty - p}{2G} \frac{r_0^2}{r} \quad .$$

Herein, r is the radius with respect to the tunnel axis. Furthermore, we assume a rigid-idealplastic relation $\tau(s)$, i.e. τ achieves immediately its maximum value τ_0. Thus, the total force transmitted by shear upon a bolt of the length l is $l\tau_0\pi d$. This force is applied via the top platen upon the tunnel wall. Assuming n bolts per m² tunnel wall we obtain thus the equivalent support pressure $p_{bolt} = nl\tau_0\pi d$. If the arrangement of bolts is given by the spacings a and b (Fig. 15.8), then $n = 1/(ab)$. Thus,

---

[2] Consider e.g. the relations used in concrete engineering: K. Zilch and A. Rogge, Grundlagen der Bemessung von Beton-, Stahlbeton- und Spannbetonbauteilen nach DIN 1045-1. In: Betonkalender 2000, BK1, 171-312, Ernst & Sohn Berlin, 2000

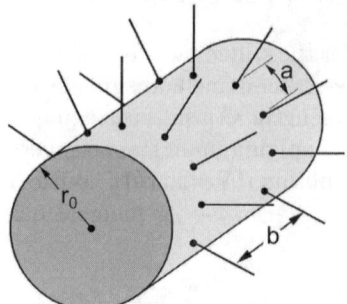

Fig. 15.8. Array of bolts

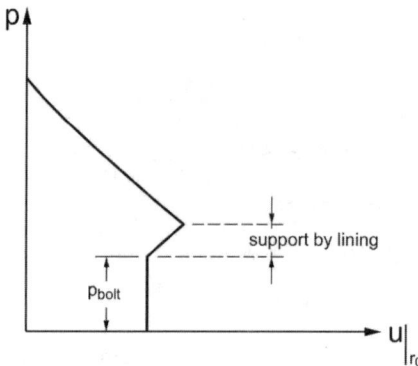

Fig. 15.9. Ground reaction line and support line affected by idealised bolts (Assumptions: rigid bolts, rigid-idealplastic shear stress transmission to the ground, ground displacement not influenced by the bolts, installation of bolts is instantaneous).

$$p_{bolt} = \frac{1}{ab}\tau_o \pi dl \qquad (15.9)$$

modifies the support line as shown in Fig. 15.9.

An alternative approach based on the multiphase model of reinforced ground is given in Appendix F.

16

Some approximate solutions for shallow tunnels

In shallow tunnels the neglection of vertical stress increase with depth due to gravity is not justified. Thus, the solutions presented so far based on hydrostatic primary stress are not applicable. In this section some approximate solutions for shallow tunnels are presented.

16.1 Janssen's silo equation

In silos (i.e. vessels filled with granular material) the vertical stress does not increase linearly with depth. Silos are, therefore, archetypes for arching. The equation of JANSSEN (1895)[1] is used for the design of silos. To derive it, we consider a slim silo with a circular cross section (Fig. 16.1).

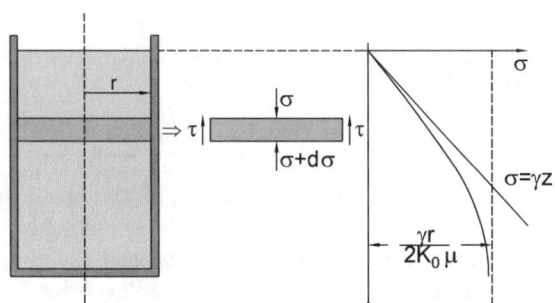

Fig. 16.1. To the derivation of the equation of JANSSEN.

Upon a disk with the radius r and the thickness dz act the own weight $\pi r^2 \gamma dz$, the stress forces $\sigma \pi r^2$ and $-(\sigma + d\sigma)\pi r^2$, as well as the shear force $-\tau 2\pi r dz$

---

[1] Jannsen, H.A. (1895), Versuche über Getreidedruck in Silozellen. Zeitschrift des Vereins deutscher Ingenieure, Band 39, No. 35

due to wall friction τ. The latter is proportional to the horizontal stress σ_H, $\tau = \mu\sigma_H$, and σ_H is assumed proportional to the vertical stress σ, i.e. $\sigma_H = K_0\sigma$. K_0 is the earth-pressure-at-rest-coefficient[2], and μ is the wall friction coefficient. Equilibrium requires that the sum of these forces vanishes. In this way one obtains the differential equation

$$\frac{d\sigma}{dz} = \gamma - \frac{2K_0\mu}{r}\sigma \quad .$$

With the boundary condition $\sigma(z=0) \stackrel{!}{=} 0$ it has the solution

$$\sigma(z) = \frac{\gamma r}{2K_0\mu}(1 - e^{-2K_0\mu z/r}) \quad . \tag{16.1}$$

Thus, the vertical stress cannot increase above the value $\gamma r/(2K_0\mu)$.
This derivation of Equ. 16.1 also applies if the silo has no circular cross section. Then, r is the so-called hydraulic radius of the cross section:

$$\frac{A}{U} = \frac{r}{2} \quad ,$$

where A is the area and U the circumference of the cross section.
If the adhesion c_a acts between silo wall and granulate (soil), then Equ. 16.1 is to be modified as follows:

$$\sigma(z) = \frac{(\gamma - 2c_a/r)r}{2K_0\mu}(1 - e^{-2K_0\mu z/r}) \quad . \tag{16.2}$$

If the surface of the granulate is loaded with the load q per unit area, then the boundary condition at $z = 0$ reads $\sigma(z=0) = q$. This leads to the equation

$$\sigma(z) = \frac{(\gamma - 2c_a/r)r}{2K_0\mu}\left(1 - e^{-2K_0\mu z/r}\right) + qe^{-2K_0\mu z/r} \quad . \tag{16.3}$$

The theory of JANSSEN points out that the granulate stored in silos 'hangs' partly at the silo walls by friction. This results in high vertical stresses in the silo walls, which may buckle. The mobilization of the shear stresses on the wall presupposes sufficiently large relative displacements between granulate and silo wall. If the granulate is moved upwards, the the sign of wall shear stress is reversed. Equation 16.1 then has to be replaced by

$$\sigma(z) = \frac{\gamma r}{2K_0\mu}(e^{2K_0\mu z/r} - 1) \quad .$$

JANSSEN's equation is often used to assess arching above tunnels:

---

[2] according to JAKY is $K_0 \approx 1 - \sin\varphi$ for un-preloaded cohesionless materials

1. TERZAGHI regarded the range ABCD represented in Fig. 16.2 as silo,[3] with the width b (for the plane deformation considered here the hydraulic radius is $r = b$), on the lower edge BC of which acts the pressure p.[4] He thus obtained the following equation for the load p acting upon the roof of a tunnel with rectangular cross section:

$$p = \frac{(\gamma - 2c/b)b}{2K \tan \varphi} \left(1 - e^{-2Kh \tan \varphi/b}\right) \tag{16.4}$$

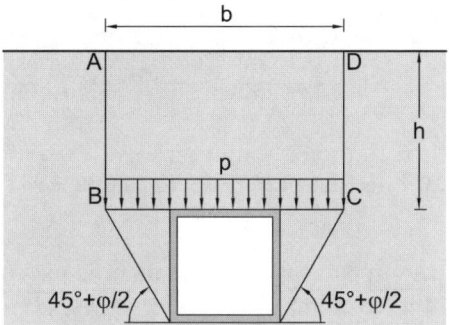

Fig. 16.2. To the derivation of the equation of TERZAGHI.

2. To estimate the pressure needed for the support of the tunnel face (e.g. of a slurry shield), JANSSEN's equation is used. The assessment of face stability is often accomplished following a collapse mechanism originally proposed by HORN (Fig. 16.3).[5] To take into account the 3D-character of the collapse mechanism, the front ABCD of the sliding wedge is taken of equal area as the one of the tunnel cross section. On the sides BDI and ACJ is set cohesion and friction (in accordance with the geostatic stress distribution $\sigma_x = K\gamma z$). The vertical force V is computed according to the silo formula. The necessary support force S is determined by equilibrium consideration of the sliding wedge, whereby the inclination angle ϑ is varied until S becomes maximum. From the consideration of the relative displacements (Fig. 16.3,c) it follows that at the sliding wedge acts also a horizontal force H, which is (erroneously!) omitted by most authors.[6] The

---

[3] The delimitation by $45° + \varphi/2$ inclined lines is not clear.

[4] K. Széchy, Tunnelbau, Springer-Verlag, Wien, 1969.

[5] J. Holzhäuser, Problematik der Standsicherheit der Ortsbrust beim TBM-Vortrieb im Betriebszustand Druckluftstützung, Mitteilungen des Institutes und der Versuchsanstalt für Geotechnik der TU Darmstadt, Heft **52**, 2000,49-62

[6] P.A. Vermeer et al., Ortsbruststabilität von Tunnelbauwerken am Beispiel des Rennsteig Tunnels, 2. Kolloquium 'Bauen in Boden und Fels', TA Esslingen, Januar 2000

silo equation presupposes a full mobilization of the shear strength at the circumference of the prism sliding downward, which implies substantial settlements at the surface.

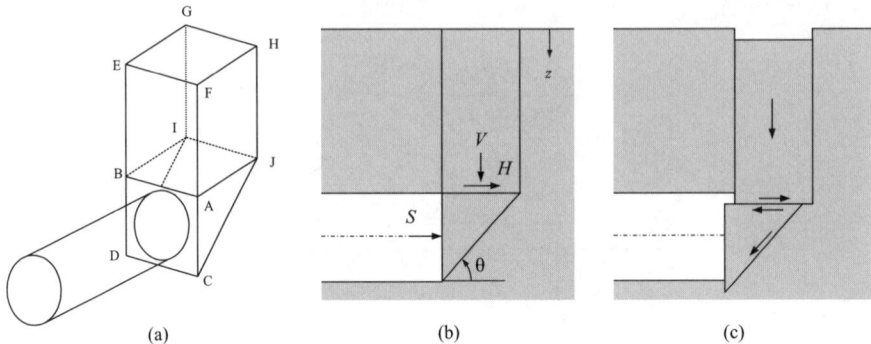

Fig. 16.3. Mechanism of HORN to analyse face stability.

3. In top heading the upper part of the tunnel is excavated first and supported with shotcrete lining. This lining constitutes a sort of arch (or bridge) whose footings must be safely founded, i.e. the vertical force F exerted by the body ABCD (Fig. 16.4) has to be introduced into the subsoil. To assess the safety against punching of the footings into the subsoil, F is estimated by means of JANSSEN's equation.[7]

16.2 Trapdoor

The link between the equation of JANSSEN and tunnelling is established by the so-called trapdoor problem (Fig. 16.5). A trapdoor is moved downwards, whereas the force Q exerted by the overburden sand is being measured and plotted over the settlement s. The similarity between a trapdoor and a discharged silo becomes obvious if we consider the part of the soil that does not

---

[7]G. Anagnostou, Standsicherheit der Ortsbrust beim Vortrieb von oberflächennahen Tunneln. Städtischer Tunnelbau: Bautechnik und funktionale Ausschreibung, Intern. Symposium Zürich, März 1999, 85-95; see also P.A. Vermeer et al., Ortsbruststabilität von Tunnelbauwerken am Beispiel des Rennsteig Tunnels, 2. Kolloquium 'Bauen in Boden und Fels', TA Esslingen, Januar 2000; J. Holzhäuser, Problematik der Standsicherheit der Ortsbrust beim TBM-Vortrieb im Betriebszustand Druckluftstützung, Mitteilungen des Institutes und der Versuchsanstalt für Geotechnik der TU Darmstadt, Heft **52**, 2000,49-62; S. Jancsecz u.a., Minimierung von Senkungen beim Schildvortrieb ..., *Tunnelbau* 2001, 165-214, Verlag Glückauf; methods of Broms & Bennemark and Tamez, cited in: M. Tanzini, Gallerie, Dario Flaccovio Editore, 2001

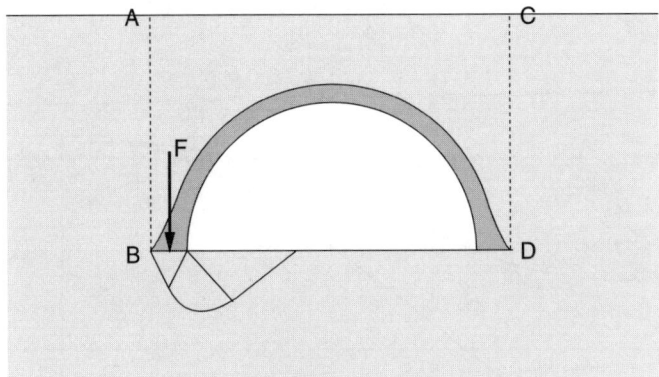

Fig. 16.4. Foundation of crown support.

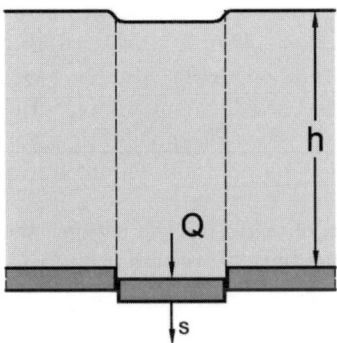

Fig. 16.5. Downwards displacement of the trapdoor.

move downwards as an equivalent to the silo wall. However, the soil is deformable, whereas the silo wall is considered as rigid. Thus, the analogy of the two problems is not complete. In fact, the stress distribution along the 'silo wall' of the trapdoor problem deviates from the one according to JANSSEN's equation. This is obtained with laboratory measurements which TERZAGHI[8] carried out with the steel-tape method (Fig. 16.6 and 16.7).

The results of the measurements are confirmed by numerical results obtained with the code FLAC. Fig. 16.9 shows the vertical stress σ_z averaged over the trapdoor width b ($\bar{\sigma}_z := \frac{1}{b}\int_0^b \sigma_z dx = \frac{1}{b}Q$, see Fig. 16.8) in dependence of the depth z. The curves have been obtained with $K = 1$. The deviations from JANSSEN's solution are obvious, especially for dilatant soil.

---

[8] K. Terzaghi: Stress distribution in dry and in saturated sand above a yielding trapdoor. Proceed. Int. Conf. Soil Mechanics, Cambridge Mass., 1936, Vol. 1, 307-311.

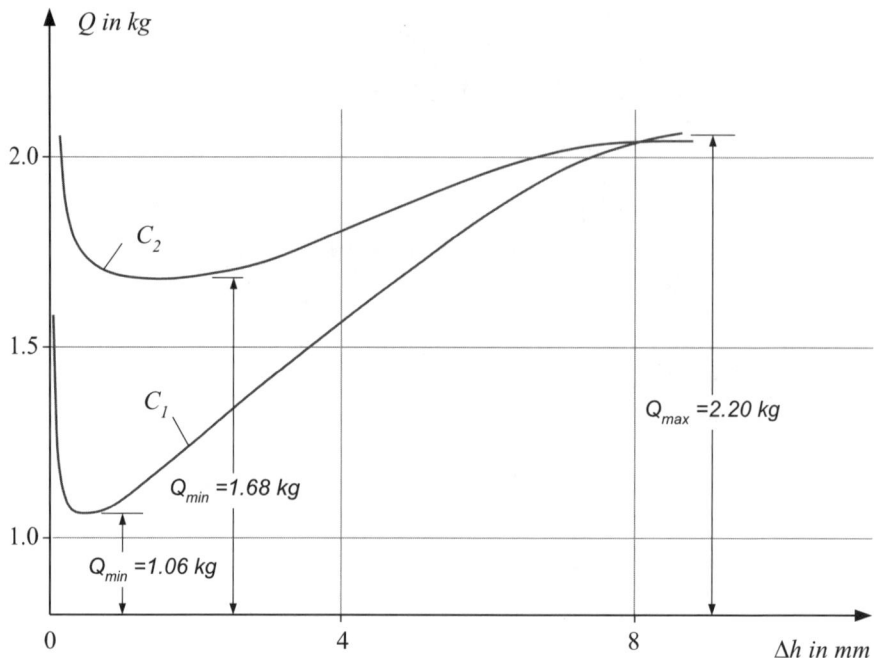

Fig. 16.6. Experimentally obtained relation between the vertical downward movement Δh of the trapdoor and the total vertical pressure Q. C_1: dense sand, C_2: loose sand (measurements reported by TERZAGHI).

The observed Q-s-curve[9] is regarded as a model of the ground reaction line. The rising branch of this curve confirms the associated concept of the NATM. Considered as a proxy of NATM and disregarding the results shown in Fig. 16.6, the rising branch has been attacked. It is interesting to note that the rising branch can also be obtained numerically using standard FEM schemes (e.g. FLAC with MOHR-COULOMB constitutive equation). However, these results are mesh-dependent. The rising branch appears only with reducing the mesh size (Fig. 16.10). This fact proves that the solution is mesh-dependent and, thus, the related numerical problem is ill-posed. Improved numerical approaches using so-called regularised approaches prove to be capable to reproduce the rising branch.

Using a constitutive law with softening (i.e. decrease of strength beyond the peak) and a non local approach, the ground reaction lines shown in Fig. 16.11 have been obtained.[10]

---

[9]Similar results are reported in E. Papamichos, I. Vardoulakis and L.K. Heil, Overburden Modelling Above a Compacting Reservoir Using a Trap Door Apparatus. *Phys. Chem. Earth (A)* Vol. 26, No. 1-2, 69-74, 2001.

[10]P.A. Vermeer, Th. Marcher, N. Ruse, On the Ground Response Curve, *Felsbau*, 20 (2002), No. 6, 19-24

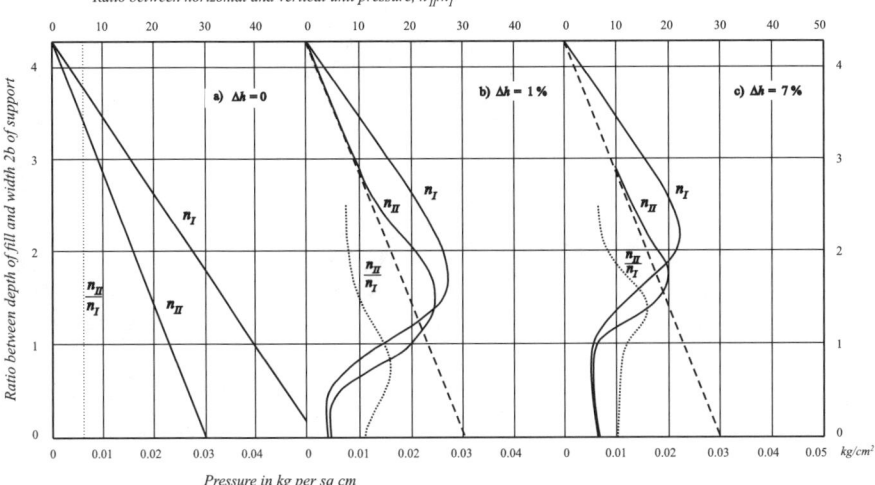

Fig. 16.7. Experimentally obtained distribution of the vertical pressures n_I and the horizontal pressures n_{II} over a plane, vertical section through the axis of the trapdoor (a) for the state preceding the downward movement of the trapdoor, (b) for the state corresponding to Q_{min}, and (c) for the state corresponding to Q_{max}.

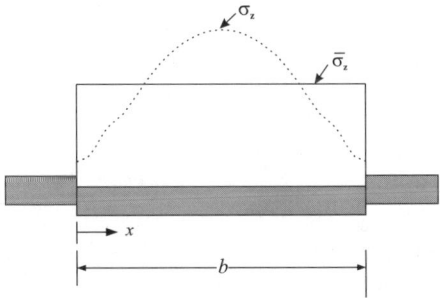

Fig. 16.8. Vertical stress σ_z averaged over the trapdoor width b. Schematic representation.

16.3 Support pressures at crown and invert

The approximate solutions presented in the sequel follow from simple assumptions and explain some principles of the New Austrian Tunnelling Method (NATM). They are based on the reasonable assumption that adjacent to the crown and the invert of a tunnel the principal stress trajectories are parallel resp. perpendicular to the tunnel boundary. Thus, at a tunnel with a circular cross section (radius r) the σ_ϑ-trajectory at the crown has the curvature radius r (cf. Fig. 14.5), provided the support pressure is sufficiently low. For a non-circular tunnel (e.g. a tunnel with a mouth profile) the radius of the σ_ϑ-trajectory at the crown coincides with the curvature radius r_c of the crown.

16 Some approximate solutions for shallow tunnels

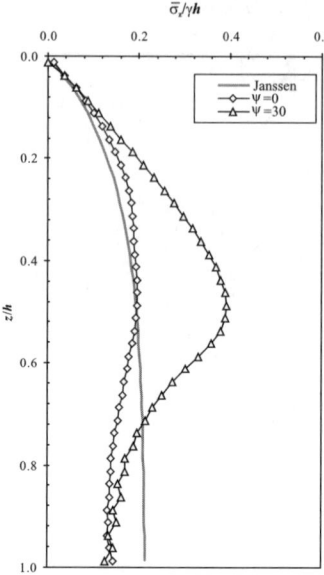

Fig. 16.9. $\bar{\sigma}_z$ vs. z for $h/b = 4$, $\varphi = 30°$, $\psi = 0/30°$

Due to the considered geometry we can express equilibrium at crown and invert using the cylindrical coordinates r and ϑ:

$$\frac{\partial \sigma_r}{\partial r} + \frac{\sigma_r - \sigma_\theta}{r} = \varrho \mathbf{g} \cdot \mathbf{e}_r . \tag{16.5}$$

Here $\mathbf{g}$ is the mass force (gravity acceleration), ϱ the density and $\mathbf{e}_r$ the radial unit vector. This equation can be expressed for point B (Fig. 16.13) in x-z-coordinates as:

$$\frac{d\sigma_z}{dz} + \frac{\sigma_x - \sigma_z}{r} = \gamma$$

with $\varrho \mathbf{g} \cdot \mathbf{e}_r = -\gamma$, $dr = -dz$, $\sigma_r = \sigma_z$, $\sigma_\theta = \sigma_x$, and for point C as:

$$\frac{d\sigma_z}{dz} - \frac{\sigma_x - \sigma_z}{r} = \gamma$$

with $\varrho \mathbf{g} \cdot \mathbf{e}_r = \gamma$ and $dr = dz$.

We regard the vertical stress σ_z in the symmetry axis ABC (Fig. 16.13). Before the excavation of the tunnel, σ_z is assumed to be linearly distributed over the depth z: $\sigma_z = \gamma z$. The primary stress state changes due to tunnel construction and we can estimate the new stress distribution: Between ground

16.3 Support pressures at crown and invert   321

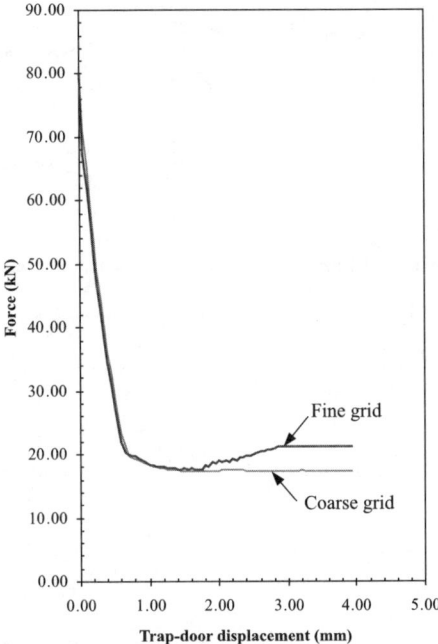

Fig. 16.10. The rising branch disappears with reducing the mesh size. The shown curves were obtained numerically with FLAC and MOHR-COLOUMB elastoplastic constitutive law.

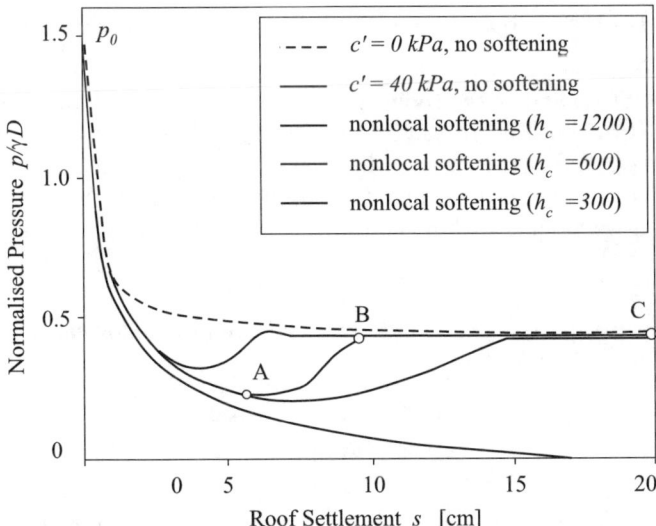

Fig. 16.11. Ground reaction lines for softening rock. Numerically obtained with a non-local approach.

322   16 Some approximate solutions for shallow tunnels

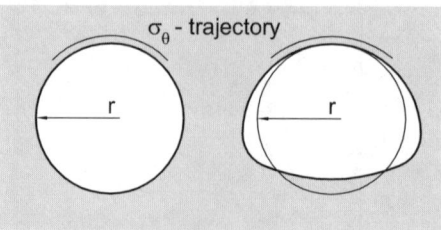

Fig. 16.12. σ_θ trajectories at the crown

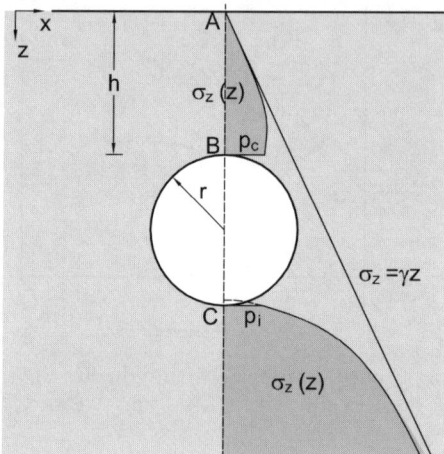

Fig. 16.13. Distribution of the vertical stress over the vertical symmetry axis

surface (point A) and crown (point B) the distribution of σ_z with depth z will be as represented in Fig. 16.13. Towards the ground surface it approaches the primary stress and at the crown it has the value p_c. p_c is the support pressure at the crown, i.e. that stress which the ground exerts upon the lining. For this distribution we assume within the range $0 \le z \le h$ a quadratic parabola:

$$\sigma_z(z) = a_1 z^2 + a_2 z + a_3 \, .$$

The coefficients a_1, a_2, a_3 can be determined from the following three requirements:

1. $\quad \sigma_z(z=0) = 0$
2. $\quad \dfrac{d\sigma_z}{dz}\bigg|_{z=0} = \gamma$

The second requirements follows from equation 16.5 and the reasonable assumption that at the point A the curvature radius of the horizontal stress trajectory is infinite ($r = \infty$). The third requirement follows from the assumption that at point B (crown) the strength of the ground is fully mobilised. For a purely cohesive material ($c \ne 0, \varphi = 0$) this relationship reads:

16.3 Support pressures at crown and invert

$$\sigma_x - \sigma_z = 2c \ . \tag{16.6}$$

Thus one obtains from equation 16.5 the third requirement:

$$3. \quad \left. \frac{d\sigma_z}{dz} \right|_{z=h} = \gamma - \frac{2c}{r_c}$$

r_c is the curvature radius of the crown. The stress distribution between the points A and B thus reads

$$\sigma_z(z) = -\frac{c}{r_c h} z^2 + \gamma z \ . \tag{16.7}$$

If we set in equation 16.7 $z = h$, then we obtain the necessary support pressure at the crown $p_c = \sigma_z(z = h)$:

$$p_c = h \left(\gamma - \frac{c}{r_c} \right) \ . \tag{16.8}$$

From equation 16.8 one sees that for

$$c \geq \gamma r_c \tag{16.9}$$

a support (at least at the crown) is not necessary. Note that according to Equ. 16.9 the overburden height h does not play a role if the cohesion exceeds the value γr_c. This apparently paradoxical result can be illustrated if one regards the simple collapse mechanism shown in Fig. 16.14: The cohesion force $2c(h+r)$ has to carry the weight $2r\gamma(h+r) - \frac{1}{2}r^2\pi\gamma$. This is obviously possible if $c \geq \gamma r$, no matter how large h is.

The relationship 16.8 can easily be generalised in the case that the ground possesses cohesion and friction ($\varphi \neq 0$). Then equation 16.6 is to be replaced by the failure condition:

$$\sigma_x - \sigma_z = \sigma_z \frac{2 \sin \varphi}{1 - \sin \varphi} + 2c \frac{\cos \varphi}{1 - \sin \varphi} \ , \tag{16.10}$$

from which one finally obtains:

$$p_c = h \frac{\gamma - \dfrac{c}{r_c} \dfrac{\cos \varphi}{1 - \sin \varphi}}{1 + \dfrac{h}{r_c} \dfrac{\sin \varphi}{1 - \sin \varphi}} \ . \tag{16.11}$$

It follows that no support is necessary if

$$c \geq \gamma r_c \frac{1 - \sin \varphi}{\cos \varphi} \ . \tag{16.12}$$

From equations 16.11 and 16.12 we infer the importance of the radius r_c. Thus, we cannot judge the stability of a large tunnel from the stability of an

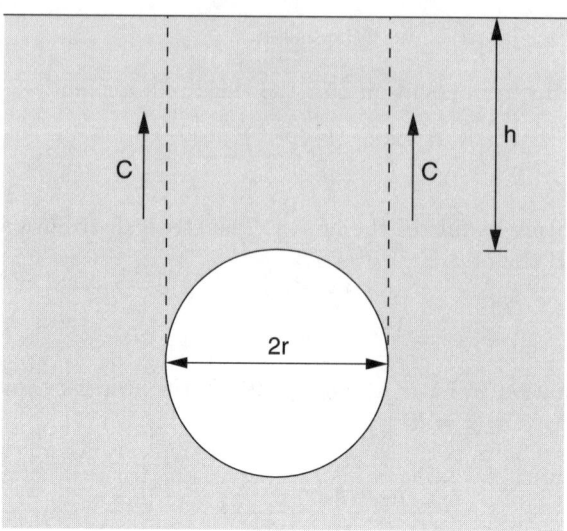

Fig. 16.14. Simple mechanism for the daylight collapse

exploration gallery with small diameter in the same ground. From equation 16.11 it is evident that the support pressure p_c increases if c decreases due to loosening (ascending branch of the ground reaction line). Therefore loosening should be avoided, which is an important principle of the NATM. If the ground surface is loaded with the constant load q per unit area, then equation 16.11 can be generalised as follows:

$$p_c = \frac{q - \dfrac{h}{r_c}\dfrac{c\cos\varphi}{1-\sin\varphi} + \gamma h}{1 + \dfrac{h}{r_c}\dfrac{\sin\varphi}{1-\sin\varphi}} \quad .$$

By laboratory model tests in the laboratory it has been shown that this equation supplies a safe estimation of the necessary support pressure.[11]

Now we use a similar consideration in order to determine the necessary support pressure p_i at the invert. We regard once again the distribution of the vertical stress σ_z along the symmetry axis ABC (Fig. 16.13). σ_z has at the invert the (still unknown) value p_i and approaches with increasing depth z asymptotically the geostatic primary stress $\sigma_z = \gamma z$. A simple analytic curve, which is sufficient for these requirements, is the hyperbola

$$\sigma_z(z) = \gamma z + \frac{a}{z} \quad . \tag{16.13}$$

with the free parameter a.

---

[11] P. Mélix, Modellversuche und Berechnungen zur Standsicherheit oberflächennaher Tunnel. Veröffentlichungen des Instituts für Boden- und Felsmechanik der Universität Karlsruhe, Heft Nr. 103, 1986.

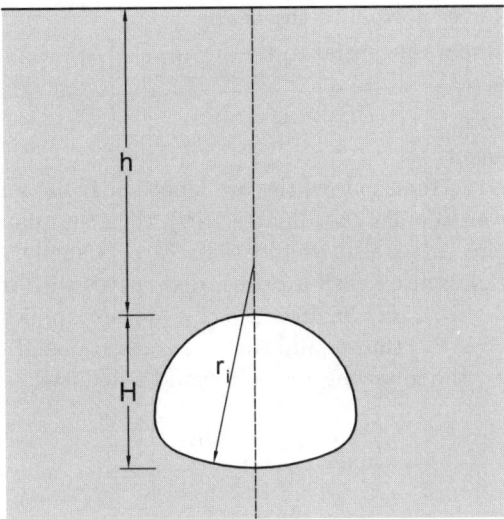

Fig. 16.15. Symbols for a mouth profile

We assume now that the strength of the ground is fully mobilised at the invert (point C in Fig. 16.13). For a frictionless material it then follows from the equation of equilibrium 16.5:

$$\left.\frac{d\sigma_z}{dz}\right|_C = \gamma + \frac{2c}{r_i} \quad . \tag{16.14}$$

r_i is the curvature radius of the invert. From equations 16.13 and 16.14 a can be determined to $-2c(H+h)^2/r_i$, such that the support pressure at $z = h+H$ is determined as:

$$p_i = (H+h)\left(\gamma - \frac{2c}{r_i}\right) \quad .$$

H is the tunnel height (Fig. 16.15). For $c < \gamma r_i/2$ is $p_i > 0$, i.e. a support of the invert is necessary. This is also an important principle of the NATM, which requires for soft grounds a rapid "ring closure". For grounds with cohesion and friction one obtains in a similar way:

$$p_i = (H+h)\frac{\gamma r_i(1-\sin\varphi) - 2c\cos\varphi}{r_i(1-\sin\varphi) + 2(H+h)\sin\varphi} \quad .$$

16.4 Forces acting upon and within the lining

The tunnel lining can be regarded as a beam with initial curvature. All relevant quantities refer to a beam width of 1 m. The quantities

326   16 Some approximate solutions for shallow tunnels

p : distributed forces normal to the beam,
q : distributed forces tangential to the beam,
N : normal force,
Q : transverse force,
M : bending moment,

can be represented as functions of the arc length s. If the form of the tunnel cross section is given in polar coordinates, $\mathbf{x}(\vartheta)$, then the quantities mentioned above can be represented also as functions of ϑ. Usually, derivatives with respect to s are represented with a prime, derivatives with respect to ϑ with a dot: $\mathbf{x}' := d\mathbf{x}/ds$, $\dot{\mathbf{x}} := d\mathbf{x}/d\vartheta$. Because of $ds = r d\vartheta$ applies (r is the radius of curvature): $\dot{\mathbf{x}} = \mathbf{x}'r$. From equilibrium considerations at a beam element with the length ds, the following relations can be deduced

$$\dot{Q} - N = -pr ,$$
$$\dot{N} + Q = -qr , \qquad (16.15)$$
$$\dot{M} = rQ ,$$

which represent a coupled system of differential equations. A simple special case results, if we accept that due to creep and cracking in a not yet completely hardened sprayed concrete all bending moments disappear, $M \equiv 0$, and that (consequently) no shear stresses act between rock and sprayed concrete lining: $q \equiv 0$. It follows then from the equations (16.15) that for sections of the lining with constant curvature ($r = $ const) must apply $p = $ const and $N = -pr = $ const.

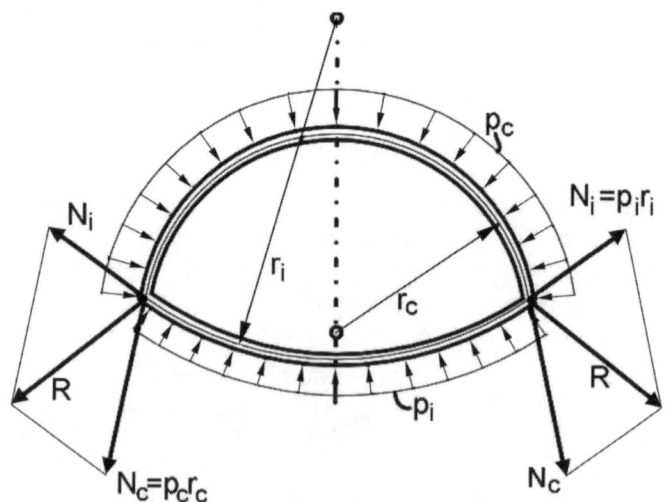

Fig. 16.16. Forces at points where the curvature of the lining changes

Let us look at a mouth profile consisting only of crown and invert arcs (cf. top heading). The resultant forces **R** represented in Fig. 16.16 exerted from the support to the rock must be taken up by suitable constructions, e.g. enlarged footings of the crown arc (so-called elephant feet) or micro piles.

The values p_c and p_i can be inferred from Section 16.3. It should be proved that the compressive stress in the sprayed concrete is permissible. If β is the compressive strength and d the thickness of the sprayed concrete lining, then

$$d > p_c r_c / \beta, \quad d > p_i r_i / \beta .$$

16.5 Estimations based on the bound theorems

16.5.1 Lower bound of the support pressure

According to the lower-bound-theorem one can obtain a safe estimation of the support pressure p in a cohesive ground ($c > 0$, $\varphi = 0$) as follows:[12] we regard a tunnel with circular cross section (radius r_0) and a support pressure p which is constant along the lining. The ground surface is loaded with the constant load q. Within a circular range which touches the surface (Fig. 16.17), we assume a stress field, which fulfils the equations of equilibrium (Section 14.2) and the limit condition $\sigma_1 - \sigma_2 = 2c$ (σ_1 and σ_2 are the principal stresses).

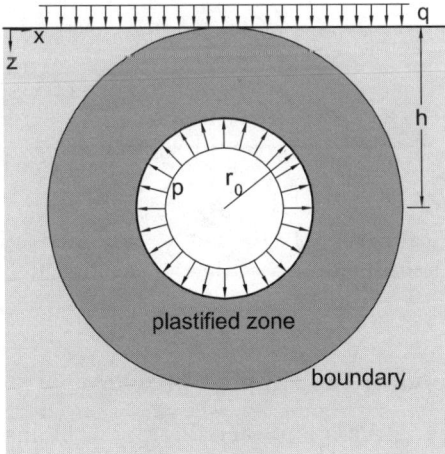

Fig. 16.17. Situation for the estimation of p according to the lower-bound-theorem by assumption of a permissible stress field

---

[12] E.H. Davis: The stability of shallow tunnels and underground openings in cohesive material. *Géotechnique* **30**, No. 4, 397-416 (1980).

16 Some approximate solutions for shallow tunnels

Outside the plastified zone the vertical stress is set $\sigma_z = q + \gamma z$ and the horizontal stress $\sigma_x = K\sigma_z$. K is specified in such a way that the normal and shear stresses which act on the boundary of the plastified zone do not suffer a jump. The results, which are represented in Fig. 16.18 in dimensionless form, are obtained numerically. For comparison, the support pressure according to Equ. 16.11 is also represented. After a re-arrangement and consideration of the load q one obtains (for $\varphi = 0$):

$$\frac{p-q}{c} = \left(\frac{\gamma r_0}{c} - 1\right)\frac{h}{r_0} \quad . \tag{16.16}$$

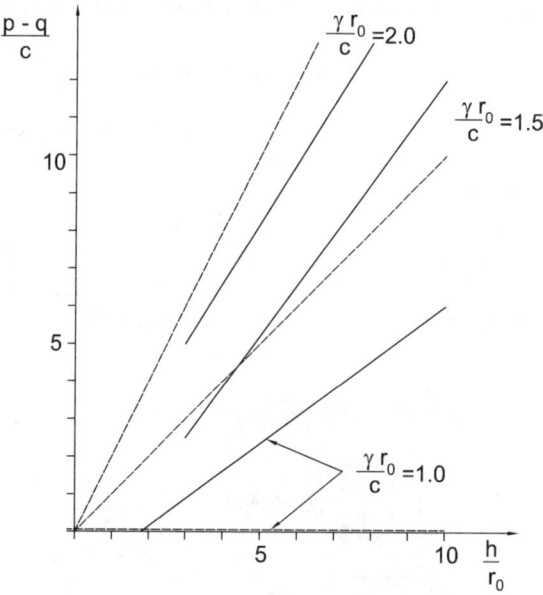

Fig. 16.18. Numerical results of DAVIS et al.: Safe estimation of the necessary support pressure p. Dashed: results according to equation 16.11

16.5.2 Upper bound of the support pressure

For $h/r \geq 1/\sin\varphi - 1$ ATKINSON and POTTS[13] obtained an upper bound (unsafe estimation) of the support pressure p:

$$\frac{p}{\gamma r} = \frac{1}{2\cos\varphi}\left(\frac{1}{\tan\varphi} + \varphi - \frac{\pi}{2}\right) \quad . \tag{16.17}$$

---

[13] J.H. Atkinson, D.M. Potts, Stability of a shallow circular tunnel in cohesionless soil, *Géotechnique* **27**,2 (1977), 203-215

17

Stability of the excavation face

In weak rock, the excavation face must be supported. It is important to estimate the necessary support pressure, in particular for slurry and EPB shields, where the pressure must be set by the operator. Several methods can be consulted for the estimation:

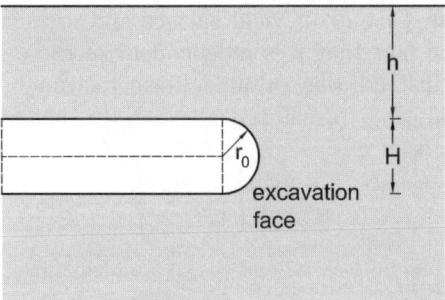

Fig. 17.1. Excavation face as hemisphere

17.1 Approximate solution for ground with own weight

We regard the distribution of the vertical stress between the ground-surface and the crown of a spherical cavity (Fig. 16.13, 17.1). We approximate this distribution by a quadratic parabola and assume that the material strength is fully mobilised at the crown. As in equation 16.11, we obtain the necessary support pressure p_c at the crown as

$$p_c = h\,\frac{\gamma - \dfrac{c}{r_f}\dfrac{\cos\varphi}{1-\sin\varphi}}{1 + \dfrac{h}{r_f}\dfrac{2\sin\varphi}{1-\sin\varphi}} \quad . \tag{17.1}$$

Thus, the crown of the unsupported excavation face is stable if:

$$c \geq \gamma r_f \frac{1-\sin\varphi}{\cos\varphi} \quad .$$

17.2 Numerical results

From numerical results obtained with the FE-code PLAXIS, VERMEER and RUSE[1] deduced the following approximation for the limit support pressure and the case $\varphi > 20°$:

$$p \approx -\frac{c}{\tan\varphi} + 2\gamma r\left(\frac{1}{9\tan\varphi} - 0.05\right) \quad . \tag{17.2}$$

The underlying results are obtained with an elastic-ideal plastic constitutive law assuming MOHR-COULOMB yield surface and associated plasticity. The authors point to the fact that p is independent of the overburden height h. For the case $\varphi = 0$, instead, they obtain a linear relation between p and h. For this relation no analytic expression is given. Note that for $\varphi = 0$ Equ. 17.1 reduces to

$$p = \gamma h\left(1 - \frac{c}{\gamma r}\right)$$

and for $h \to \infty$ Equ. 17.1 reduces to

$$p = -\frac{c}{2\tan\varphi} + 2\gamma r\frac{1-\sin\varphi}{4\sin\varphi} \quad . \tag{17.3}$$

Despite the striking similarity between (17.2) and (17.3), Equ. 17.1 provides much higher p-values than (17.2). Also the bound theorems (Section 17.3) provide higher p-values than Equ. 17.2.

17.3 Stability of the excavation face according to the bound theorems

The lower-bound-theorem supplies a simple but very conservative estimation of the necessary support pressure p at the excavation face, for which the form

---
[1] P.A. Vermeer and N. Ruse, Die Stabilität der Tunnelortsbrust im homogenen Baugrund, *Geotechnik* **24** (2001), Nr. 3, 186-193

17.3 Stability according to the bound theorems

of a hemisphere is assumed (Fig. 17.2). We first consider the case $\gamma = 0$. Within a spherical zone (radius $r = r_0 + h$) around the excavation face we assume that the limit condition $\sigma_\theta - \sigma_r = 2c$ is fulfilled. For the spherically symmetric case regarded here, the equation of equilibrium in radial direction reads:

$$\frac{d\sigma_r}{dr} + \frac{2}{r}(\sigma_r - \sigma_\theta) = 0 \ .$$

Using the limit condition and integration of the differential equation with consideration of the boundary condition $\sigma_r(r = r_0) = p$ gives

$$p = \sigma_r - 4c \ln \frac{r}{r_0} \ .$$

Outside the spherical plastified zone we assume a constant hydrostatic stress $\sigma_r = \sigma_\theta = q$. Equilibrium at the boundary of the two zones $(r = r_0 + h)$ requires $\sigma_r = q$. Thus we obtain the necessary support pressure p as

$$p = q - 4c \ln \left(1 + \frac{h}{r_0}\right) \ .$$

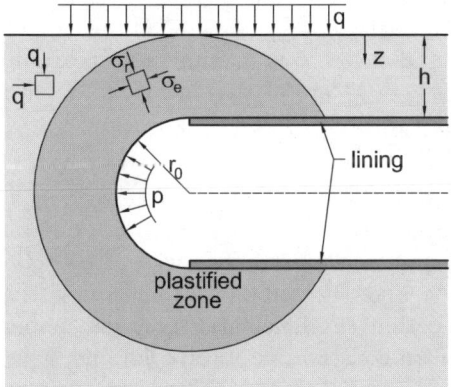

Fig. 17.2. Layout plan for the derivation of a lower bound for the support pressure at the excavation face. Case $\gamma = 0$

Now we consider the case $\gamma > 0$ by overlaying the hydrostatic stress $\sigma_z = \sigma_x = \sigma_y = \gamma z$ to the above mentioned stress field.[2] We obtain, thus, a support pressure that increases linearly with depth (Fig. 17.3):

$$p = \gamma z + q - 4c \ln \left(1 + \frac{h}{r_0}\right) \ .$$

---

[2] The limit condition is not violated by the overlay of a hydrostatic stress

17 Stability of the excavation face

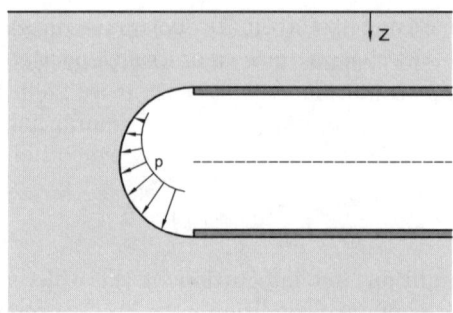

Fig. 17.3. Support pressure increasing linearly with depth z

For plane deformation ('infinitely' long tunnel with circular cross section) the necessary support pressure can be estimated in a similar way[3] as:

$$p = \gamma z + q - 2c \ln\left(1 + \frac{h}{r_0}\right) \quad .$$

The necessary support pressure p at the excavation face can also be estimated (on the unsafe side) by the upper-bound-theorem, where we look at the sliding of two cylindrical rigid blocks made of rock (Fig. 17.4). By variation of the geometry (i.e. of the angle shown in Fig. 17.4) the support pressure obtained from the upper-bound-theorem is maximised. The results of the numerical computation of DAVIS et al. are plotted in Fig. 17.5. On the y-axis is plotted the so-called stability ratio N:

$$N := \frac{q - p + \gamma(h + r_0)}{c} \quad .$$

More complex collapse mechanisms for rocks with friction and cohesion are considered by LECA and DORMIEUX.[4] From comparison with model tests it can be concluded that the kinematic solutions (upper bounds) are more realistic than the ultra conservative static solutions (lower bounds).

The assessment of excavation face stability is often accomplished following the collapse mechanism proposed by HORN (Section 16.1).

---

[3] A. Caquot: Équilibre des massifs à frottement interne. Gauthier-Villars, Paris, 1934, p. 37

[4] E. Leca and L. Dormieux, Upper and lower bound solutions for the face stability of shallow circular tunnel in frictional material. *Géotechnique* **40**, No. 4, 581–606 (1990)

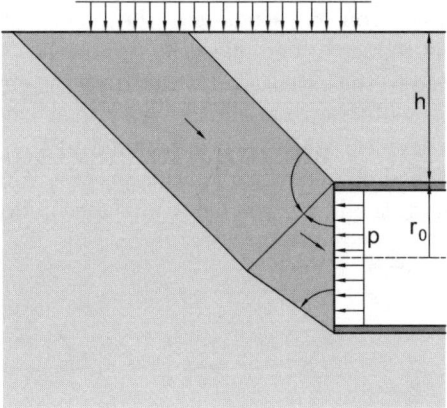

Fig. 17.4. Failure mechanism at the excavation face

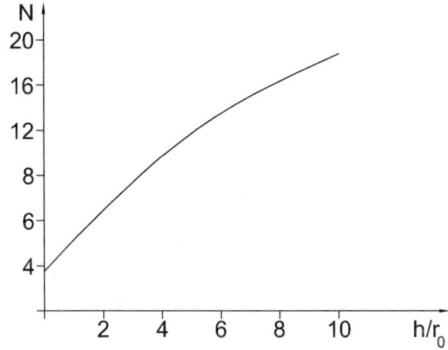

Fig. 17.5. Estimation of the support pressure at the excavation face with the upper-bound-theorem (according to DAVIS et al.)

17.4 Stand-up time of the excavation face

The excavation face is stable for a certain stand-up time. The delay of collapse is attributable partly to creep of the ground and partly to pore water pressure.[5,6] The latter effect can be explained as follows:

According to TERZAGHI's consolidation theory, a load suddenly applied on a water-saturated cohesive soil acts, in the first instance, only upon the pore water. It is gradually transmitted to the grain skeleton, to the extent that the pore water is squeezed out. Exactly the same procedure occurs at unloading (for instance due to the construction of a cut or the excavation of a tunnel):

---

[5] P.R. Vaughan and H.J. Walbancke: Pore pressure changes and the delayed failure of cutting slopes in overconsolidated clay. *Géotechnique* **23**, 4, 1973, 531-539

[6] J.H. Atkinson and R.J. Mair: Soil mechanics aspects of soft ground tunnelling. *Ground Engineering* 1981

Initially, the grain skeleton 'does not feel' the unloading, and the pressure in the pore water is reduced. The effective stresses are thus increased and, subsequently, reduced to the extent that water from the environment is sucked into the voids. This reduction can finally lead to a cave-in. The so-called consolidation coefficient c_v, which is proportional to the permeability of the material, controls the time necessary for this process. Consequently the less permeable the ground, the larger the delay of the cave-in is.

18

Earthquake effects on tunnels

18.1 General remarks

Experience shows that underground structures, especially deep ones, are far less vulnerable to earthquakes than superficial ones. The latter are endangered by earthquakes due to the fact that the motion of the ground can be amplified by the response of the structure to such an extent that the induced strains damage the structure. The earthquake waves can also be amplified within soft superficial strata. In addition, loose water-saturated soil may loose its strength (so-called liquefaction), and this can lead to landslides or failure of foundations and retaining walls. In contrast, deep buried structures, especially flexible ones, are not expected to oscillate independently of the surrounding ground, i.e. amplification of the ground motion can be excluded. This is manifested by the relatively low earthquake damage of tunnels.[1] Of course, the portals may be damaged by earthquake-induced landslides. Very revealing on earthquake effects is the report of what happened to the driving of a 7 m diameter tunnel in the underground of Los Angeles during the San Fernando M 6.7 earthquake in 1971:[2]

> 'The earthquake caused an outage of electrical power that caused the tunnel pumps to stop. Amid the attendant confusion and anxiety, the miners made their way to the locomotive and drove 5 miles out of the tunnel in pitch darkness. This means that the rails were not significantly distorted to cause a derailment. However, Southern Pacific Railroad tracks on the surface were distorted and broken.'

Earthquakes can endanger tunnel and other structures if they are buried in loose watersaturated soil that can be liquefied by dynamic excitation. Lique-

---

[1] Y.M.A. Hashash, J.J. Hook, B. Schmidt, I. I-Chiag Yao, Seismic design and analysis of underground structures. *Tunnelling and Underground Space Technology* 16 (2001) 247-293

[2] R.J. Proctor, The San Fernando Tunnel Explosion, California. *Engineering Geology* 67 (2002) 1-3

faction implies a drastic loss of shear strength, the consequence of which can be large displacements of structures. Countermeasures are stone columns or soil improvement by densification or grouting.

A point of concern is when a tunnel has to cross an active fault. In this case, the tunnel cross section can be enlarged to accommodate the expected displacement. The latter, however, can hardly be predicted nor can it be easily judged whether a fault is active or not.

18.2 Imposed deformation

The main loading of deep tunnels results from their deformation, which may be assumed to be identical with the deformation of the surrounding ground. This assumption implies that the tunnel is infinitely thin and flexible, so that it can be considered as a material line. The distortion of this line results completely from the wave motion of the embedding continuum. For design purposes, this motion must be somehow predicted. This is a very difficult task called seismic hazard analysis. Predictions can be attempted based on deterministic or probabilistic analysis.[3] In either case the results are highly uncertain but possibly still the best achievable assumption.

Assume now that the wave motion is given. We consider harmonic waves with the circular frequency ω. Non-harmonic waves can be decomposed into harmonic ones. Let the unit vector $\mathbf{l}$ denote the direction of wave propagation and let $\mathbf{a}$ be the amplitude of oscillation. Then the displacement $\mathbf{u}$ of a point with spatial coordinates $\mathbf{x}$ is given by the following expressions. Herein, $\mathbf{u}_p$ denotes the p-wave and $\mathbf{u}_s$ denotes the s-wave displacement vectors.

$$\mathbf{u}_p = a_p \, \mathbf{l} \, \exp\left[i\omega\left(t - \frac{\mathbf{l} \cdot \mathbf{x}}{c_p}\right)\right]$$

$$\mathbf{u}_s = \mathbf{a}_s \times \mathbf{l} \, \exp\left[i\omega\left(t - \frac{\mathbf{l} \cdot \mathbf{x}}{c_s}\right)\right]$$

c_p, c_s, a_p and a_s are the corresponding propagation speeds and amplitudes, respectively. Let $\mathbf{t}$ be the unit tangential vector at a particular point P of the tunnel axis. Then, the earthquake-induced longitudinal strain and the change of curvature of tunnel can be obtained as:

$$\varepsilon_{\max} = \frac{\omega}{c} \, a \, \cos^2 \alpha$$

$$\kappa_{\max} = \left(\frac{\omega}{c}\right)^2 a \, \cos^2 \alpha$$

with α being the angle between $\mathbf{l}$ and $\mathbf{t}$ (see Appendix G).

---

[3] S.L. Kramer, Geotechnical Earthquake Engineering. Prentice Hall, 1996

18.2 Imposed deformation

Thus, a joint between two rigid tunnel elements will suffer the elongation $s = L\varepsilon$ and the rotation $\vartheta = \kappa L$ (Fig. 18.1). Herein, L is the length of each tunnel element.

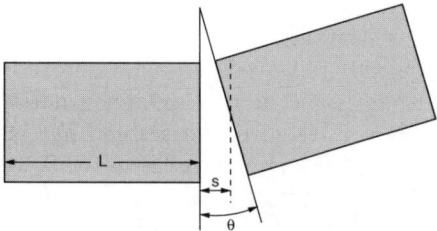

Fig. 18.1. Distortion between two rigid tunnel elements

19

Settlement of the surface

19.1 Estimation of settlement

Apart from the assessment of stability, the determination of settlements at the surface is very important in tunnelling. However, in geotechnical engineering, deformations can be forecast with less accuracy than stability. This is mainly because the ground has a nonlinear stress-strain-relationship, so that one hardly knows the distribution of the stiffnesses. We consider here some rough estimations of the settlement of the ground surface due to the excavation of a tunnel. One should be aware of their limited accuracy.

For the determination of the distribution of the surface settlements let us take first LAMÉ's solution (see Equ. 14.8) of the problem of a cylindrical cavity in a weightless elastic space, loaded by the hydrostatic stress σ_∞. We regard the vertical displacement u_v of the ground-surface shown in Fig. 19.1.

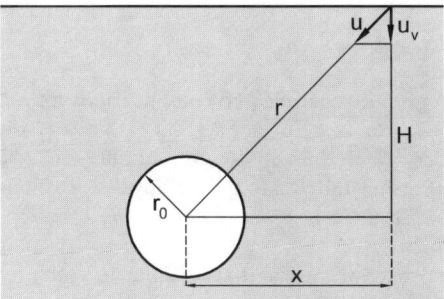

Fig. 19.1. Vertical displacement at the ground surface

The vertical component u_v of the displacement u reads

$$u_v = \frac{H}{r} \cdot u \quad . \tag{19.1}$$

With $r^2 = H^2 + x^2$ we obtain from LAMÉ's solution

$$u_v = \frac{\sigma_\infty - p}{2G} \cdot \frac{r_0^2 H}{H^2 + x^2} \quad . \tag{19.2}$$

The maximum settlement $u_{v,\max}$ is obtained at $x = 0$, and the distribution of the settlement reads:

$$u_v = \frac{u_{v,\max}}{1 + (x/H)^2} \quad .$$

This distribution[1] is not realistic, when compared with measurements.[2] It is also inconsistent, because it uses a solution for the full space for a problem of the halfspace. A more realistic description of the measured settlement is obtained according to PECK by the GAUSS-distribution

$$u_v = u_{v,\max} \cdot e^{-x^2/2a^2} \quad .$$

The parameter a (standard deviation) is to be determined by adjustment to measurements. It equals the x-coordinate at the inflection point of the GAUSS-curve. It can be estimated with the diagram of PECK (Fig. 19.2)[3] or according to the empirical formula:[4]

$$2a/D = (H/D)^{0.8} \quad . \tag{19.3}$$

D is the diameter of the tunnel and H is the depth of the tunnel axis (Fig. 19.3). For clay soils is $a \approx (0,4\ldots 0,6)H$, for non-cohesive soils is $a \approx (0,25\ldots 0,45)H$.

Another estimation of a is given in table 19.1.[5]

---

[1] It also follows from more complicated computations for a linear-elastic material, see A. Verruijt and J.R. Booker: Surface settlements due to deformation of a tunnel in an elastic half space. *Géotechnique* **46**, No. 4 (1996), 753-756

[2] see e.g. J.H. Atkinson and D.M. Potts: Subsidence above shallow tunnels in soft ground. *Journal of the Geotechnical Engineering Division*, ASCE, Volume **103**, No. GT4, 1977, 307-325

[3] Peck, R.B., Deep excavations and tunnelling in soft ground. State-of-the-Art report. In Proceedings of the 7th International Conference on Soil Mechanics and Foundation Engineering, Mexico City, State-of-the-Art Volume, 1969, 225-290

[4] M.J. Gunn: The prediction of surface settlement profiles due to tunnelling. In 'Predictive Soil Mechanics', Proceedings Wroth Memorial Symposium, Oxford, 1992

[5] J.B. Burland et al., Assessing the risk of building damage due to tunnelling - lessons from the Jubilee Line Extension, London. In: Proceed. 2nd Int. Conf. on Soil Structure Interaction in Urban Civil Engineering, Zürich 2002, ETH Zürich, ISBN 3-00-009169-6, Vol. 1, 11 -38.

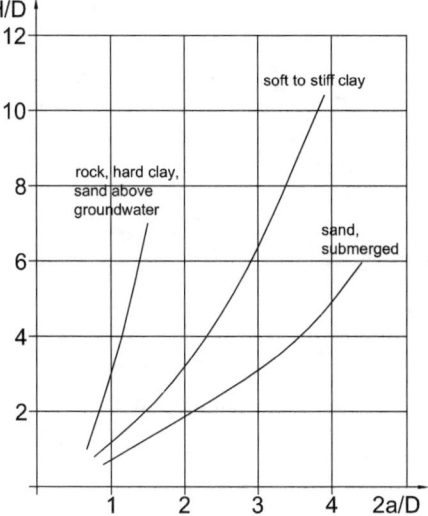

Fig. 19.2. Estimation of A by PECK

Soil	a/H
granular	0.2 - 0.3
stiff clay	0.4 - 0.5
soft silty clay	0.7

Table 19.1. Estimation of a

The horizontal displacements u_h of the ground-surface follow from the observation that the resultant displacement vectors are directed towards the tunnel axis (as shown in Fig. 19.1) i.e.

$$u_h = \frac{x}{H} u_v$$

The distribution of the settlements in longitudinal direction of a tunnel under construction is represented in Fig. 19.3.
The volume of the settlement trough (per current tunnel meter) results from the GAUSS-distribution to

$$V_u = \sqrt{2\pi} \cdot a \cdot u_{v,\max} \qquad (19.4)$$

and is usually designated as volume loss[6] (ground loss). The volume loss amounts to some percent of the tunnel cross-section area per current meter. If

---

[6]This designation is based on the conception that the soil volume V_u is dug additionally to the theoretical tunnel volume

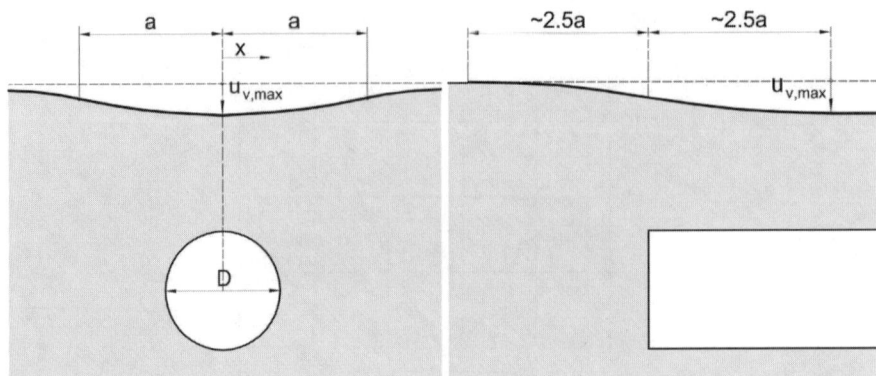

Fig. 19.3. Settlement trough over a tunnel (left); Approximate distribution of the surface settlements in tunnel longitudinal direction. The shown curve coincides reasonably with the function $y = \mathrm{erf}\, x = \dfrac{1}{\sqrt{2\pi}} \int_0^x e^{-y^2/2} dy$ (right).

this ratio is known by experience for a given soil type, then the maximum settlement $u_{v,\max}$ can be estimated with (19.3) and (19.4). MAIR and TAYLOR[7] give the following estimated values for V_u/A:
Unsupported excavation face in stiff clay: 1-2%
Supported excavation face (slurry or earth mash), sand: 0.5%
Supported excavation face (slurry or earth mash), soft clay: 1-2%
Conventional excavation with sprayed concrete in London clay: ... 0.5-1.5%

The volume loss depends on the skill of tunnelling. Due to improved technology the volume loss has been halved over the last years.
The evaluation of numerous field surveys and lab tests with centrifuge leads to an empirical relationship[8] between the volume loss V_u related to the area A of the tunnel cross section and the stability number $N := (\sigma_v - \sigma_t)/c_u$. Herein, σ_v is the vertical stress at depth of the tunnel axis, σ_t is the supporting pressure (if any) at the excavation face, and c_u is the undrained cohesion. If N_L is the value of N at collapse, then:

$$V_u/A \approx 0.23\ e^{4.4N/N_L}\ .$$

The estimations represented here refer to the so-called greenfield. If the surface is covered by a stiff building, then the settlements are smaller[9].

---

[7] R.J. Mair and R.N. Taylor, Bored tunnelling in the urban environment. 14th Int. Conf. SMFE, Hamburg 1997

[8] S.R. Macklin, The prediction of volume loss due to tunnelling in overconsolidated clay based on heading geometry and stability number. *Ground Engineering*, April 1999

[9] 'Recent advances into the modelling of ground movements due to tunnelling', *Ground Engineering*, September 1995, 40-43

19.1 Estimation of settlement

The maximum settlement $u_{v,\max}$ can also be roughly estimated by the following consideration: Let ε_{r0} and ε_{v0} be the radial strain and the volume strain at the crown, respectively. We can determine these values by a triaxial or biaxial extension test in the laboratory. Then we have $\varepsilon_{\vartheta 0} = \varepsilon_{v0} - \varepsilon_{r0} = u_0/r_0$, whereby u_0 is the displacement (settlement) of the crown and r_0 is the radius of the tunnel. We now assume

$$u = u_0 \left(\frac{r_0}{r}\right)^d \qquad (19.5)$$

for the distribution of the displacement above the crown[10] (Fig. 19.4). With $\frac{du}{dr}|_{r_0} = \varepsilon_{r0}$ it follows from Equ. 19.5

$$\varepsilon_{r0} = -d\frac{u_0}{r_0} \quad .$$

With

$$u_0 = \varepsilon_{\vartheta 0} r_0 = (\varepsilon_{v0} - \varepsilon_{r0}) r_0$$

it follows

$$d = -\frac{\varepsilon_{r0}}{\varepsilon_{v0} - \varepsilon_{r0}} \quad .$$

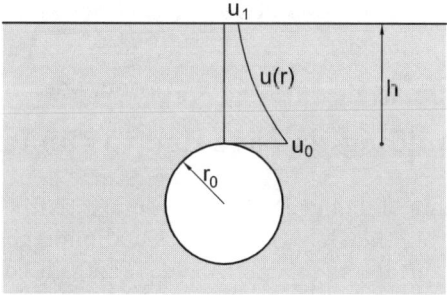

Fig. 19.4. Distribution of the vertical displacement u above the crown

The settlement of the surface ($r = r_0 + h$) results then via

$$\varepsilon_{r0} = \left.\frac{du}{dr}\right|_{r_0} = -u_0 d \left(\frac{r_0}{r}\right)^{d-1} \left.\frac{r_0}{r^2}\right|_{r=r_0} = -u_0 \frac{d}{r_0} = -d\,\varepsilon_{\vartheta 0}$$

to

---
[10] cf. C. Sagaseta: Analysis of undrained soil deformation due to ground loss. *Géotechnique* **37**, No. 3 (1987), 301-320; R. Kerry Rowe and K.M. Lee: Subsidence owing to tunnelling. II. Evaluation of a prediction technique. *Can. Geotech. J.* Vol. **29**, 1992, 941-954

$$u_1 = (\varepsilon_{v0} - \varepsilon_{r0})r_0 \left(\frac{r_0}{r_0+h}\right)^d .$$

The surface settlement is therefore smaller, the larger h is and the smaller the crown displacement u_0 is. One can keep the surface settlement small, if one keeps the strain ε_{r0} (and, consequently ε_{v0}) at the crown small. This can be obtained by rapid ring closure. MÜLLER-SALZBURG reported that he could always keep the crown displacement u_0 between 3 and 5 cm.[11]

19.2 Reversal of settlements with grouting

With shield driving the surface settlements result mainly from the tail gap, if the excavation face is suitably supported (e.g. by pressurised slurry). Grouting of the tail gap is expected to reverse the surface settlement. However, it is observed that even if the grouted mass exceeds the theoretical gap volume, the surface settlement is not reversed.[12] This fact can be explained in terms of soil mechanics: A cycle of loading and unloading leaves behind a net volume change, usually a compaction (Fig. 19.5). The effect of soil compaction due to a loading-unloading cycle in shield tunnelling is shown in Fig. 19.6 which represents the surface settlement due to closure of a 7 cm thick tail gap (curve a) and the one obtained after the grouting of the gap (i.e. reversing of the convergence of 7 cm). This result is obtained with the FEM programme ABAQUS and use of the hypoplastic constitutive equation calibrated for medium dense sand.[13]

19.3 Risk of building damage due to tunnelling

For rough assessments of damage risk it is assumed that surface buildings are completely flexible, i.e. they have no stiffness and undergo the same deformation as the ground surface of the 'greenfield'. This is a conservative assumption, because the real deformation will be reduced due to the stiffness of the building as compared to the one of the greenfield.[14] Evaluating the predicted

---

[11] L. Müller-Salzburg und E. Fecker: Grundgedanken und Grundsätze der 'Neuen Österreichischen Tunnelbauweise'. In: Grundlagen und Anwendungen der Felsmechanik. Felsmechanik Kolloquium Karlsruhe 1978, Trans Tech Publications, Clausthal 1978, 247-262

[12] S. Jancsecz et al., Minimierung von Senkungen beim Schildvortrieb am Beispiel der U-Bahn Düsseldorf. Tunnelbau 2001, VGE Essen, 165-214

[13] M. Mähr, Settlements from tail gap grouting due to contractancy of soil, *Felsbau*, in print, 2004

[14] This section is based mainly on the book "Building Response to Tunnelling. Case Studies from Construction of the Jubilee Line Extension, London", Vol. 1, edited by J. B. Burland et al, Telford, London, 2001

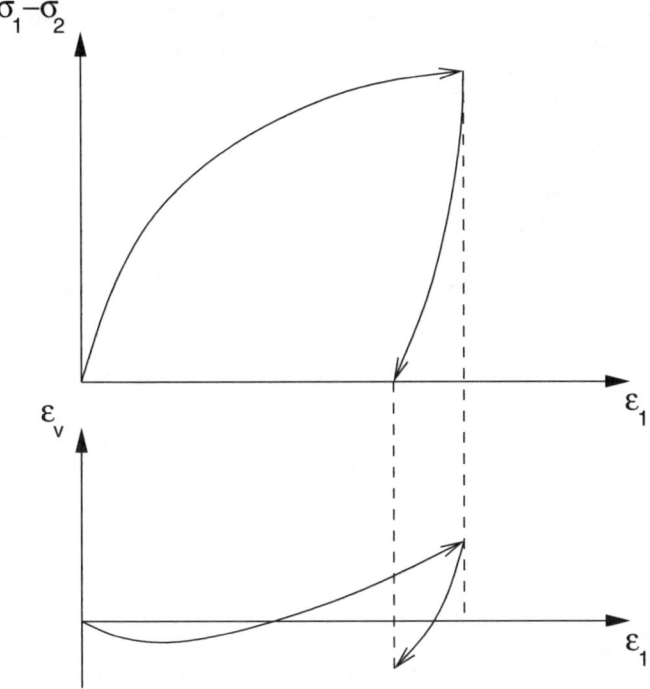

Fig. 19.5. A loading-unloading cycle (here shown for the example of triaxial test) leaves behind a permanent densification

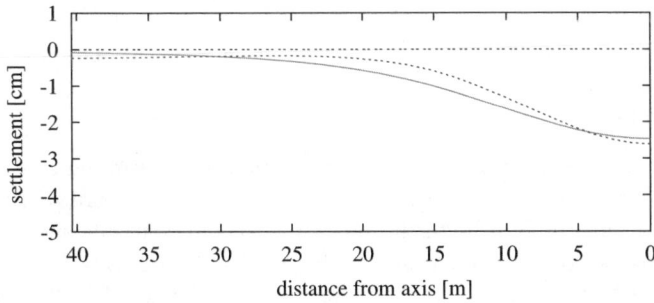

Fig. 19.6. Surface settlement due to gap closure (full line) and after grouting of the gap (dashed). Numerically obtained results.

settlements of the greenfield, we can assess that buildings with a maximum tilt of 1/500 and a settlement of less than 10 mm have negligible risk of damage. For the remaining buildings, a risk assessment must be undertaken which is still based on the greenfield deformation and, therefore, is quite conservative (because settlements are over-estimated). Damage of buildings is assessed in terms of tensile strain ε according to Table 19.2. The strain of the building is to be inferred from the settlement trough. BURLAND and his co-authors consider a building as TIMOSHENKO beam[15] and derive its strains from the deflection Δ (Fig. 19.7). Possible cracks (due to shear and due to bending, as shown in Fig. 19.8) are perpendicular to maximum tensile strains.

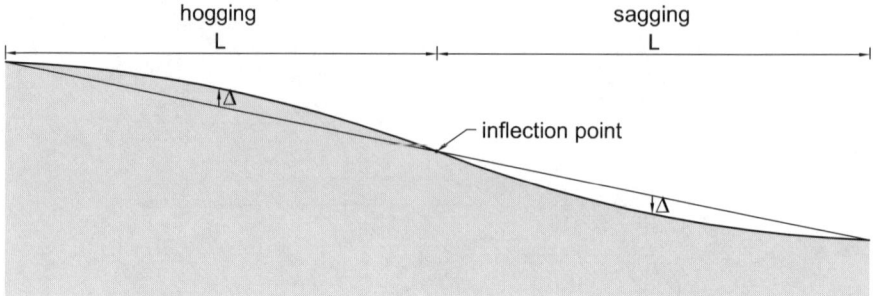

Fig. 19.7. Deflections Δ in sagging and hogging zones.

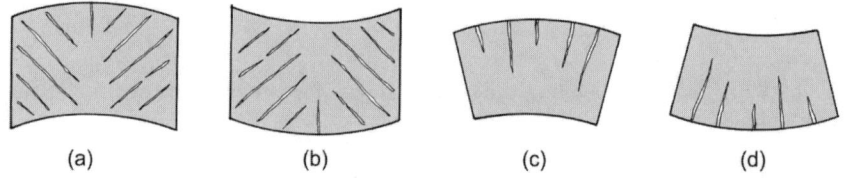

Fig. 19.8. Deformation and cracks due to pure shear (a, b) and pure bending (c, d).

However, the estimation of the strain in the building, assuming that it behaves like a beam is more or less academic, because (i) it contradicts the starting assumption that the building is infinitely flexible, and (ii) the shear and bending stiffnesses of this beam can hardly be assessed, especially for old masonry buildings with vaults, timbering etc. For a rough estimation it appears reasonable to use table 19.2 with the assumption that the maximum tensile strain in a building situated over the inflexion point has the order of magnitude of $u_{v,max}/a$. For more elaborate estimations, a method is used in the cited book that attempts to take the stiffness of the building into account. This method

---

[15] Contrary to the 'classical' or EULER-BERNOULLI beam, where shear forces are recovered from equilibrium but their effect on beam deformation is neglected, in the TIMOSHENKO beam cross sections remain plane but do not remain normal to the deformed longitudinal axis. The deviation from normality is produced by transverse shear.

is based on the results of FEM computations. It should be added that time-dependent settlements may be observed several years after the damage of the tunnel. Thereby, the settlement through expands laterally.

Degree of severity	Description of typical damage	Tensile strain ε (%)
Negligible	Hairline cracks less than about 0.1 mm.	0 - 0.05
Very slight	Damage generally restricted to internal wall finishes. Close inspection may reveal some cracks in external brickwork or masonry. Cracks up to 1 mm. Easily treated during normal decoration.	0.05 - 0.075
Slight	Cracks may be visible externally and some repointing may be required to ensure weather-tightness. Doors and windows may stick slightly. Cracks up to 5 mm.	0.075 - 0.15
Moderate	Doors and windows sticking. Service pipes may fracture. Weather-tightness often impaired. Cracks from 5 to 15 mm.	0.15 - 0.30
Severe	Windows and door frames distorted, floor sloping noticeably. Walls leaning or bulging noticeably, some loss of bearing in beams. Service pipes disrupted. Cracks 15...25 mm. Extensive repair work involving breaking-out and replacing sections of walls, especially over doors and windows.	> 0.30
Very severe	Beams lose bearing, walls lean badly and require shoring. Windows broken with distortion. Danger of instability. Cracks > 25 mm. Major repair is required involving partial or complete rebuilding.	

Table 19.2. Relation between damage and tensile strain (according to BURLAND et al.)

20

Stability problems in tunnelling

Leaving aside a precise definition of mechanical stability we only need to mention that a loss of stability occurs if in the course of a loading process a mechanical system becomes suddenly (often the term 'spontaneously' is used) softer, so that large deformations appear. These can cause serious damage. The most widespread known stability problem is the buckling of a rod. Here we consider the buckling of a tunnel lining or, equivalently, a pipe.

20.1 Rockburst

Rockburst can occur in deep tunnelling and mining and is manifested by spalling of the cavity walls. Large amounts of stored elastic energy are released and transformed into kinetic energy so that rock plates of several dm thickness are accelerated into the cavity causing worldwide many casualties per year. The mechanism of rockburst is not yet completely understood. The thickness d_f of the spalled rock can be estimated according to the following empirical formula:[1]

$$\frac{d_f}{r} = 1.25 \frac{\sigma_{max}}{q_u} - 0.51 \pm 0.1 \qquad (20.1)$$

where r is the tunnel radius, σ_{max} is the maximum circumferential stress at the tunnel wall and q_u is the unconfined strength of the rock.

20.2 Buckling of buried pipes

The buckling under consideration is caused by external forces acting upon the pipe (or tunnel lining). We should distinguish between forces exerted by a fluid and forces exerted by the ground. The latter depend on the deformation

---

[1]P.K. Kaiser et al., Underground works in hard rock tunnelling and mining. GeoEng 2000, Melbourne

of the pipe, whereas fluid loads are independent of deformation and always normal to the pipe surface.

20.2.1 Buckling of pipes loaded by fluid

We consider the pipe as a beam with initial curvature, where the differential equations of bending can be applied. Plane deformation (i.e. displacements $u_r \neq 0$, $u_\theta \neq 0$, $u_z \equiv 0$) implies that the YOUNG's modulus E usually applied in bending theory of beams has to be replaced by $E^* := E/(1-\nu^2)$. The differential equation follows from the known relation $\Delta M = EJ \cdot \Delta\kappa$, where $\Delta\kappa$ is the change of the beam curvature and J is the moment of inertia of the area shown in Fig. 20.3. With $r := r_0 + u$, $\dot{r} := dr/d\theta = r_0 dr/ds$ and $r' := dr/ds$ we can express the curvature as

$$\kappa = \frac{r^2 + 2\dot{r}^2 - r\ddot{r}}{(r^2 + \dot{r}^2)^{3/2}}$$

Introducing $r = r_0 + u$ and neglecting terms quadratic in u and u' as well as the product uu' we obtain:

$$\kappa \approx \frac{1}{r_0} - \frac{u}{r_0^2} - u''\ .$$

With $\kappa_0 = 1/r_0$ for the initially prevailing circular form we finally obtain

$$\Delta\kappa = \kappa - \kappa_0 \approx -\frac{u}{r_0^2} - u''\ ,$$

and the differential equation

$$u'' + \frac{u}{r^2} = -\frac{M}{E^*J} \tag{20.2}$$

or

$$\ddot{u} + u = -r^2 \frac{M}{E^*J}\ . \tag{20.3}$$

We now assume that the buckled shape of the pipe is symmetric with respect to the x and y axes, i.e.

$$u(\vartheta) = u(-\vartheta) \tag{20.4}$$
$$u\left(\frac{\pi}{2} - \vartheta\right) = u\left(\frac{\pi}{2} + \vartheta\right) \tag{20.5}$$

Referring to Fig. 20.2 we can express the bending moment M in dependence of ϑ:

$$M(\vartheta) = M_0 - N_0 \left[r_0 + u_0 - (r_0 + u)\cos\vartheta\right]$$
$$- \frac{1}{2}p\left[r_0 + u_0 - (r_0 + u)\cos\vartheta\right]^2 - \frac{1}{2}p(r_0 + u)^2 \sin^2\vartheta\ . \tag{20.6}$$

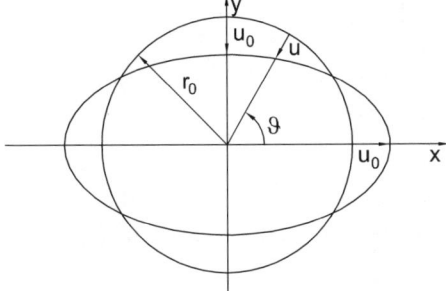

Fig. 20.1. Buckled pipe

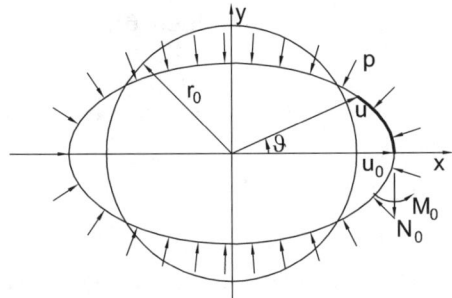

Fig. 20.2. Deriving the bending moment as function of ϑ

Taking into account that $N_0 = p(r_0 + u_0)$ and neglecting small terms[2] yields from (20.6):

$$M \approx M_0 - pr_0(u_0 - u). \tag{20.7}$$

Introducing (20.7) into (20.3) yields

$$\ddot{u} + k^2 u = -\frac{r_0^2}{E^*J}(M_0 - pr_0 u_0) \tag{20.8}$$

with

$$k^2 := 1 + \frac{r_0^2}{E^*J}p \quad . \tag{20.9}$$

The solution of (20.8) reads

$$u = A\cos k\vartheta + B\sin k\vartheta - \frac{(M_0 - pr_0 u_0)r_0^2}{E^*Jk^2} \quad . \tag{20.10}$$

From (20.4) follows $B = 0$ and from (20.5) follow

---

[2] i.e. terms of the order of u^2

$$\cos\left[k(\frac{\pi}{2}-\vartheta)\right]=\cos\left[k(\frac{\pi}{2}+\vartheta)\right] \quad , \quad \sin\left(k\frac{\pi}{2}\right)\sin(k\vartheta)=0$$

or

$$\sin\left(k\frac{\pi}{2}\right)=0 \quad .$$

It follows $k = 2n$ with $n = 1, 2, 3, \ldots$. Thus, the smallest (critical) buckling load p_{cr} is obtained (GRASHOF, 1859) from (20.10) for $k = 2$ as

$$p_{cr} = 3\frac{E^*J}{r_0^3} \quad .$$

20.2.2 Buckling of elastically embedded pipes

We consider a pipe embedded within a material in such a way that the interaction is governed by the subgrade modulus K_r. According to NIKOLAI[3], the buckling load is obtained as

$$p = (k^2 - 1)\frac{E^*I}{r^3} + K_r\frac{r}{k^2 - 1} \quad , \tag{20.11}$$

with $k = 2, 3, 4, \ldots$. If we introduce into Equ. 20.11 the number k which minimises p [4], we obtain the critical buckling load according to DOMKE and TIMOSHENKO:

$$p_{cr} = \frac{2}{r}\sqrt{K_r E^* J} \quad . \tag{20.12}$$

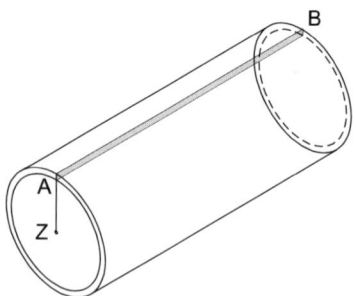

Fig. 20.3. Area referring to moment of inertia J

---

[3]Cited in N.S. Bulitschew: Mekhanika podsemnych sooruženij, Moscow, Nedra Publishing House, 1994

[4]This can be obtained by formal differentiation, $\frac{dp}{dk} = 0$.

21
Monitoring

Measurements are indispensable for various kinds of feedback in tunnelling:

Verification of design: Computational predictions of the behaviour of the ground during tunnelling have to be continuously verified by measurements (this is the central idea of PECK's observational method). If the deviations between predicted and measured values are too large, then computations should be repeated with revised input parameters.

Indication of imminent damage: Measurements can indicate imminent collapse and, thus, make it possible to take countermeasures.

Measuring devices[1] are expected to register a quantity without influencing it. This is hardly to achieve in case of stress measurements, where the stiffness of the measuring device alters the stress field.

The names of various measuring devices are not unique. They vary depending on the manufacturer[2], the country as well as between soil mechanics and rock mechanics.

Clearly, online measurements are preferable, as they allow a fast evaluation and feedback.

---

[1]H.-P. Götz und E. Fecker: Verformungs- und Spannungsmessungen im Tunnelbau; ein unentbehrliches Mittel zur Überprüfung des Vortriebs- und Ausbaugeschehens, VDI-Berichte Nr. 472/1983. W. Schuck und E. Fecker: Geotechnische Messungen in bestehenden Eisenbahntunneln, *Tunnelbau* 1998, Verlag Glückauf, 44-84. H. Bock, European practice in performance monitoring for tunnel design verification, *Tunnels and Tunnelling International,* July 2000, 35-37, H. Bock, European practice in geotechnical instrumentation for tunnel construction control, *Tunnels & Tunnelling International,* April 2001, 51-53 and May 2001, 48-50

[2]such as DMT, GeoConcept, Glötzl, Interfels, Reflex Instrument, Slope Indicator, Spacetec, Tunnel Consult, SolExperts and others (see *Tunnels & Tunnelling International,* December 2002, 57).

21.1 Levelling

The settlement trough due to tunnelling is monitored by levelling. It is recommended to plot not only the settlement troughs of several sections but also contour maps. Levelling is also carried out inside a tunnel to determine e.g. the crown settlement (the levelling rod can be suspended from the tunnel crown). A method for online monitoring of surface settlements (especially valuable in case of compensation grouting) is to use electronic liquid level gauges, mounted e.g. in the cellars of buildings. A vertical displacement of the gauges can be registered either by LVDT float sensors or with pressure transducers. The accuracy is ±0.3 mm. Note that temperature compensation is very important for this method.

21.2 Monitoring of displacements and convergence

The monitoring of convergence takes place with the help of pins (target plates) mounted to the tunnel wall as early as possible after the excavation. The placement and monitoring of pins at the invert lining is difficult (in particular if the invert is covered by a temporary fill to provide a carriageway) but, in some cases, also very important. The distance between the pins is measured with invar wires or steel tapes. Nowadays optical methods are applied more and more, e.g. laser beams (so-called tunnel scanner) or photogrammetric methods (i.e. evaluation of stereo images). The obtained images are used to check the excavation profile. Comparison of consecutive images provides information on convergence and also on the thickness of sprayed concrete.

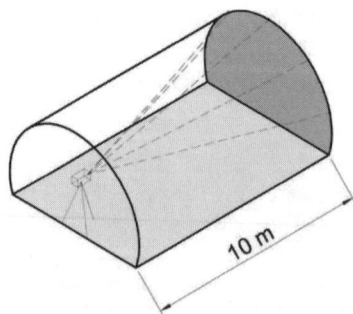

Fig. 21.1. Monitoring of convergence with tachymeter

Methods for monitoring of convergence[3]			
Method	Manufacturer	Resolution (mm)	System accuracy (mm)
Pins and invar wire	SolExperts	0.001	±0.003
Steel tapes	Interfels	0.01	±0.05
Geodetic tachymeter	e.g. Leica	1	±(2 − 3)
Tunnel scanner	DIBIT, GEODATA	1	±10

The data should be graphically represented as shown in Fig. 21.2. To detect convergence, the differential displacements between several pins of the same cross section should be plotted over the time.

21.3 Extensometers and inclinometers

Deformations of the ground can be registered by measuring the deformation of embedded pipes. Changes of length (longitudinal extensions) are measured by registering the extension of so-called extensometers, whereas deflections are measured by inclinometers and deflectometers. What is actually measured is the shape of a borehole. Comparison of the shapes at two different times yields the deformation.

Several measuring principles have been applied to determine the location of a probe moving within a borehole, such as gyroscopic positioning and measuring of earth's magnetic field. In geotechnical engineering and tunnelling the following instruments are mainly used:

Extensometer: A wire or a rod is fixed at the lower end of a borehole (the latter is cased, if necessary; however, the casing should be flexible enough to follow the deformation of the surrounding ground). A dynamometer maintains a constant tension within the wire, such that the displacement Δl at the fixed end is registered at the top of the borehole. The accuracy can reach 0.01 mm per 10 m. The ratio of the displacement to the length of the wire equals the mean strain $\varepsilon = \Delta l/l$ between the two points.

Multiple extensometer: A more detailed picture of the strain distribution along a borehole can be obtained if several (up to 6) wires or rods are fixed at various points within the borehole (Fig. 21.4). The wires are fixed with cement grout pumped into the space between two packers. In case of jointed rock, the grout is pumped into a fleece sack. Alternatively, the

---

[3] No responsibility is taken for the correctness and/or completeness of this information.

[4] The collapse of NATM tunnels at Heathrow Airport, HSE Books, Crown copyright 2000, Her Majesty's Stationery Office, St Clemens House, 2-16 Colegate, Norwich NR3 1BQ, ISBN 0 7176 1792 0

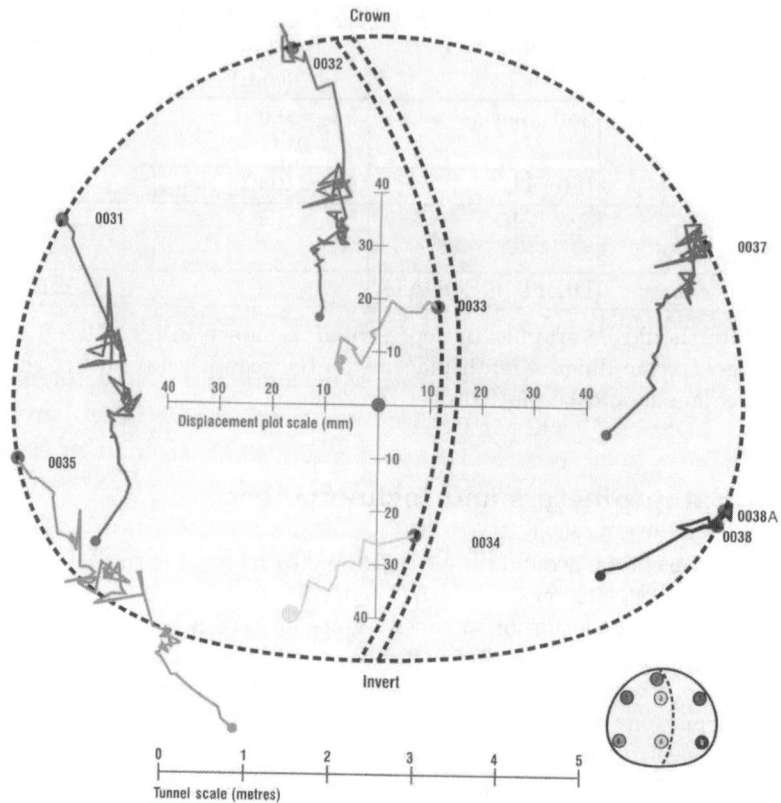

Fig. 21.2. Vector plots of displacements of target points at the tunnel wall.[4]

fixation occurs with the principle of Swellex anchors. The borehole can be subsequently filled with a deformable material to protect the borehole walls from cave-in.

Sliding micrometer (Gleitmikrometer): This is a probe introduced into a borehole. The borehole contains a PVC tube equipped with ring-shaped measuring marks with spacing of (initially) 1 m length. The PVC tube is fixed with the surrounding rock by grout. The probe is consecutively placed between two adjacent marks and extended in such a way that its two spherical shaped tips touch the conical marks (Fig. 21.5). Their actual distance is measured by means of a linear displacement transducer (LVDT) with an accuracy of $\pm 2\,\mu$m. The range of measurement is 1 cm per 1 m. If larger deformations are expected the so-called sliding deformeter (Gleitdeformeter) can be used, which has the same principle, a larger range and a smaller accuracy.

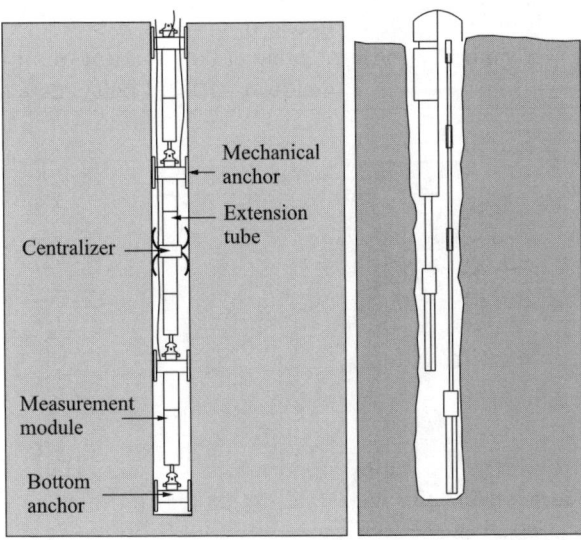

Fig. 21.3. Extensometer

The probe is submersible in water up to a pressure of 1.5 MPa and is temperature compensated. The time needed to measure a 40 m deep borehole (one measurement downwards and one upwards) amounts to ca 1/2 hour. Repeated measurements allow to detect the elongations between two consecutive measurements.

Inclinometer: A flexible tube is fixed within a vertical or horizontal borehole by means of grout. The inclinometer (Fig. 21.7) is moved within this tube and its inclination is being electronically measured at various positions (usually every 1 m). A traversing is thus obtained, the numerical integration of which yields the 3D form of the borehole (provided its lower end is fixed, a condition fulfilled e.g. if the lower end is sufficiently deep). Measurements at two different times yield the displacements perpendicular to the borehole axis that occurred in this time interval.

Deflectometer: The deflectometer consists of rigid rods interconnected with articulated joints in such a way that the relative twist can be measured. A traversing can be thus obtained which corresponds to the numerical integration of the differential equation $y' = f(x)$. To obtain a 3D picture, one has to measure two twist angles in two perpendicular planes.

21.4 Monitoring stresses within the lining

Stresses are measured via deformations of calibrated pressure cells. Flat pressure cells consist of two stainless steel plates with a thin fluid-filled cavity

---
[5] *SolExperts* Ltd.

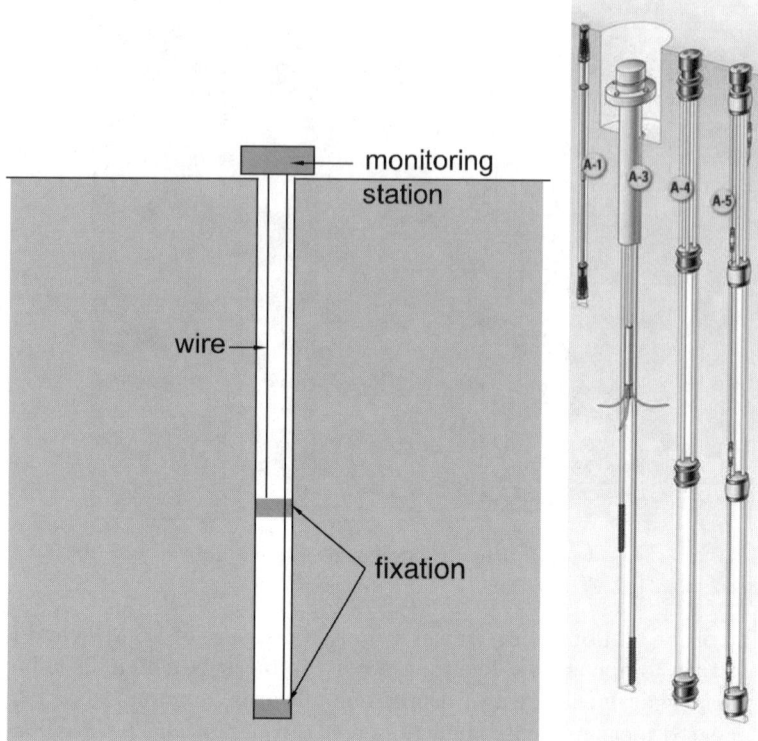

Fig. 21.4. Multiple extensometer (principle), *GEOKON Inc.*: Model A-1 Single Point Mechanical, Model A-3 Multiple Point Groutable Anchor, Model A-4 Multiple Point Snap-Ring Anchor, Model A-5 Multiple Point Hydraulic Anchor

between them. They are embedded within the shotcrete lining or between lining and adjacent rock. The pressure cells are filled with a fluid (usually mercury, which is relatively stiff but also hazardous for health. In recent time is used oil), the pressure of which is measured. The measuring principle of non-electric pressure cells is based on a reflux valve that is activated when the external pressure equals the internal one set by the operator (Fig. 21.8). Pressure cells are typically installed in one of two orientations: radial (to record σ_ϑ) and tangential (to record σ_r).[6] To provide reliable measurements, pressure cells should be stiff and flat. Temperature changes of the order of 20°C (e.g. due to setting of shotcrete) may simulate changes of up to 3 MPa. Also shrinkage of shotcrete may lead to significant pressure increases.

---

[6]C.R.I. Clayton et al., The performance of pressure cells for sprayed concrete tunnel linings. *Géotechnique* **52**, No. 2, 107-115 (2002).

21.4 Monitoring stresses within the lining

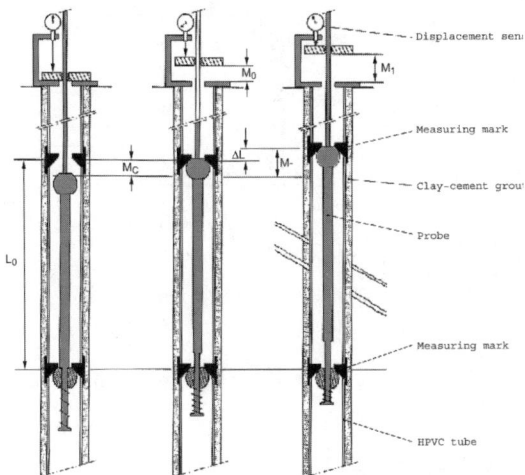

Fig. 21.5. Sliding deformeter[5]

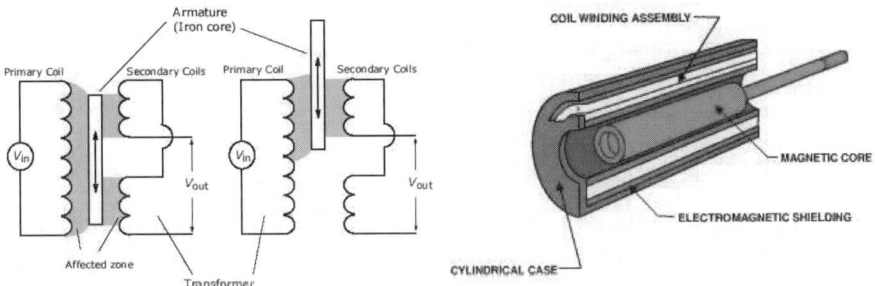

Fig. 21.6. LVDT

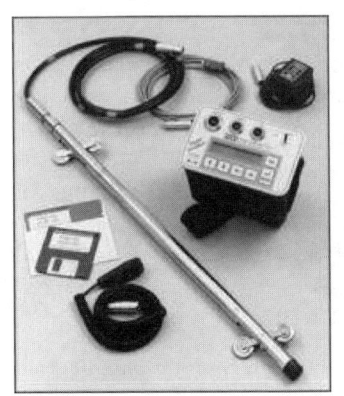
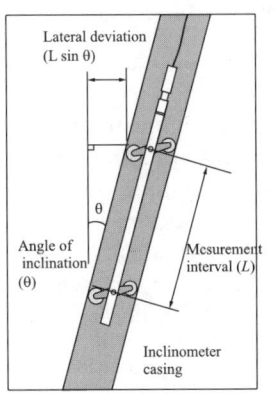

Fig. 21.7. Inclinometer

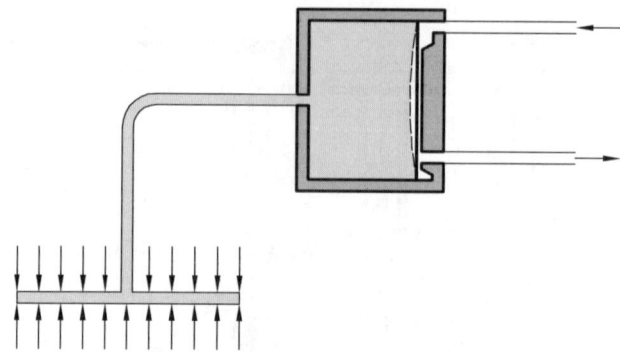

Fig. 21.8. GLÖTZL-pressure cell

21.5 Measurement of primary stress

The primary stress is used in the structural analysis of the tunnel lining. The applied measurement methods are based on:[7]

- Hydraulic fracturing by means of pressurising a fluid.
- Measurement of deformation of rock samples or slots upon unloading.
- Compensation methods: The deformations due to unloading are reversed by measurable mechanical loadings. It is assumed that the required stress is equal to the primary stress.

All these methods presuppose linear elastic behaviour. In other words, they are only applicable in good rock. The obtained results are, often, not very reliable and hardly deserve the related effort.

21.5.1 Hydraulic fracturing

A part of an uncased borehole, limited between two inflatable packers, is pressurised with water.[8] As soon as the water pressure reaches the value p_1, fracturing of the rock sets on, indicated by a sharp drop of the water pressure (Fig. 21.9). After stopping the pump, pressure is reduced to p_2, which means that the fissure is closed. Pumping anew leads eventually to re-opening of fissures, as indicated by the pressure p_3.
For the evaluation, it is assumed that the vertical stress σ_z is a principal stress[9], consequently the two other principal stresses σ_x and σ_y are perpendicular to the vertical axis of the borehole (Fig. 21.10).

---

[7]Several articles on rock stress estimation can be found in *International Journal of Rock Mechanics & Mining Sciences*, **40** (2003)

[8]B. Bjarnason, B. Lijon, O. Stephansson 'The Bolmen Project: Rock Stress Measurements Using Hydraulic Fracturing and Overcoring Techniques', *Tunnelling and Underground Space Technology*, Vol. 3, No. 3, 305-316, 1988

[9]At the wall of the borehole σ_z needs not be equal to γz

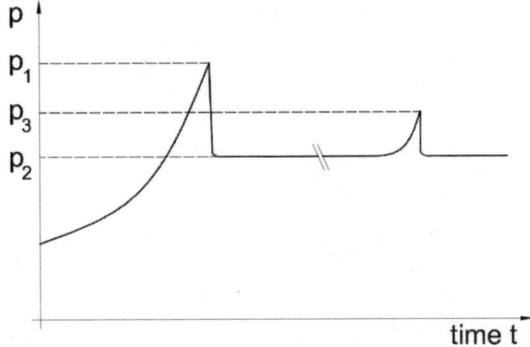

Fig. 21.9. Pressure-time record at hydraulic fracturing (schematic)

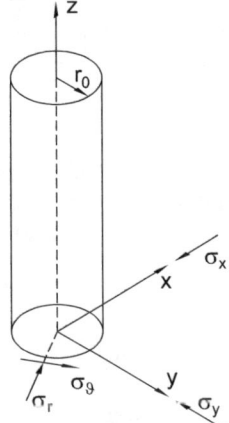

Fig. 21.10. Primary stress field σ_x, σ_y, σ_z and stresses σ_r and σ_ϑ at the wall of the borehole

σ_x, σ_y and σ_z are the principal stresses to be measured. They are considered as primary stresses or as stresses of a homogeneous far-field. These stresses are changed by the construction of the uncased borehole. We consider now the new stresses that prevail at the wall of the borehole. The circumferential stress σ_ϑ can be obtained under the assumption of linear elasticity from Equ. 14.5 with $\sigma_y := K\sigma_x$, $K < 1$, as $\sigma_\vartheta = (\sigma_x + \sigma_y) - 2(\sigma_x - \sigma_y)\cos 2\vartheta$. In particular, we have: $\min \sigma_\vartheta = 3\sigma_y - \sigma_x$. It is assumed that the first opening of fractures occurs as soon as the water pressure reaches the sum of $\min \sigma_\vartheta$ (due to stress) and the tensile strength σ_f of the rock:

$$p_1 = 3\sigma_y - \sigma_x + \sigma_f \ .$$

σ_ϑ is a minimum for $\vartheta = 0$. Hence, the fracture will be perpendicular to the minimum principal stress σ_y (Fig. 21.11).

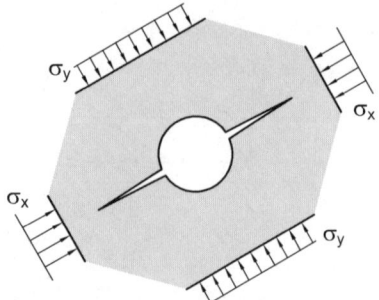

Fig. 21.11. Location of the fracture and directions of the principal stresses

For the second opening of the fractures the strength of the material needs no more to be overcome. We thus have:

$$p_3 = 3\sigma_y - \sigma_x \quad , \tag{21.1}$$

i.e. $\sigma_f = p_1 - p_3$. If the fracture is open, the geometry does not correspond to the one laid down for derivation of Equ. 14.5. It is however assumed that the fracture is closed if p has been reduced to the value of the minimum principal stress σ_y:

$$p_2 = \sigma_y \quad . \tag{21.2}$$

From (21.1), (21.2) and σ_x can be determined σ_y.

This derivation presupposes that the minimum principal stress is horizontal. If the minimum principal stress is σ_z, then the fracture will not open as shown in Fig. 21.11. In this case the pressure records cannot be evaluated. Note that for depths larger than 1,000 m the needed pressure cannot be supplied by usual hydraulic equipment.

The method described above requires sections in the borehole free from fractures. In case of pre-existing fractures, the HTPF ('hydraulic tests on pre-existing fractures')-method is applied. It consists in re-opening of the fractures.

21.5.2 Unloading and compensation methods

Most of these methods are based on the overcoring technique: the bottom of a borehole is overcored and thus released from the horizontal stresses. The accompanied deformation is measured. The corresponding stresses follow if one assumes elastic behaviour. The measurements exhibit large scatter. Thus, one needs a large number (at least 10) of repeated measurements to obtain reliable results.

If σ_1 is the maximum primary stress and σ_c is the unconfined strength, the results of stress measurement up to $\sigma_1/\sigma_c > 0.2$ are very difficult to interpret, and for $\sigma_1/\sigma_c > 0.3$ they are useless.[10] For the Canadian plate these values correspond to depths of 1,000 to 1,500 m.

The measurement of the unloading deformation is accomplished either by measuring the expansion of an interior, smaller borehole (Fig. 21.12) by means of a special probe[11] or by means of a glued strain gauge rosette[12] or with other methods[13].

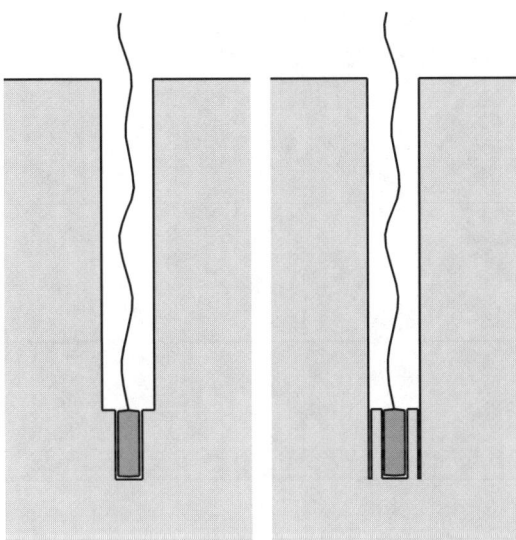

Fig. 21.12. Overcoring to measure the unloading deformation

Borehole slotting releases tangential stresses within the borehole wall. The accompanied strains are measured. They are related with the stresses according to theory of elasticity. With compensation methods the stress needed to re-open holes or slots back to their original size is measured.

21.6 Cross sections for monitoring

Monitoring cross sections of tunnels are distributed along the tunnel axis. About every 300 m principal monitoring cross sections and every 30 to 50 m

---
[10]P.K. Kaiser et al., Underground works in hard rock tunnelling and mining. Geo-Eng 2000, Melbourne

[11]deformation meter of the U.S. Bureau of Mines

[12]so-called doorstopper of the South African Council for Scientific and Industrial Research

[13]see J.A. Franklin, M.B. Dusseault : Rock Engineering. McGraw-Hill, 1989

secondary monitoring cross sections are installed.[14] Primary monitoring cross sections are equipped with devices to measure surface settlement, convergence, extensometers or gliding micrometers, load cells at or in the shotcrete lining. Secondary monitoring cross sections are equipped with devices to measure surface settlement and convergence.

---

[14]RVS 9.32, Blatt 7

22
Numerical analysis of tunnels

22.1 General remarks

Numerical analysis (also called numerical simulation) aims at predictions of the behaviour of the support (lining, rockbolts etc.) and the deformations of the ground and buildings. By means of such predictions it is tried to assess the safety and to optimise the construction. With a posteriori 'predictions' (so-called class B predictions) it is tried to improve the understanding, to adjust the involved parameters and to analyse failures.

The application of numerical simulation is nowadays standard in tunnelling. However, its results are not always convincing and, therefore, usually not appropriately integrated in the decision and construction process. It is thus interesting to look at the capabilities and limitations of numerical simulation. The basic ingredients of numerical simulations are the balance equations (for mass, energy and momentum) and constitutive equations. The first ones are beyond any doubt, whereas constitutive equations describe the behaviour of the involved materials and are always approximate. Simple constitutive laws (such as HOOKE's law) are often not realistic enough, and more realistic ones exhibit a complexity, which is sometimes prohibitive. In addition to the balance and constitutive equations, initial and boundary conditions are needed. The resulting system of equations determines the deformation and stress fields within the considered body. In this relation one should ask whether the sought-after fields can be uniquely determined. This is not necessarily the case. Loss of uniqueness, bifurcation and multiple solutions can be implied by the very nature of the involved equations and may have a physical background. At any rate, they impose severe difficulties onto the numerical procedures. To determine the deformation and stress fields at every point of the continuous body means to determine a multiple infinity of unknowns by solving the involved equations, which are differential ones. To overcome this difficulty, we restrict our attention to a finite number of points ('nodal points') embedded within the body and require that equilibrium is fulfilled only at these points. The deformation between the nodal points is appropriately assumed (interpolated).

This approach enables to replace the differential equations by algebraic ones. Upon the considered points may act external forces (e.g. due to gravitation) and internal forces due to the deformation. These forces are not the real but approximated ones. They result from the interpolated displacements between the nodal points. Equilibrium at the nodal points is often addressed as the 'week formulation' of the equilibrium differential equations. It is formulated by means of the principle of virtual displacements, which turns out to express GALERKIN's principle, which states that the approximate solution is 'orthogonal' to the defect (= difference between real and approximate solutions).

At present, a multiplicity of finite element codes are available. Their quality is often unproven and sometimes questioned. A main reason could be found in the fact that finite element codes are often used as black boxes without understanding of the theory in the background. It turns out that the quality of predictions can hardly be checked against measurements. It seems, therefore, that the expectations posed on computability should be moderated.

It is interesting to mention a benchmark test carried out by the working group 'Numerical Methods in Geotechnics' of the German Society for Geotechnics.[1] A tunnel with prescribed geometry and boundary conditions, prescribed constitutive equations and prescribed material parameters has been computed by several participants of the test. The obtained surface settlements did not considerably scatter. However, the obtained loads in the shotcrete lining (normal force and bending moments) exhibited a remarkable scatter of up to 300 %. This means that prescription of the above items does not guarantee uniqueness of results.

Thus, one should take carefully into account the several limitations of computability. By no means, however, one should decline computations. Besides predictions, numerical simulations can be used for sensitivity analyses, which provide an insight into the roles of the various involved parameters.

22.1.1 Initial and boundary conditions

The consideration of initial and boundary conditions turns out to be particularly difficult in geotechnical (and, thus, tunnelling) problems.

Initial conditions Constitutive equations are *evolution* equations, i.e. they try to predict stress changes due to deformations starting from a known stress state. In geotechnical engineering, however, the initial stress field is hardly known, as it results from a very complex and little known geological history. Even if the geological deformation history were known, we can hardly compute the resulting stress field, as we lack constitutive equations that describe the material behaviour of rock at very slow processes lasting thousands or millions of years. An inspection of present geological

---

[1] H.F. Schweiger, Results from two geotechnical benchmark problems. In: Proceed. 4th Europ. Conf. Num. Methods in Geotechnical Engineering, Udine, 1998. A. Cividini (ed.), Springer, 1998, 645-654

situations (especially of folds) reveals that brittle rock can be very ductile when deformed extremely slow.

The only case where the initial stress field can be easily assessed is the normally consolidated (i.e. non pre-loaded) halfspace bounded with a horizontal plane. In such a situation the soil results from sedimentation and the stress field is a geostatic one, i.e. the vertical stress increases linearly with depth, $\sigma_z = \gamma \cdot z$, and the horizontal stresses are $\sigma_x = \sigma_y = K\sigma_z$. The shear stresses vanish: $\sigma_{xy} = \sigma_{yz} = \sigma_{xz} = 0$. In absence of tectonic compression or expansion in horizontal direction we have $K = K_0$.

In case of linear elasticity, the initial stress field does — by definition — not play any role. This is why the difficulty related with the initial stress field is often overlooked.

Boundary conditions Traditionally, the finite element method is applied to bodies $\mathcal{B}$, on the boundary $\partial \mathcal{B}$ of which apply prescribed displacements or tractions. Usually the motion of the body is 'topological', i.e. the boundary consists always of the same particles. In contrast, non-topological motions are characterised by the fact that internal particles become boundary ones and vice versa. This is the case e.g. when the considered body is teared or penetrated by other objects. Exactly this is often the case in geotechnical engineering and especially in tunnelling. All sorts of excavation, drilling, driving piles and shields etc. are non-topological motions and pose severe difficulties to the application of the finite element method. The standard approach is to reduce (gradually or at one step) the stiffness and the weight of the elements to be removed (excavated). This is reasonable in the case e.g. of the excavation of a pit in soft soil. If, however, the excavated ground is cohesive, then a surplus force has to be applied to overcome cohesion. This force can leave behind a self-equilibrated stress field in the ground. Excavation by blasting most probably deteriorates the nearby rock. Penetration (e.g. of a shield) cannot be modelled by merely replacing the material properties of some elements, i.e. 'transforming' them from soil to steel. One should take into account that penetration implies also the application of shear and compressive stresses to the surrounding ground. If the support of the face is accomplished by a pressurised slurry, then the applied pressure has to be considered as a static boundary condition. On the same time, the motion of the surrounding soil into the shield face has to be taken into account.

Tunnelling constitutes a complicated soil-structure-interaction problem. The main difficulty lies in the fact that the construction (here: the tunnel including its support) disturbs the primary stress field in a hardly detectable way. We regard an intermediate construction level in Fig. 22.1. The total convergence, i.e. the reduction of the diameter from d_0 to d_2 implies a reduction of the radial stress on the tunnel boundary from the primary stress σ_0 to σ_2. In order to determine d_2, one must consider also the deformation behaviour of the support, which depends on the deformation of the ground occurred since the application of support. The

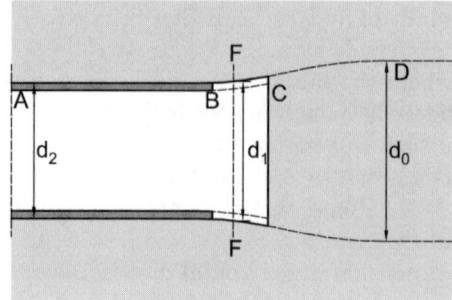

Fig. 22.1. Intermediate construction level (with exaggerated deformations). Sprayed concrete within the range AB is already installed, installation within the range CB is imminent

still unsupported range BC owes its stability to the supporting effect of the excavation face in longitudinal direction. This is assessed (in case of plane deformation) to $(1-\beta)\sigma_0$, so that the sprayed concrete lining [2] experiences at the time of its application only the load $\beta\sigma_0$.[3] An alternative procedure to assess the supporting effect of the excavation face consists in assuming a reduced stiffness for the rock within the tunnel cross section. ($E \rightarrow \alpha E, \alpha \leq 1$). In this way the stress reduction $\sigma_0 \rightarrow (1-\beta)\sigma_0$ is obtained from the interaction between unaltered and weakened rock. It must be added that the values of α and β have to be more or less arbitrarily guessed.

22.1.2 Coping with non-linearity

Treatment of non-linear problems requires high-level mathematics. Here, an engineer's approach is attempted, without mathematical rigour. It aims at physical understanding of the applied methods, which are indispensable in numerical simulations in tunnelling. Non-linearity arises from large deformations ('geometrical non-linearity') and from the material behaviour. Geometrical non-linearity can be avoided in most cases by applying the deformation in sufficiently small portions and updating the positions of the material points at each step. The material non-linearity has to be distinguished in elastic and inelastic one. Non-linear elasticity means that the stress-strain curve is curved but reversible, i.e. the same for loading and unloading. This type of non-linearity can, equally, be treated by applying sufficiently small deformation increments. The inelastic non-linearity is, however, more severe, as it implies different stress-strain curves (and, thus, different stiffnesses) for loading and

---

[2] An instantaneous setting is presupposed.

[3] See to bibliography cited in D. Härle, P.M. Mayer, B. Spuler: Stability Investigations for outer lining and rock with driving the Engelberg Base tunnel, *Tunnel* 3 (1997), 12-22.

unloading. This non-linearity cannot be treated by applying small increments and is, therefore, called 'incremental non-linearity' or 'non-linearity in the small'. This non-linearity is immanent in elasto-plasticity and hypoplasticity. The methods to treat non-linearity will be demonstrated on a system with one degree of freedom, the generalisation to many degrees of freedom will be shown later. The considered body can be conceived as a (non-linear) spring loaded by the weight P of a mass m (Fig. 22.2). The spring force I depends on the spring elongation/compression x in a non-linear way. The deformation x_F of the spring can be determined by solving the non-linear equation $y(x) := P - I(x) = 0$. P is the external force, I is the internal (spring) force, and y represents a sort of defect or excess force, that is to be annihilated. Usually, the equation $y(x) = 0$ is solved iteratively, i.e. an initial guess x_0 is improved to x_1 and so on. The iteration scheme has to improve the solution x_i by adding $\Delta x_i := x_{i+1} - x_i$. A possible iteration scheme is to set $\Delta x_i = \kappa y(x_i)$. Clearly, $\Delta x_i = 0$ if $y(x_i) = 0$. This iteration process is graphically shown in Fig. 22.3 left. It converges if $\kappa \leq 1/y'(x_i)$ for every i.

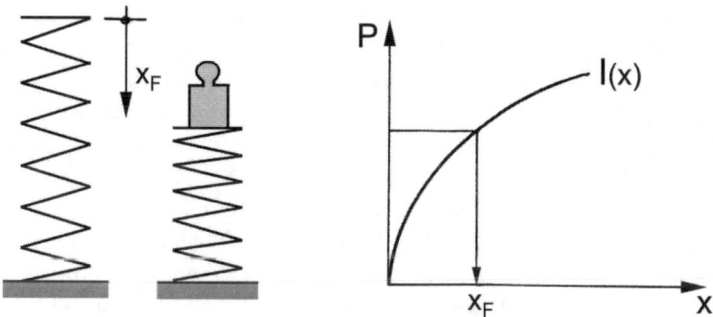

Fig. 22.2. Deformation of a non-linear spring.

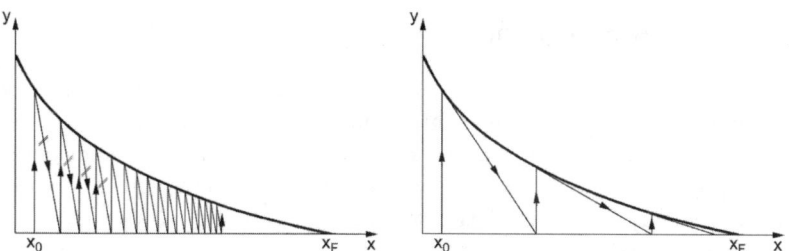

Fig. 22.3. Iteration schemes

Much faster convergence is achieved by NEWTON's scheme: $\Delta x_i = y(x_i)/y'(x_i)$. This scheme requires the evaluation of the (non-vanishing!) derivative y' at every point x_i.

If we consider systems with n degrees of freedom, then x and y have to be replaced by n-dimensional vectors $\mathbf{x} = \{x_1, x_2, ..., x_n\}$, $\mathbf{y} = \{y_1, y_2, ..., y_n\}$ and $y' = dy/dx$ has to be replaced by the (stiffness) matrix $\partial \mathbf{y}/\partial \mathbf{x} = \partial y_k/\partial x_l$. NEWTON's scheme now reads: $(\frac{\partial \mathbf{y}}{\partial \mathbf{x}})_i (\Delta \mathbf{x})_i = (\mathbf{y})_i$. This is a system of linear equations the solution of which yields $(\Delta \mathbf{x})_i$. At each iteration step, the matrix $\partial \mathbf{y}/\partial \mathbf{x}$ has to be determined, either analytically or numerically. To the vanishing of y' in the one-dimensional case corresponds the vanishing of the determinant $|\partial \mathbf{y}/\partial \mathbf{x}|$, i.e. the appearance of vanishing eigenvalues or the appearance of so-called bifurcations. Usual programmes stop execution whenever an eigenvalue becomes negative.

The fast iteration of NEWTON's scheme is partly counterbalanced by the time-consuming determination of the stiffness matrix. This is avoided by the aforementioned scheme $\Delta x_i = \kappa y(x_i)$ (going back to JACOBI), which however converges much slower. This scheme can be given a physical illustration: If the force P is suddenly applied, the mass will oscillate. The oscillation will be perpetual unless energy is extracted by some sort of damping. If the damping is very large compared with the inertia forces, the latter can be neglected. Assuming linear viscous damping we then have the excess force $y = P - I$ counteracted by the viscous drag $\eta \dot{x}$. The equation $y(x) = \eta \dot{x}$ can be numerically solved, say with $(x_{i+1} - x_i)/\Delta t = y(x_i)/\eta$ or $\Delta x_i = \frac{\Delta t}{\eta} y(x_i)$, which is the aforementioned scheme of JACOBI.

If inertia is not neglected, the numerical schemes are called 'dynamic relaxation' (such as FLAC or ABAQUS/EXPLICIT). With sufficiently small time steps they take into account that loading of the boundary propagates with waves into the interior of the body. They can be applied to dynamic problems. For quasi-static problems a numerical damping is applied, which can be fictitious and provides a fast convergence. However, the damping must be 'manually' adapted to the applied problem and constitutive equations, otherwise spurious oscillations of the solution may occur.

22.1.3 Constitutive equation

In the early days of FEM-applications in tunnelling, computers had a very limited memory. Thus, management of data was the most important issue, whilst the choice of the appropriate constitutive equation was considered of minor importance. This attitude survived until our days and one can observe cases of numerical simulation where the used constitutive equation is not even mentioned. This is by no means justified. The (proper!) use of the proper constitutive equation is of decisive importance. True, the behaviour of soil and rock is extremely complex and, therefore, realistic constitutive equations can be complex to such an extend that they cannot be used. Of course, everybody seeks to avoid unnecessary complexity[4] and tries to choose the simplest possible constitutive equation that can be applied in each case. No doubt, the most simple constitutive equation is HOOKE's law and is, in fact, broadly used in geotechnical engineering. One should, however, be aware that HOOKE's law does not comprise any sort of dilatancy, yield or collapse. The next step of complexity is generally considered to be elastoplasticity with MOHR-COULOMB yield condition and associated ($\psi = \varphi$) or non-associated ($\psi < \varphi$) flow rule. Herein, φ denotes the friction angle and ψ denotes the dilatancy angle. This approach, too, has shortcomings. E.g., it predicts dilatancy at unloading instead of the drastic contractancy which occurs in reality. Thus, it is incapable to realistically describe volume loss due to tunnelling (cf. Section 19.2).[5]

It turns out that the law of parsimony is difficult to be applied, because (in view of true predictions) we often do not know in advance which effects are relevant for the considered problem and which ones are irrelevant. HOOKE's law, e.g., is characterised by constant stiffness and constant POISSON's-ratio, whereas soil is characterised by a stiffness that depends on (i) the actual stress level and (ii) the direction of deformation. It is often believed that this extreme variability of stiffness can be sufficiently well modelled by HOOKE's law if the constant stiffnesses E and G are appropriately chosen for each element and for each load step. However, this procedure virtually presupposes that the stress and deformation fields are more or less known and need not be computed. Moreover, it should be emphasised that dilatancy and contractancy can *not* be modelled by HOOKE's law, no matter which value has been assigned to the POISSON's-ratio. It should also be taken into account that small effects are not always negligible. There are situations, especially near to bifurcation or critical points and also at cyclic loading, where small disturbances may have large consequences.

22.2 Method of subgrade reaction

A simple approach to the analysis of problems of soil-structure interaction is to assume that the ground consists of infinitely thin and independent (uncoupled) springs. Linear or elastic subgrade reaction is given if the springs are linear ($p = Ku$). Here, p is the pressure between building and ground, u is the displacement of the building and K is the subgrade reaction modulus or the subgrade reaction number. Note that the building is considered as flexible (deformable) and, thus, the relation $p = Ku$ holds locally, i.e. for every point of the interface. The subgrade reaction approach $p = Ku$, going back to WINKLER, is applied for mat foundations, piles, retaining walls and tunnels (Fig. 22.4).

---

[4]This general principle is called 'Ockham's razor' or 'law of parsimony' in science.
[5]See I. Herle, Constitutive models for numerical simulations. In: Rational Tunnelling, D. Kolymbas (ed.), Logos, Berlin 2003, 27-60.

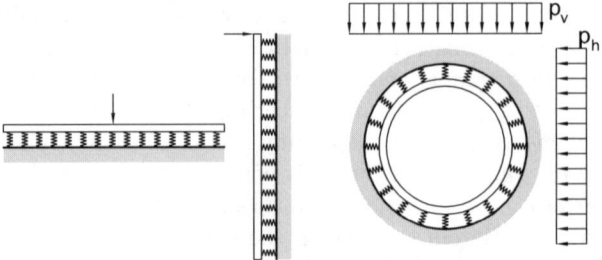

Fig. 22.4. Soil-structure interaction with subgrade reaction approach

The subgrade reaction approach permits the application of elegant mathematical procedures. Thus, the differential equation of a straight beam, $EJu^{(4)} = -p$, leads to the linear differential equation $EJu^{(4)} = -Ku$ with the general solution

$$u(x) = \sinh\frac{x}{L}\left(C_1\sin\frac{x}{L} + C_2\cos\frac{x}{L}\right) + \cosh\frac{x}{L}\left(C_3\sin\frac{x}{L} + C_4\cos\frac{x}{L}\right)$$

with $L := \sqrt[4]{4EJ/K}$.

However, it should be taken into account that the unjustified assumption of uncoupled springs implies that the subgrade reaction modulus K is *not measurable*. Note that a measurement is misleading, if the underlying physical concept is not realistic. The practically hardly attainable measurement of pressure *distributions* implies that the assumption of the subgrade reaction is not examinable. Pressure distributions are hardly measurable, because the load cells, built into the bearing construction, usually exhibit a different stiffness than the surrounding body and, therefore, affect the stress distribution. A method for the measurement of the subgrade reaction modulus is to strut a hydraulic piston (press) against the wall of a gallery (Fig. 22.5). The force and the extension of the hydraulic piston are measured. Since the stress distribution at the end plates is not constant (it depends among other criteria on the disk diameter), the subgrade reaction modulus cannot be inferred.

The often undertaken refinements of the subgrade reaction approach, like the nonlinear subgrade reaction $K = K(u)$ or the locally variable subgrade reaction modulus $K = K(x)$, resp. $K = K(u,x)$, are not real improvements as they enhance the above mentioned difficulties.

In tunnelling, the subgrade reaction approach $\sigma_r = K_r u_r$ is used[6] with the subgrade reaction number K_r obtained from the elastic ground line: $K_r = 2G = E/(1+\nu)$, which presupposes, however, a uniform expansion resp. convergence of a circular cross section with hydrostatic primary stress.

---

[6]if the support is sufficiently stiff relative to the surrounding rock, i.e. if $K_r r_0^3/(EJ) < 200$ m^{-1}

Subsequently, a load is assumed to act on the lining according to Fig. 22.4, which is, however, arbitrary.

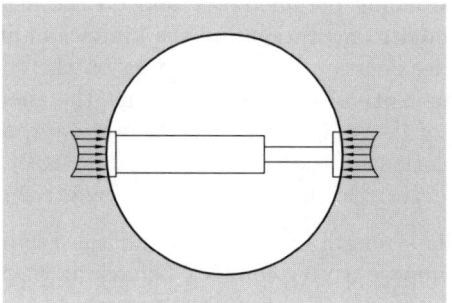

Fig. 22.5. Plate load test

22.3 Difficulties related to the design of shotcrete lining

The shotcrete lining is expected to carry the loads exerted by the surrounding rock. This task is either temporary (if a permanent lining is subsequently installed) or permanent. To properly design the shotcrete lining one needs to know the loads acting upon it. This is, however, an extremely difficult question that is still not satisfactorily answered. The reason is that the loads are not given a priori but depend on the interaction between ground and shotcrete lining. So-called soil-structure interaction problems are known to be very difficult, and in tunnelling the difficulties are increased by the following facts:

- The initial stress field and the mechanical behaviour of the ground are poorly known, especially in the case of squeezing or swelling rock.
- The initial conditions are poorly known. The shotcrete lining is installed piecemeal upon circumferential strips of freshly excavated rock. There, the radial stress upon the rock surface has reduced to zero and, consequently, a — hardly measurable — convergence has taken place. With advancement of the face, this convergence increases transferring thus load upon the shotcrete lining.
- The mechanical properties of shotcrete vary with time due to ageing of the hardening concrete, which is usually taken into account with variable YOUNG's modulus and compressive strengths, e.g. both varying according to time functions of the type $\exp(\alpha/t^\beta)$. In addition to ageing, shotcrete is a rate-dependent material, the stiffness of which depends on the time history and rate of loading. Creep (i.e. deformation at constant stress) and relaxation (i.e. stress change at constant deformation) are but only two

manifestations of rate dependence. According to some new developments, shotcrete is considered as a visco-elastoplastic material. The elastic regime is limited by a yield surface of the Drucker-Prager type that depends on elastoplastic hardening parameters χ and on the degree of hydration ξ. Creep is attributed to hydration and the kinetics of dislocations with the corresponding deformation rates depending on the actual stress. In view of the high complexity of the faced relations, the enormous experimental difficulties and of the scarcity of experimental data, one should take into account that relations of the above type are, virtually, mere assumptions, which are, however, welcome in absence of any other information.

The reduced tensile strength of shotcrete and the related high creep rates at tensile stresses impose a very complex behaviour whenever normal forces within the lining interact with bending moments. Even the mere existence of bending moments renders the confinement-convergence approach futile. As known, this approach assigns to a given convergence u a pressure $p_R(u)$ that is exerted by the rock upon the lining and a pressure $p_L(u)$, with which the lining reacts to the convergence. The convergence of the lining is obtained from the equation $p_R = p_L$. In presence of bending moments, neither u nor p are constant along the circumference of the cavity. In this case one has to take into account the known equilibrium equations for curved beams:[7]

$$Q' - N/r = -p$$
$$N' + Q/r = -q$$
$$M' = Q$$

Herein, the prime denotes derivative with respect to the arc length s, r is the radius of curvature (for $r \to \infty$ we recover the equations for a linear beam) and q is the tangential force per unit length acting upon the beam/lining. Even for $q \equiv 0$ the above equations imply that M and N are interrelated:

$$M'' - N/r = -p \ . \tag{22.1}$$

To relate the deformation of the lining with M, N and Q, we note that a change of curvature $\kappa = 1/r$ can be approximated as

$$\Delta\kappa \approx \frac{u}{r^2} + u'' \ .$$

Obviously, the first term denotes the curvature change due to the reduction of $r : r \to r - u$. The longitudinal strain within the lining reads

$$\varepsilon \approx -\frac{u}{r} \ .$$

It appears thus reasonable to assign N to ε, and M to u'':

---
[7]cf. Equ. 16.15

22.3 Difficulties related to the design of shotcrete lining

$$N = -AE\varepsilon = AEu/r; \quad M = f(u'') \quad . \tag{22.2}$$

Note, however, that there is no way to control u and u'' independently. In the linear-elastic case, Equ. 22.2 reads $M = EJu''$ but it becomes strongly non-linear as soon as tensile stresses appear within the lining. Because of Equ. 22.1 and because u cannot be separated from u'', the convergence-confinement approach is no more applicable. As N influences the appearance of tensile stresses, Equ. 22.2 is to be re-written as $M = f(N, u'')$. Note that the function $f(N, u'')$ also depends on the position of reinforcement. However, the present practise in tunnelling does not allow to precisely keep a prescribed position of the reinforcement. Thus, the design of shotcrete reinforcement bars is burdened with many simplifying assumptions. The situation is better for fibre reinforced shotcrete because, in this case, the reinforcement is homogeneously distributed within the lining.

A proper design of the lining has to assure that the actual combinations of M and N are admissible. To this end, a sort of yield function $g(M, N, t)$ has to be introduced in such a way that the inequality $g(M, N, t) < 0$ indicates safe states. The variable t (time) is included to take ageing and creep into account. In the present state of the art, the function $g(M, N, t)$ is missing.

A step towards simplification could be to assume that creep occurs only if the shotcrete is cracked. So, the design has either to assure that shotcrete does not crack (in which case no reinforcement is needed) or that the time dependent ('creep') deformations do not exceed some acceptable limit. Note that time dependent deformations are not creep deformations, as — strictly speaking — creep refers only to deformation under constant stress. In tunnels, however, the rock behaviour implies that the stresses acting upon the lining depend on its deformation (and also on time, in case of squeezing or swelling rock).

The remarks presented here demonstrate that tunnelling is demanding from an engineering point of view and still needs intensive research.

Part III

Appendices

A

Physics of detonation

This appendix[1] is intended to shed some light on the nature of explosions and their effects upon soil/rock.

A.1 Detonation

A detonation front is a shockwave which breaks the molecules of the explosive (which is a mixture of oxygen and combustible substances) into pieces. This takes place within a zone whose thickness is comparable to the free flight of the molecules (approx. 10^{-5} to 10^{-6} cm). Subsequently an exotherm reaction (oxidation) takes place. The result of this oxidation is a mixture of gases. Fig. A.1 shows a snapshot of the various zones involved in a detonating cylindrical charge.

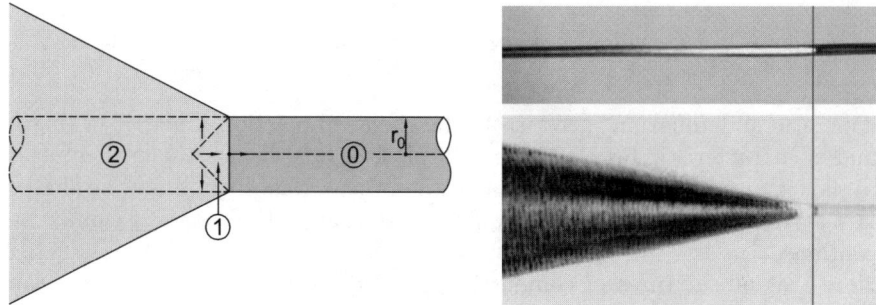

Fig. A.1. Detonation zones within a cylindrical charge

---

[1] It is based mainly on B.N. Kutusow's book 'Rasruscheniye gornikh porod vsrivom' (Rock Blasting), Publishing House of the Moscow Institute of Mining, Moscow 1992.

The detonation front propagates with the supersonic speed u_D. In zone 0 the explosive has not yet 'felt' the detonation and is still in its initial state. In zone 1 the chemical reaction takes place, but the substances involved have not as yet expanded. The expansion takes place within zone 2.

The boundary between zones 1 and 2 is a moving discontinuity surface. In the case of a stable (or stationary, if registered by an observer moving with the velocity u_D) detonation, the discontinuity moves with the speed u_D. The combustion in zone 1 is the driving force for this process.

The energy released by the combustion propagates to the detonation front and keeps it working. This is the mechanism of the so-called homogeneous detonation, which occurs within homogeneous explosives and propagates with speeds $u_D = 6$ to $7\,\text{km/s}$. Within commercial explosives, where mixtures contain certain inert components, the processes are more complex.

The duration (speed) of the chemical reaction within zone 1 depends on the radius r_0 of the charge. Note that for each type of explosive there is a characteristic (critical) radius r_c and that the detonation expires if the charge radius is smaller than the critical one. In the range $r_c < r_0 < r_g$ the detonation speed u_D increases with r_0, Fig. A.2.

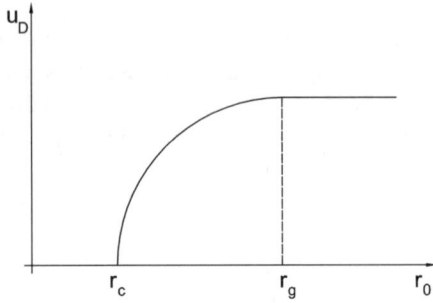

Fig. A.2. Detonation speed u_D in dependence of the size of the charge

The ratio of detonation speed u_D to the chemical reaction time t_c is a length and is characteristic for a particular explosive. It indicates the critical mass of the explosive. If the mass is smaller than the critical mass, then the charge is scattered by the previous detonation, so that the detonation cannot be continued.[2]).

Note that all mixtures of combustible substances and oxygen suppliers can detonate provided that the exotherm reaction releases sufficiently energy, that the size of the mass is sufficiently large and that there is an appropriate ignition.

---

[2] According to YU. KHARITON, see Ia.B. Zeldovich and A.S. Kompaneets, Theory of Detonation, Academic Press, 1960

Explosive	critical diameter (mm)
Lead acid	0.01-0.02
Hexogen	1.0-1.5
Trotyl	8.0-10
Ammonite	10-12

The critical mass of an explosive also depends on its containment. It is reduced if the explosive is encased in such a way that the expansion of the resulting gases is impeded. Therefore, explosive cartridges should abut to the boreholes.

A.2 Underground explosions

Charges are either lengthy or short ('spherical') and can be ignited either on the surface or inside the rock. The action of buried blasts depends on their depth, Fig. A.3. Deep blasting does not produce any permanent deformation of the ground surface, whereas blasting at reduced depths produce spalling (b), loosening (c) and throw-off (d).

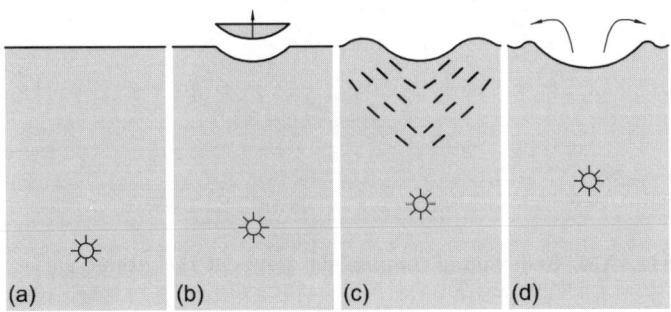

Fig. A.3. Actions of buried blastings in dependence of depth

In addition, the action of blasting upon rock depends on whether it is jointed or not.

Explosions in soil: The high pressure of the gases produces a cavity in the soil which can subsequently collapse. Cratering occurs if the depth of the explosion is small. Note that a loose cohesionless soil can be compacted by explosions, especially if the soil is water-saturated.

Explosions in unjointed rock: The detonation front of an explosion in unjointed rock hits the containing rock with a velocity of 4 to 6 km/s. There it releases a strong shock, which propagates with a speed of 3 to 5 km/s in the rock mass. The accompanying stresses exceed the strength of the rock, so that it is decomposed into small grains. The related dumping reduces the amplitude of the shock. This reduction is approximately proportional to $1/r^5$, where r is the distance from the charge. In a distance of $r \approx 5$ to $6\ r_0$ (r_0 = radius of the charge) the propagation speed is reduced to the speed of elastic compression waves, but the associated stresses still exceed the strength of the rock, which is destroyed up to a distance of 10 to 12 r_0, mainly by radial fissures.

Beyond this zone elastic waves propagate and may disturb or damage nearby structures. Besides the direct action of the propagating shocks there are also some effects due to the pressure (4 to $7 \cdot 10^3$ MPa) of the released gases. The proportion of fumes penetrating into the rock fissures approx. amounts to 30 to 40 vol % for uncontained explosions, and to 70 vol % for contained ones.

If the charge is placed near the ground surface, then the compression waves are reflected on the surface and come back as extension waves (Fig. A.4). One can imagine that the source of these reflected waves is the mirror location of the charges. As the tensile strength of rock is much lower than its compressive strength, the reflected waves are much more destructive.

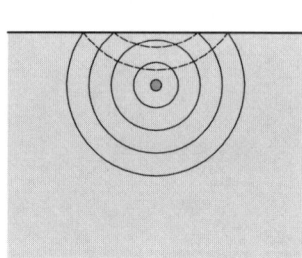

Fig. A.4. Reflexion of compression waves at the ground surface

Explosions in jointed rock: Reflections at the joints are responsible for the much stronger attenuation of the waves at increasing distance from the charge. Thus, the destroyed zone is much smaller in case of jointed rock.

The above statements only refer to chemical explosions. For the effects of nuclear underground explosions see the article of FAIRHURST.[3]

A.3 Interaction of charges

The ignition of solitary charges is not usual in mining and tunnelling. Instead, many charges are ignited and the interaction of the detonations produces some

---

[3]Ch. Fairhurst, Rock Mechanics of Underground Nuclear Explosions, *ISRM News Journal*, **6**, 3/2001, 21-25

beneficial effects. In Fig. A.5 we see that the action of two simultaneous detonations is enhanced along their connecting line, whereas it is attenuated outside this line. To understand this effect one has to note that the radial stresses are compressive ones, whereas the circumferential stresses are extensive. Thus, at point A we have an enhancement of stress, whereas the stress is reduced at point B. This effect is exploited to produce smooth surfaces of blast excavation.

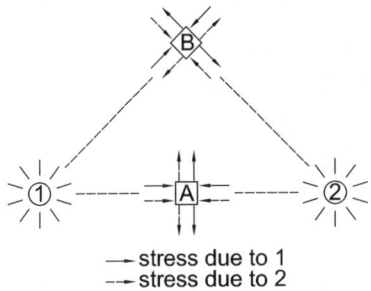

Fig. A.5. Stress fields due to simultaneous detonation of the charges 1 and 2

So-called millisecond blasting is applied to improve the destruction of rock and to reduce the disturbance in the neighbourhood. An ignition delay by a certain time Δt has the following effects:

1. Interference of compression waves emanating from adjacent charges ($\Delta t < 5$ ms)
2. Formation of additional surfaces ($15 < \Delta t < 200$ ms)
3. Additional demolition due to collision of individual blocks ($\Delta t > 200$ ms)

These effects can be roughly explained as follows (the underlying theory is still incomplete):

Interference: The compression wave starting from charge 1 reaches the ground surface at speed v (Fig. A.6) and is reflected there. As it is now an extension wave, it reaches after the time lapse $\Delta t = \sqrt{a^2 + 4y^2}/v$ the charge 2, which is ignited exactly at this time. Such delays can be achieved by loops in the ignition string (the detonation propagates within the ignition string with a speed of 6.5 m per millisecond). Delays can also be achieved if one uses detonators that are triggered by the compression wave. Utilisation of interference presupposes very precise ignition times. They can hardly be calculated, especially in the case of jointed rock.

Formation of additional surfaces: Free surfaces are of importance because they reflect compression waves transforming them to extension waves. The effect is shown in Fig. A.7.

A Physics of detonation

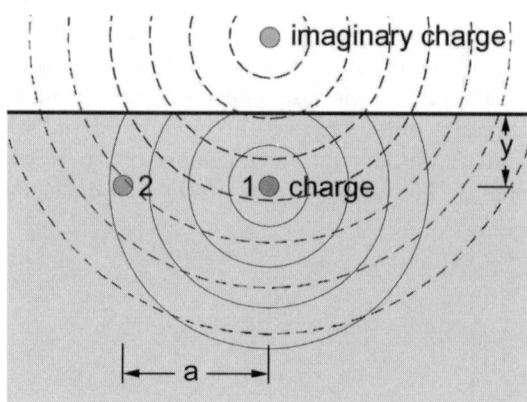

Fig. A.6. Reflexion of compression wave at free surface

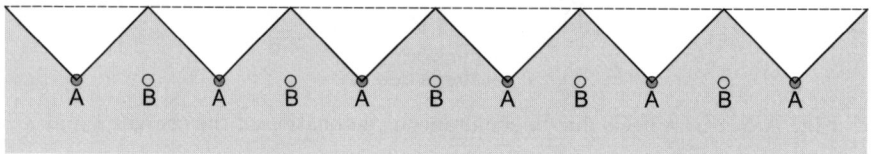

Fig. A.7. The ignition of the charges A forms additional free surfaces

Collision: Rock blocks from the second ignition fly with a velocity of 20 to 60 m/s and impinge the blocks from the first ignition, the velocity of which has reduced to approx. 3 to 6 m/s. The maximum demolition is achieved if the two blocks have perpendicular orbits and the relative velocity amounts at least 15 m/s.

B

Support of soil with a pressurised fluid

Consider the interface of soil with a fluid. The soil is fluid-saturated and there is no separating membrane between soil and fluid. Applying a pressure upon the fluid gives rise to percolation of the fluid through the soil. The hydrodynamic drag force exerted by the fluid upon the grain skeleton is $-dp/dx$ for a flow in horizontal direction. This is a volume force (i.e. a force exerted upon a unit volume) that stabilises the grain skeleton. Thus, it is important to note that the stabilisation is not due to the applied pressure p but due to the pressure gradient dp/dx. In other words, the stabilising action is not of hydrostatic but of hydrodynamic nature. Of course, the distribution of p (and, consequently, of dp/dx) depends on the hydraulic boundary conditions and, in general, the pressure gradient at the interface will not be particularly high. This will be e.g. the case if the considered fluid is water. If, however, the fluid is a bentonite suspension then the soil pores adjacent to the interface will be clogged very soon after the beginning of the percolation. This is the case because a bentonite suspension is a so-called BINGHAM-fluid that will no flow if the applying shear stress is lower than a limit τ_f.[1] The bentonite penetrates into the soil and forms a so-called filter cake of the thickness l. To estimate l, we assume that the pore is a cylinder with diameter d. Equilibrium of forces yields

$$l\pi d\tau_f = p\pi d^2/4 \quad ,$$

hence $l = pd/(4\tau_f)$. From experience we may set $d \approx 2d_{10}$, where d_{10} is the grain diameter of soil not exceeded by 10 % of the soil mass. Thus,

$$l \approx \frac{d_{10} \cdot p}{2\tau_f} \quad .$$

Obviously, for large τ_f and small d_{10} the obtained penetration length (or cake thickness) l is very small. Thus, the filter cake can be considered as an impermeable membrane applied upon the interface.

---

[1] In other words, the bentonite suspension is a clay with a very high liquid limit w_L and a very low undrained cohesion $c_u \equiv \tau_f$.

C

A simple analytical approximation for frost propagation

As is known, the propagation of heat occurs with radiation and/or conduction. Here we neglect radiation. The heat flux $\mathbf{q}$ is proportional to the temperature gradient: $\mathbf{q} = -\lambda \nabla T$. The factor of proportionality λ is called thermal conductivity. Thus, the amount of heat flowing into a volume element within a time unit reads $-\text{div}\,\mathbf{q} = \lambda \text{div}(\nabla T) = \lambda \triangle T$. Herein, $\triangle$ is the LAPLACE-operator. The influx of heat causes a rise of temperature per time unit: $\dfrac{\partial T}{\partial t} = -\dfrac{1}{c\varrho}\text{div}\,\mathbf{q}$. ϱ is the density, c is the specific heat of the considered material. From these equations follows the differential equation of FOURIER

$$\frac{\partial T}{\partial t} = \frac{\lambda}{c\varrho} \triangle T \quad . \tag{C.1}$$

$\alpha := \dfrac{\lambda}{c\varrho}$ is the diffusivity.

Some data follow, which are useful for thermodynamic computations:

heat conductivity	λ (kJ/(m·h·°K))
Sand frozen	18.4
Sand unfrozen	9.2
clay frozen	9.6
clay unfrozen	6.3

specific heat	c (kJ/(kg·°K))
water	4.18
ice	1.80
soil	0.80 depends strongly on water content

Latent heat of water: 334.5 kJ/kg

C A simple analytical approximation for frost propagation

Let us consider a long freezing pipe with an external diameter $2r_0$, which is surrounded by frozen soil of radius R (Fig. C.1). We consider a plane problem with axial symmetry, i.e. there is no temperature flux in z-direction. Due to permanent loss of heat the radius of the frozen soil increases with the rate $u = \dot{R}$. The surface $R=$const is thus a moving surface of discontinuity. Referring to this discontinuity the balance equations have to be expressed as jump relations. Let $[x]$ denote the jump of a quantity x across the discontinuity. The jump relation for energy balance reads (for a non-moving medium)[1]

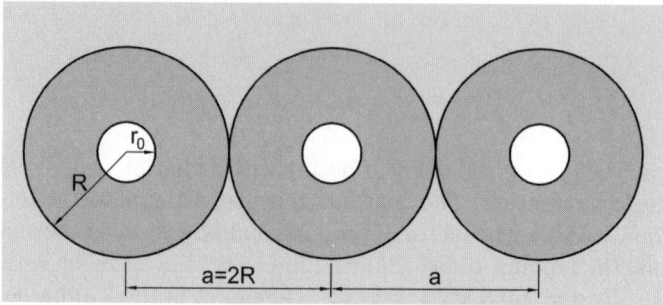

Fig. C.1. Propagation of frost around freezing pipes

$$[\varrho e u - q] = 0 \ . \tag{C.2}$$

e is the specific internal energy, i.e. the internal energy per volume unit. As is known, internal energy is that part of the energy of a body which does not depend on the motion of the observer. If we denote the frozen soil with the subscript b and the unfrozen soil with a, Equ. C.2 reads: $\varrho_a e_a u - q_a = \varrho_b e_b u - q_b$ or

$$u(\varrho_a e_a - \varrho_b e_b) = q_a - q_b = \lambda_b \left(\frac{\partial T}{\partial r}\right)_b - \lambda_a \left(\frac{\partial T}{\partial r}\right)_a \tag{C.3}$$

$\left(\frac{\partial T}{\partial r}\right)_a$ and $\left(\frac{\partial T}{\partial r}\right)_b$ denote the right and the left limits at $r = R$, respectively. We set $\varrho := \varrho_a \approx \varrho_b$. $h := e_a - e_b$ is the latent heat, i.e. the heat that must be extracted from the soil to freeze it. To determine the propagation of the freezing front with time we have to solve the following initial value problem:

$$\begin{aligned}
&\text{For } t = 0: T = T_\infty &&\text{in } r_0 < r < \infty \\
&\text{For } t > 0: T = T_0 &&\text{for } r = r_0 \\
&\phantom{\text{For } t > 0: }T = T_1 (= 0°C) \text{ for } &&r = R \\
&\phantom{\text{For } t > 0: }T = T_\infty &&\text{for } r \to \infty
\end{aligned}$$

---

[1] See e.g. E. Becker and W. Bürger: Kontinuumsmechanik (equation 5.22), Teubner, 1975

C A simple analytical approximation for frost propagation

In the ranges $r_0 < r < R$ and $r > R$ the differential equation C.1 is valid and for $r = R$ the jump relation C.3 is valid. As the boundary $r = R$ varies with time, we have a so-called STEFAN-problem.

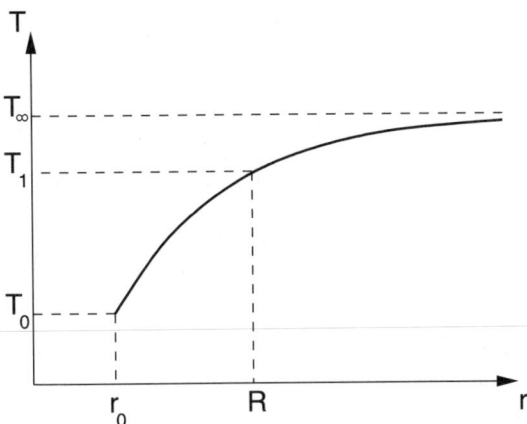

Fig. C.2. Instantaneous temperature distribution around a freezing pipe

To obtain an approximate solution of this complicated initial-boundary-value problem, we assume the following distributions of temperature:

$$T = T_b = T_0 + (T_1 - T_0)\left(\frac{r - r_0}{R - r_0}\right)^\mu \quad \text{for} \quad r_0 < r < R \tag{C.4}$$

and

$$T = T_a = T_\infty - (T_\infty - T_1)\left(\frac{r}{R}\right)^\nu \quad \text{for} \quad R < r < \infty \quad , \tag{C.5}$$

where the parameters μ and ν are not yet determined. To use the LAPLACE-Operator in cylindrical coordinates (with $\frac{\partial}{\partial z} \equiv 0$)

$$\Delta T = \frac{1}{r} \cdot \frac{\partial T}{\partial r} + \frac{\partial^2 T}{\partial r^2} \quad ,$$

we differentiate the expressions (C.4) and (C.5)

$$\frac{\partial T_b}{\partial r} = (T_1 - T_0)\mu \left(\frac{r - r_0}{R - r_0}\right)^{\mu - 1} \frac{1}{R - r_0} \quad , \tag{C.6}$$

$$\frac{\partial^2 T_b}{\partial r^2} = (T_1 - T_0)\mu(\mu - 2)\left(\frac{r - r_0}{R - r_0}\right)^{\mu - 1} \frac{1}{R - r_0}^2 \quad , \tag{C.7}$$

$$\frac{\partial T_a}{\partial r} = -\nu(T_\infty - T_1)\left(\frac{r}{R}\right)^{\nu-1}\frac{1}{R}, \tag{C.8}$$

$$\frac{\partial^2 T_a}{\partial r^2} = -\nu(\nu-1)(T_\infty - T_1)\left(\frac{r}{R}\right)^{\nu-2}\frac{1}{R^2}. \tag{C.9}$$

The derivatives with respect to time t are:

$$\frac{\partial T_b}{\partial t} = -\dot{R}(T_1 - T_0)\mu\left(\frac{r - r_0}{R - r_0}\right)^{\mu-1}\frac{r - r_0}{R - r_0}^2, \tag{C.10}$$

$$\frac{\partial T_a}{\partial t} = \dot{R}\nu(T_\infty - T_1)\left(\frac{r}{R}\right)^{\nu-1}\frac{r}{R^2}. \tag{C.11}$$

Using equations C.8, C.9 and C.11 we write the differential equation C.1 at $r = R + 0$ (i.e. to the right adjacent to the frost boundary). It then follows

$$\nu = -\frac{\dot{R}}{\alpha_a} \cdot R \tag{C.12}$$

with $\alpha_a = (\frac{\lambda}{c\varrho})_a$. Using (C.6), (C.7), (C.10) we write Equ. C.1 at $r = R - 0$ (i.e. to the left, adjacent to the frost boundary). It follows

$$\mu = 1 - \left(\frac{\dot{R}}{\alpha_b} + \frac{1}{R}\right) \cdot (R - r_0). \tag{C.13}$$

We introduce the obtained expressions into the jump relation (C.2) and obtain:

$$\dot{R}\varrho h = \lambda_b(T_1 - T_0)\frac{1}{R - r_0}\left[1 - \left(\frac{\dot{R}}{\alpha_b} + \frac{1}{R}\right)(R - r_0)\right] - \lambda_a(T_\infty - T_1)\frac{\dot{R}}{\alpha_a}.$$

This is differential equation for $R(t)$. With the abbreviations

$$A := \varrho h + \varrho c_a(T_\infty - T_1) + \varrho c_b(T_1 - T_0)$$
$$B := \lambda_b(T_1 - T_0)$$

it reads:

$$\dot{R}A = B\left(\frac{1}{R - r_0} - \frac{1}{R}\right) = B\frac{r_0}{(R - r_0)R}.$$

Separation of variables leads to

C A simple analytical approximation for frost propagation

$$(R^2 - Rr_0)dR = \frac{B}{A}r_0 dt$$

from which follows the solution (taking into account the initial condition $R = r_0$ for $t = 0$)

$$\frac{1}{3}(R^3 - r_0^3) - \frac{1}{2}r_0(R^2 - r_0^2) = \frac{B}{A}r_0 t \quad . \qquad (C.14)$$

By means of Equ. C.14 we can approximately determine the closure time t_s, after the lapse of which the frozen soil cylinder obtains the radius $a/2$. With the abbreviation

$$C = \frac{r_0}{4A}\left(\sqrt{-6Bt(r_0^2 A - 6Bt)} - \frac{1}{4}(r_0^2 A - 12Bt)\right)$$

the solution of equation C.14 reads:

$$R = C^{\frac{1}{3}} + \frac{1}{4} \cdot \frac{r_0^2}{C^{\frac{1}{3}}} + \frac{r_0}{2} \quad .$$

If we set R equal to $a/2$, where a is the distance between two adjacent freezing pipes, we obtain the closure time t_s. Taking into account that $r_0 \ll R = a/2$, we can simplify Equ. C.14 as follows: Dividing by r_0^3 we obtain

$$\frac{1}{3}\left[\left(\frac{R}{r_0}\right)^3 - 1\right] - \frac{1}{2}\left[\left(\frac{R}{r_0}\right)^2 - 1\right] = \frac{B}{Ar_0^2}t_s \quad .$$

As $\left(\frac{R}{r_0}\right)^3 \gg \left(\frac{R}{r_0}\right)^2 \gg 1$, we obtain with $R = \frac{a}{2}$:

$$t_s \approx \frac{1}{3} \cdot \frac{A}{Br_0} \cdot \left(\frac{a}{2}\right)^3 \quad .$$

Herein, t_s is the closure time i.e. the time needed for two adjacent freezing fronts to get in touch (Fig. C.1). The heat $\dot{Q}$ extracted from the freezing pipe per time and length units reads:

$$\dot{Q} = 2\pi r_0 q|_{r_0} \approx 2\pi r_0 \lambda_b \frac{T_1 - T_0}{R - r_0} \quad .$$

D

Rigorous solution for the steady water inflow to a circular tunnel

The rigorous solution, obtained with complex analysis, is presented here in full length, as it can be hardly found in the literature. We consider a tunnel with circular cross section within a soil/rock with homogeneous and isotropic permeability. With an appropriate drainage system, the circumference of the tunnel is kept at a constant hydraulic head h_a.

The water ingress into the tunnel can be determined by conformal mapping. The circle and the horizontal line in Fig. D.1 can be transformed to two concentric circles by the MÖBIUS transformation

$$w = f(z) = \frac{r(z - cr)}{cz + r}$$

with $c = i \dfrac{h - \sqrt{h^2 - r^2}}{r} = ib$ and $i = \sqrt{-1}$.

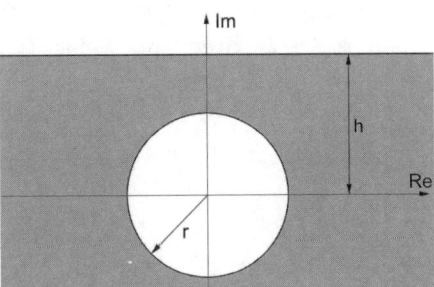

Fig. D.1. Circle and line in the complex plane

$f(z)$ maps the circle $z\bar{z} = r^2$ into the same circle: With $\bar{f}(z) = \frac{r(\bar{z} + cr)}{-c\bar{z} + r}$ we obtain $f\bar{f} = r^2$. As usual in complex analysis, the bar denotes the conjugate complex number.

D Rigorous solution for the steady water inflow to a circular tunnel

The horizontal line $z = ih + a$, $a \in \mathcal{R}$, is mapped to a circle with radius R:

$$f(z) = \frac{r\,(ih + a - cr)}{c\,(ih + a) + r} = r^2\,\frac{a + i\,\sqrt{h^2 - r^2}}{(h - \sqrt{h^2 - r^2})(ai + \sqrt{h^2 - r^2})}$$

$$\rightsquigarrow |f\bar{f}| = \frac{r^2}{h - \sqrt{h^2 - r^2}} =: R \ .$$

We introduce the complex potential $F(z) = \log f(z) = \log r + \log(z - cr) - \log(cz + r)$. With $\log w = \underbrace{\log|w|}_{\Phi} + i\Psi$ we obtain Φ as the potential to describe flow into the tunnel. At the circumference of the tunnel it obtains the value

$$\Phi_T = \log r$$

and at the straight line $\mathcal{I}m(z) = h$ it obtains the value

$$\Phi_S = \log R.$$

In order to attain there the prescribed values h_a and H (Fig. 8.7), respectively, it has to be re-scaled:[1]

$$\Phi^* := \frac{R - \rho}{R - r}\frac{\Phi}{\Phi_T} h_a + \frac{\rho - r}{R - r}\frac{\Phi}{\Phi_S} H = \Phi Q$$

with

$$Q = \frac{1}{R - r}\left[\frac{h_a}{\log r}(R - \rho) + \frac{H}{\log R}(\rho - r)\right].$$

Noting, however, that $Q = Q(\rho)$ we infer that this re-scaling is not admissible, since Φ^* is no more a potential ($\Delta\Phi^* \neq 0$, Δ is the Laplacian operator). Thus, we have to proceed as follows: By appropriate choice of the datum of geodetic head we may set $H = 0$. In this case, h_a obtains the meaning of Δh, i.e. the head difference between the line $\mathcal{I}m(z) = h$ and the tunnel circumference, $\Delta h := H - h_a$. The re-scaled potential now reads $\Phi^* = \Delta h \cdot \frac{\Phi - \Phi_S}{\Phi_T - \Phi_S}$.

With $z = \rho\,e^{i\varphi} = \rho\,(\cos\varphi + i\,\sin\varphi)$ and $\frac{\partial}{\partial\rho}\log|f| = \frac{\partial}{\partial\rho}\log(f\bar{f})^{1/2} = \frac{1}{2}\frac{\partial}{\partial\rho}\log f\bar{f}$ we obtain:

$$\frac{\partial\Phi}{\partial\rho} = \frac{\partial}{\partial\rho}\Big[\log|\rho\,e^{i\varphi} - cr| - \log|c\rho\,e^{i\varphi} + r|\Big]$$

$$= \frac{\partial}{\partial\rho}\Big[\log|\rho\,\cos\varphi + i(\rho\,\sin\varphi - br)|$$

$$- \log|-b\rho\,\sin\varphi + r + ib\rho\,\cos\varphi|\Big]$$

$$= \frac{\rho - rb\,\sin\varphi}{\rho^2 - 2\rho rb\,\sin\varphi + b^2 r^2} - \frac{b^2\rho - br\,\sin\varphi}{b^2\rho^2 - br\rho\,\sin\varphi + r^2}$$

$$\left.\frac{\partial\Phi}{\partial\rho}\right|_{\rho=r} = \frac{1}{r}\frac{1 - b^2}{1 - 2b\,\sin\varphi + b^2}$$

---

[1] In this section ρ denotes the radial coordinate

D Rigorous solution for the steady water inflow to a circular tunnel

Thus, the radial velocity at the tunnel circumference reads

$$v = -k\frac{\partial \Phi^*}{\partial \rho} = -k\frac{\Delta h}{\Phi_T - \Phi_S}\left.\frac{\partial \Phi}{\partial \rho}\right|_{\rho=r} = \frac{k\Delta h}{\log(R/r)\cdot r} \cdot \frac{1-b^2}{1-2b\sin\varphi + b^2}$$

To integrate along the circumference we use the expression

$$\int \frac{d\varphi}{A - B\sin\varphi} = -\frac{\tan^{-1}\left[\frac{B\cos\varphi/2 - A\sin\varphi/2}{\cos\varphi/2 \cdot \sqrt{A^2-B^2}}\right]}{\sqrt{A^2 - B^2}}$$

with $A := 1 + b^2$, $B := 2b$, $\sqrt{A^2 - B^2} = 1 - b^2$. Thus the water ingress

$$q = \int_0^{2\pi} vr\,d\varphi = -\frac{k\Delta h(1-b^2)}{\log(R/r)}\int_0^{2\pi}\frac{d\varphi}{1+b^2-2b\sin\varphi}$$

reads:

$$q = \frac{2k\Delta h}{\log(R/r)}\tan^{-1}\frac{2b\cos\varphi/2 - (1+b^2)\sin\varphi/2}{\cos\varphi/2 \cdot (1-b^2)}\bigg|_{-\pi/2}^{\pi/2}$$

$$= \frac{2k\Delta h}{\log(R/r)}\left[\tan^{-1}\frac{2b-(1+b^2)}{1-b^2} - \tan^{-1}\frac{2b+(1+b^2)}{1-b^2}\right]$$

$$= \frac{2k\Delta h}{\log(R/r)}\left[\tan^{-1}\left(-\frac{(1-b)^2}{1-b^2}\right) - \tan^{-1}\frac{(1+b)^2}{1-b^2}\right]$$

Using $\tan^{-1}x + \tan^{-1}y = \tan^{-1}\frac{x+y}{1-xy}$ we finally obtain

$$q = \frac{\pi k\Delta h}{\log(R/r)} = \frac{\pi k(H - h_a)}{\log\left(\frac{r}{h - \sqrt{h^2 - r^2}}\right)} \quad \text{(D.1)}$$

Equ. (D.1) shows that q depends non-linearly on the tunnel radius. E.g. for $h = 30$ m, an increase of r from 3 m to 6 m (i.e. by 100%) causes an increase of q of only 30%.

The merit of equation D.1 lies in its consistency and validity not only for deep but also for shallow tunnels. It presupposes homogeneous and isotropic permeability.

E

Aerodynamic pressure rise in tunnels

We consider the case of a piston (train) entering with velocity V into a tube (tunnel). For simplicity, we neglect the gap between piston and tube. As shown in Fig. E.1, a shock front is released and moves with the propagation speed c. In region 1 the velocity v, pressure p and density of air are V, p_1, ρ_1, respectively, whereas the corresponding values in region 0 are: 0, p_0, ρ_0.

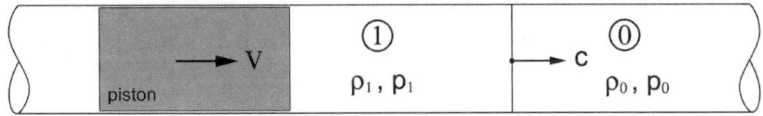

Fig. E.1. Piston moving within a tube

Mass balance across the shock requires

$$[\rho(c - v)] := \rho_1(c - v_1) - \rho_0(c - v_0) = 0 ,$$

hence

$$c = \frac{\rho_1}{\rho_1 - \rho_0} V . \tag{E.1}$$

Momentum balance across the shock requires:

$$[\rho v(c - v) - p] := \rho_1 v_1(c - v_1) - p_1 - \rho_0 v_0(c - v_0) + p_0 = 0 , \tag{E.2}$$

hence

$$\rho_1 V(c - V) = p_1 - p_0 . \tag{E.3}$$

Adiabatic compression of a gas is described by $p/\rho^\kappa = \text{const} \rightsquigarrow dp = \frac{p}{\rho}\kappa d\rho$ with κ being the adiabatic exponent ($\kappa \approx 1.4$ for air). Thus,

398   E Aerodynamic pressure rise in tunnels

$$p_1 - p_0 \approx \frac{p_0}{\rho_0} \kappa \cdot (\rho_1 - \rho_0) \ . \tag{E.4}$$

Combining equations E.1, E.3 and E.4 yields a quadratic equation for c, the solution of which reads

$$c = \frac{1}{2}\left(V + \sqrt{V^2 + 4\frac{p_0}{\rho_0}\kappa}\right) . \tag{E.5}$$

With the known values for atmospheric pressure and density, i.e. with $\rho_0 = 1{,}293$ g/cm^3 and $p_0 = 10^5$ Pa $= 10^5$ N/m^2, we obtain $4\frac{p_0}{\rho_0}\kappa = (659$ km/h$)^2$. Thus, for a train with $V = 200$ km/h we obtain $c = 444$ km/h and $\Delta p = p_1 - p_0 = p_0 \cdot \frac{\kappa}{c/V-1} = 10^5$ Pa$\cdot\frac{1.4}{444/200-1} = 114$ kPa.

This value is far too high, because the air flow in the gap has been neglected. If we take the gap into account, we have to consider the ratio of the cross section areas $n := A_{tunnel}/A_{train}$. Neglecting air compressibility and assuming that the velocities are averaged over the corresponding cross sections, continuity yields (Fig. E.2) $vA_{tunnel} = VA_{train} - v_g(A_{tunnel} - A_{train})$ or

$$v = nV - (1-n)v_g \ . \tag{E.6}$$

In the equations E.1, E.3 and E.5 V has now to be replaced by v. The gap velocity v_g is related to the pressure rise Δp via Equ. 2.2 and decreases with increasing length l . Thus, it depends on the travel time of the train in the tunnel.

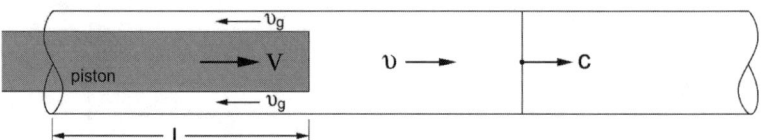

Fig. E.2. Piston with gap

For any time t, v must be determined iteratively from

$$\Delta p = \lambda \frac{l}{d} \frac{g}{2} v_g^2 = p_0 \frac{\kappa}{c/v - 1} .$$

Herein, d is an appropriate hydraulic radius for the flow in the gap.

F

Multiphase model of reinforced ground

For multiphase media (such as composite materials) it is assumed that each point is occupied by all constituents, which in our case are ground and reinforcement. The corresponding field quantities are denoted by the indices g and r, respectively. Let us consider a representative volume element (REV) with the volume V. The included volumes of ground and reinforcement are V_g and V_r, respectively. The corresponding volume fractions are $\alpha_g := V_g/V$ and $\alpha_r = V_r/V$ with $\alpha_g + \alpha_r = 1$. It can be shown that the volume fractions are equal to the area fractions, e.g. $\alpha_g = A_g/A$ with A_g being the cross section occupied by ground, and A being the total cross section. Referring to multiphase media and quantities (such as density and stress) of the several constituents, it has to be distinguished between the 'real' (or 'effective') quantities, that prevail (or are averaged) over the individual phases and the 'partial' quantities, that are averaged over the entire REV. Thus, we have the real densities of ground and reinforcement, ρ^g and ρ^r, respectively. The corresponding partial densities are $\rho_g = \alpha_g \rho^g$ and $\rho_r = \alpha_r \rho^r$. Similarly, it has to be distinguished between the stresses σ^g and σ^r on the one hand, and $\sigma_g = \alpha_g \sigma^g$ and $\sigma_r = \alpha_r \sigma^r$ on the other.[1] In the quasi-static case (i.e. accelerations are negligible) the equations of equilibrium for the two phases read:

$$\nabla \cdot \sigma_g + \mathbf{P}_{rg} + \rho_g \mathbf{g} = 0 \qquad (F.1)$$
$$\nabla \cdot \sigma_r + \mathbf{P}_{gr} + \rho_r \mathbf{g} = 0$$

$\mathbf{g}$ is the gravity acceleration, $\mathbf{P}_{rg} = -\mathbf{P}_{gr}$ is a vector that characterises the interaction of the two phases. $\mathbf{P}_{rg}$ is the force per unit volume exerted by the reinforcement upon the ground. For the case of the fully mobilised rigid-idealplastic shear stress τ_0 the interaction force can be determined as follows: The shadowed volume (Fig. F.1)

---

[1] In the notation of multiphase media, σ^g is the 'effective' stress in the ground. This quantity should *not* be confused with the effective stress in the usual sense of soil mechanics.

$$V_0 = \int_0^{\theta_0} \int_0^b \int_{r_0}^{r_0+l} r\,dr\,dz\,d\theta \quad,$$

with $\theta_0 = a/r_0$, corresponds to one bolt. Thus, the interaction force $\pi d\tau_0 l$ is obtained as:

$$\pi dl\tau_0 = \int_0^{\theta_0} \int_0^b \int_{r_0}^{r_0+l} P_{rg} r\,dr\,dz\,d\theta \quad. \tag{F.2}$$

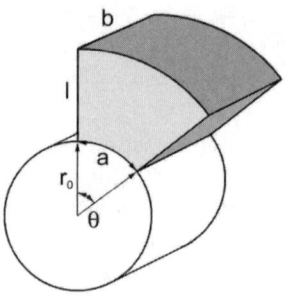

Fig. F.1. Volume corresponding to one bolt

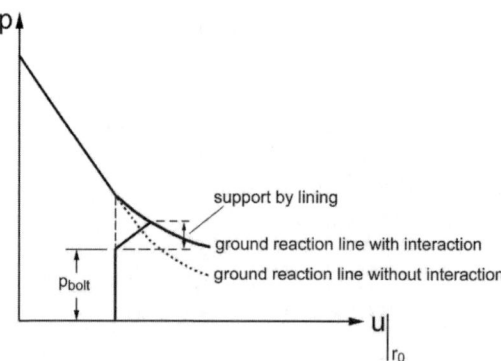

Fig. F.2. Ground reaction line altered by the action of bolts and corresponding support line.

Knowing that P_{rg} depends on r, we set $P_{rg} = \text{const}/r$ and obtain from (F.2):

$$P_{rg} = \frac{\pi d\tau_0 r_0}{ab} \cdot \frac{1}{r} \quad.$$

F Multiphase model of reinforced ground

Obviously, for very thin bolts ($d \to 0$) the interaction force P_{rg} becomes negligible and, thus, its neglection in Sect. 15.1.3 becomes justified. A non-vanishing interaction force can be taken into account as follows: As usual for deep tunnels, we neglect gravity ($\mathbf{g} \approx \mathbf{0}$). Then, equation F.1 written in cylindrical coordinates reads:

$$\frac{d\sigma_r}{dr} + \frac{\sigma_r - \sigma_\theta}{\theta} = -\frac{\pi d\tau_0 r_0}{ab} \cdot \frac{1}{r} \tag{F.3}$$

Introducing the MOHR-COULOMB yield condition leads to

$$\frac{d\sigma_r}{dr} - \frac{(\sigma_\theta + \sigma_r)\sin\varphi}{r} - \frac{2c\cot\varphi}{r} = -\frac{\pi d\tau_0 r_0}{ab} \cdot \frac{1}{r} . \tag{F.4}$$

We can reduce Equ. F.4 to the original equation if we replace the cohesion c by $\hat{c}$, where

$$\hat{c} := c - \tau_0 \frac{\pi d r_0}{2ab} \cdot \tan\varphi. \tag{F.5}$$

Thus, the action of the bolts upon the ground is equivalent to a reduction of cohesion according to Equ. F.5. Note that Equ. F.5 holds for $\varphi > 0$. For $\varphi = 0$ and $\sigma_\theta - \sigma_r = 2c$ the equilibrium equation F.4 has to be replaced by

$$\frac{d\sigma_r}{dr} - \frac{2c}{r} = -\frac{\pi d\tau_0 r_0}{ab} \cdot \frac{1}{r} .$$

The solution (14.30) will be valid if we replace c by $\hat{c}$, where

$$\hat{c} := c - \tau_0 \frac{\pi d r_0}{2ab} .$$

The volume force due to the action of the bolts alters the ground reaction line as shown in (Fig. F.2).

G

Deformation of a tunnel due to seismic waves

We omit here the indices s and p; the operations stated below should be executed twice, once for the p-wave and once for the s-wave.
From the displacement field $\mathbf{u}(\mathbf{x},t)$ we obtain the (geometrically linearised) deformation field:

$$\boldsymbol{\varepsilon} = \frac{1}{2}\left[\frac{\partial \mathbf{u}}{\partial \mathbf{x}} + \left(\frac{\partial \mathbf{u}}{\partial \mathbf{x}}\right)^T\right]$$

In index notation we obtain:

$$u_i = a\, l_i\, \exp\left[i\omega\left(t - \frac{l_k x_k}{c}\right)\right]$$

$$\rightsquigarrow \qquad \frac{\partial u_i}{\partial x_j} = -i\,\frac{\omega}{c}\, a\, l_i\, l_j\, \exp\left[i\omega\left(t - \frac{l_k x_k}{c}\right)\right] = -i\,\frac{\omega}{c}\, l_j u_i$$

hence

$$\varepsilon_{ij} = -i\,\frac{\omega}{2c}(l_i u_j + l_j u_i) = -i\,\frac{\omega}{c}\, a\, l_i l_j\, \exp\left[i\omega\left(t - \frac{l_k x_k}{c}\right)\right]\ .$$

The strain ε of the tunnel axis is given as:

$$\varepsilon = \boldsymbol{\varepsilon} : \mathbf{t} \otimes \mathbf{t} = \varepsilon_{ij}\, t_i\, t_j\ ,$$

$$\varepsilon_{\max} = \frac{\omega}{c}\, a\, l_i\, l_j\, t_i\, t_j\ .$$

Let α denote the angle between the tunnel axis $\mathbf{t}$ and the wave propagation direction $\mathbf{l}$. With $l_i t_i = \cos\alpha$ we finally have

$$\varepsilon_{\max} = \frac{\omega}{c}\, a\, \cos^2\alpha$$

Now let us determine the change of the curvature κ of the tunnel axis imposed by the displacement field $\mathbf{u}(\mathbf{x},t)$. We assume that the initial curvature is small and, thus, negligible. Let the vectorial representation of the tunnel axis be $\mathbf{r}(s) = \mathbf{r}_0(s) + \mathbf{u}(s)$, where s is the arc length. The curvature is then given by $\kappa \approx |\mathbf{r}''|$, where the prime denotes derivation with respect to s. We thus have

$$\Delta \kappa = \kappa - \kappa_0 \approx \kappa \approx |\mathbf{u}''| \;.$$

$\mathbf{u}''$ can be determined as follows:

$$\mathbf{u}' = u_i' = \frac{\partial u_i}{\partial x_j} \frac{\partial x_j}{\partial s} = \frac{\partial u_i}{\partial x_j} t_j \;.$$

Similarly

$$(u_i')' = \frac{\partial}{\partial x_k}\left(\frac{\partial u_i}{\partial x_j} t_j\right) t_k$$

$$= \frac{\partial^2 u_i}{\partial x_k \partial x_j} t_j t_k + \frac{\partial u_i}{\partial x_j} \frac{\partial t_j}{\partial x_k} t_k = \frac{\partial^2 u_i}{\partial x_k \partial x_j} t_j t_k + \frac{\partial u_i}{\partial x_j} \frac{\partial t_j}{\partial s}$$

For an initially straight tunnel ($\partial t_i/\partial s = 0$) we have:

$$u_i'' = \frac{\partial^2 u_i}{\partial x_k \partial x_j} t_k t_j$$

hence

$$\kappa = \left| \frac{\partial^2 u_i}{\partial x_k \partial x_j} t_k t_j \right|$$

$$\rightsquigarrow \quad \kappa = \frac{\omega^2}{c^2} l_j l_k t_j t_k |u_i|$$

and

$$\kappa_{max} = \left(\frac{\omega}{c}\right)^2 a \, l_j l_k t_j t_k$$

or

$$\kappa_{max} = \left(\frac{\omega}{c} \cos\alpha\right)^2 a \;.$$

H
A rational approach to swelling

In mechanical terms, swelling can be described as follows: We decompose the strain ε into a part due to swelling (ε_s) and a part due to mechanical loading/unloading (ε_b):

$$\varepsilon = \varepsilon_s + \varepsilon_b \quad . \tag{H.1}$$

To allow for immediate access of water to the swelling minerals, we consider an infinitely thin layer. By lack of experimental results, we assume the most simple relation, namely a linear one, between ε_s and w:

$$\varepsilon_s = \varepsilon_{s,\max} \frac{w}{w_{\max}} \quad . \tag{H.2}$$

Of course, this relation can be replaced by a more realistic one as soon as corresponding test results are available. w is the water content. It refers not only to the free water within the pores (as usual in soil mechanics) but also to the water responsible for swelling.
$w_{\max}$ is the water content at saturation, i.e. when no more swelling occurs. The propagation of the water within the ground is a diffusion process governed by the known diffusion equation, also known from the consolidation theory:

$$\frac{\partial w}{\partial t} = c \frac{\partial^2 w}{\partial z^2} \tag{H.3}$$

for problems with one spatial dimension. For simplicity, we have assumed that c is constant. Obviously, the same solution as for 1D-consolidation applies. For the mechanical part of the strain we use the known logarithmic relation of soil mechanics:

$$\varepsilon_b = -C_b \ln \frac{\sigma}{\sigma_0} \quad , \tag{H.4}$$

where compression is taken as negative. σ_0 is the initial stress. Again, instead of equation H.4 a more appropriate relation can be used if available.

We consider now a layer with thickness l. The access of water to its lower and upper boundaries is given at $t = 0$. The instantaneous distribution of water content within this layer is given by the same isochrones known from the consolidation theory (Fig. H.1).

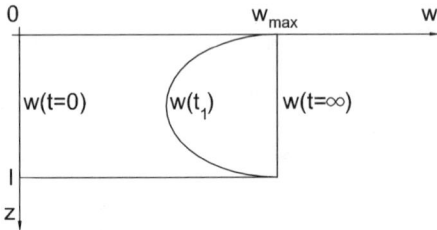

Fig. H.1. Distribution of w across the thickness $0 \leq z \leq l$ of a swelling layer at several times

If w is known for every depth z and for every time t, the upheaval due to swelling can be obtained:

$$(\Delta l)_s = \int_0^l \varepsilon_s \, dz = \varepsilon_{s,\max} \int_0^l \frac{w}{w_{\max}} \, dz = \varepsilon_{s,\max} \int_0^l \mu \, dz = \varepsilon_{s,\max} \bar{\mu} \, l \quad (H.5)$$

The solution of the diffusion equation H.3 yields $\bar{\mu}$ as function of the dimensionless time $\tau := 4ct/l^2$. In the theory of consolidation, $\bar{\mu}$ is the ratio of the actual settlement at time t to the final settlement s_∞. Here, $\bar{\mu}$ denotes the ratio of the (not inhibited) expansion due to swelling at time t to the final expansion for $t \to \infty$. As known, the relation between $\bar{\mu}$ and τ can be approximated by

$$\tau = \begin{cases} \frac{\pi}{4}\bar{\mu}^2 & \text{for } \bar{\mu} < 0,6 \\ -0,933 \log_{10}(1 - \bar{\mu}) - 0,085 & \text{for } \bar{\mu} > 0,6 \end{cases}$$

From equation H.5 follows the average expansion due to swelling

$$\bar{\varepsilon}_s := \left(\frac{\Delta l}{l}\right)_s = \bar{\mu} \varepsilon_{s,\max} \quad . \quad (H.6)$$

From (H.1), (H.4) and (H.6) follows the total expansion of the considered layer:

$$\bar{\varepsilon} = -C_b \ln \frac{\sigma}{\sigma_0} + \varepsilon_{s,\max} \bar{\mu}$$

In a semilogarithmic diagram (Fig. H.2) this relation between $\bar{\varepsilon}$ and σ is represented by straight lines.

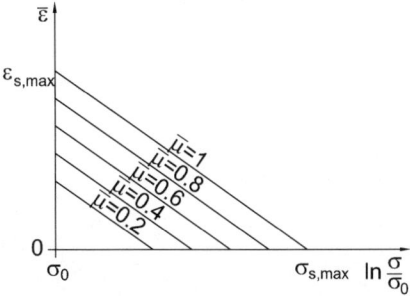

Fig. H.2. Relation between $\bar{\varepsilon}$ and σ. The access of water to the boundaries occurs at $t = 0$ for $\varepsilon = 0$ and $\sigma = \sigma_0$. The case $\bar{\mu} = 1$ is obtained for $t \to \infty$.

The family of straight lines in Fig. H.2 is parametrised by $\bar{\mu}$ (resp. the time t). Adding a compressible layer, consisting e.g. of soft grains, prescribes to the swelling layer a particular σ-$\bar{\varepsilon}$-relation, e.g. an elastic - idealplastic relation. If we plot this relation into Fig. H.2 we obtain the development of stress and strain with time (Fig. H.3).

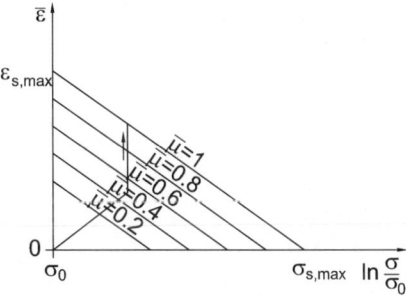

Fig. H.3. If the relation between $\bar{\varepsilon}$ and σ is prescribed, as shown here by the elastic-idealplastic curve, then the development of strain with time can be inferred from the shown parametrisation.

9
Glossary

9 Glossary

9.1 English - German

abutment	Auflager
adit	Fensterstollen
admission	Zulassung
advance grouting	Vorausinjektion
advance investigation	Vorerkundung
aggressiveness	Aggressivität
anchor length	Ankerlänge
angle of dilatancy	Dilatanzwinkel
approximation	Näherung
approximative solution	Näherungsformel
arc-length	Bogenlänge
articulation	Gelenk
axial symmetry	Zylindersymmetrie
back-up	Nachläufer
bar spacer	Abstandhalter
basin	Mulde
beam	Träger
bearing construction	Tragsystem
bearing	Lager
bear	tragen
bench	Strosse
bentonite	Bentonit
bifurcation	Verzweigung
bitumen	Bitumen
blow-out	Ausbläser
body force	Massenkraft
boom	Lafette
borehole	Bohrloch
boulder	Findling
bound theorem	Schrankentheorem
boundary condition	Randbedingung
boundary value problem	Randwertproblem
boundary	Grenze, Rand
bound	Schranke
break through	Durchbruch
breakdown bay	Pannenbucht
breakthrough point	Durchstoßpunkt
brittle	spröde
buckle	beulen
building splice	Bauwerksfuge
bulkhead	Abschottung
buoyancy	Auftrieb
cable duct	Kabelkanal
cake	Filterkuchen
cantilever	Auskragung
carriageway	Fahrbahn
cased	verrohrt

casing	Verrohrung
cast concrete	Ortbeton
catchpit	Auffangbecken
cave-in	Verbruch, Niederbruch
cavity wall	Lochrand
cement grouted rock dowel	SN-Anker
cement mortar	Zementmörtel
chainage	Station
chemical composition	Chemismus
chisel	Meissel
circular cross section	Kreisquerschnitt
circumference	Umfang
clay	Ton
clearance	lichte Breite
coefficient	Beiwert, Koeffizient
cohesion	Kohäsion
cohesive	bindig
collapse load	Bruchlast
collapse mechanism	Bruchmechanismus
collapse	Versagen
conductivity	Leitfähigkeit
consistency	Konsistenz
consolidation	Konsolidierung
construction design	Ausführungsplanung
construction level	Bauzustand
construction lot	Baulos
construction site	Baustelle
construction	Baumaßnahme
contaminated material	Altlast
continuum mechanics	Kontinuumsmechanik
contractancy	Kontraktanz
contractor	Auftragnehmer
convergence	Konvergenz
core drilling	Kernbohrung
core	Bohrkern, Kern
crossover	Querschlag
crown displacement	Firstsenkung
crown settlement	Firstsetzung
crown	Firste, Kalotte
crusher	Steinbrecher
curvature radius	Krümmungsradius
damping	Dämpfung
daylight collapse	Tagbruch
deformability	Verformbarkeit
deformable	verformbar
design load	Lastannahme
design-build contract	funktionale Ausschreibung
design	planen
diaphragm wall	Schlitzwand

diffusivity	Temperaturleitzahl
digging bucket	Baggerschaufel
dilatancy	Dilatanz
discontinuous	diskontinuierlich
disk cutter	Diskenrolle
displacement	Verschiebung
distortion	Verzerrung
diving plate	Tauchwand
drained	dräniert
drawback	Rückschlag
dray bit	Schrämmeissel
drive	auffahren
ductile	duktil
elastic	elastisch
emergency lay-by	Anhaltenische
end configuration	Endkonfiguration
entry cut	Voreinschnitt
envelope	Umhüllende
equation of equilibrium	Gleichgewichtsbedingung
evaluation	Auswertung
excavation face	Ortsbrust
excavation radius	Ausbruchradius
excavation	Auffahrung, Ausbruch, Ausheben, Vortrieb
excavator	Bagger
exhaust	Abluft
expansion joint	Dehnungsfuge
exploration gallery	Probestollen
extension test	Extensionsversuch
face support	Bruststützung
fading	Ausklingen
failure	Ausfall, Bruch
far field	Fernfeld
fault	Verwerfung
favorable	günstig
fill	auffüllen
fissure	Riss
flat jack	Druckkissen
flow	Strömung
force	Kraft
foreman	Polier
friction angle	Reibungswinkel
frictional material	Reibungsmaterial
from technical point of view	technisch
full face advance	Vollausbruch
functional iteration	Fixpunktiteration
gallery	Stollen
gang	Drittel
geostatic	geostatisch
geotechnical engineering	Geotechnik

grain skeleton	Korngerüst
granulate	Schüttgut
ground reaction line	Gebirgskennlinie
ground surface	Geländeoberfläche
groundwater	Bergwasser
ground	Gebirge
groutability	Injizierbarkeit
grouted ground	Injektionskörper
grouting	Injektion
halfspace	Halbraum
hard hut	Schutzhelm
heading	Frontbereich, Vortrieb
hemisphere	Halbkugel
hydraulic piston	Hydraulikzylinder
hydrocyclone	Zyklon
hydrogeological	hydrogeologisch
immersed tunnel	eingeschwommener Tunnel
impermeability	Dichtigkeit
inclination angle	Neigungswinkel
inclined	geneigt
increment of displacement	Verschiebungsinkrement
inertia force	Trägheitskraft
inflection point	Wendepunkt
initial settlement	Sofortsetzung
inlet	Einlage
installation	Einrichtung
interaction	Wechselwirkung
interlocked length	Haftlänge
invert	Sohle
investigation	Erfassen, Untersuchung
irreversible	irreversibel
iteration scheme	Iterationsschema
jointed	geklüftet
joint	Kluft
kinematic	kinematisch
lateral stress	Seitendruck
lattice girder	Gitterträger
leaking	Austritt
levelling bar	Abziehbohle
lighting	Beleuchtung
limit condition	Grenzbedingung
limit state	Grenzzustand
line	Kennlinie
lining	Auskleidung, Schale
loading process	Belastungsprozeß
load	Beanspruchung, Last
locked	kraftschlüssig
lock	Schleuse
long distance train	Fernzug

longitudinal section	Längsschnitt
loose	weitmaschig
lowering of groundwater table	Grundwasserabsenkung
marl	Mergel
mass force	Massenkraft
material constant	Stoffkonstante
matrix	Matrix
measurement	Messung
method of excavation	Ausbruchsart, Vortriebsverfahren
method of subgrade reaction	Bettungszahlverfahren
mining worker	Bergmann
mobilized	mobilisiert
momentum balance	Impulserhaltungssatz
mouth profile	Maulprofil
muck spoil	Bohrgut, Abraummaterial
mucking	Schuttern
nail	Nagel
NATM	NÖT
neglection	Vernachlässigung
nodal force	Knotenkraft
nodal load	Knotenlast
nodal point	Knotenpunkt
non-cohesive	kohäsionslos
normal consolidated	normalkonsolidiert
numerical	numerisch
observation method	Beobachtungsmethode
one-dimensional	eindimensional
overburden, cover	Überdeckung, Überlagerungshöhe
overconsolidated	überkonsolidiert
overexcavation	Mehrausbruch
owner	Bauherr
partial face advance	Teilausbruch
pattern of failure	Bruchmuster
pattern bolting	Systemankerung
penetrate	durchörtern
permeability	Durchlässigkeit
perpetuation of evidence	Beweissicherung
pile	Pfahl
pilot bore	Pilotstollen
pin	Meßbolzen
pipe arch	Rohrschirm
pipe screen cover	Rohrschirmdecke
pipe spile	Rohrspieß
plane	Ebene
plasticity	Plastizität
plastic	plastisch
plastification	Plastifizierung
plastified	plastifiziert
plate load test	Lastplattenversuch

plot	Verlauf
polar coordinate	Polarkoordinate
pore pressure	Porenwasserdruck
power law	Potenzgesetz
power of dissipation	Dissipationsleistung
power station	Kraftwerk
precise	genau
preliminary design	Vordimensionierung
preloaded	vorbelastet
pressure bulkhead	Druckwand
pressure distribution	Druckverteilung
pressure relief	Druckentlastung
pressuremeter	Pressiometer
primary stress	Primärspannung
principal stress	Hauptspannung
proof of operability	Lagesicherheitsnachweis
proof	Nachweis
propagation velocity	Ausbreitungsgeschwindigkeit
pump-in-test	Abpressversuch
quadratic	quadratisch
quantity	Größe
quasi-static	quasistatisch
radial stress	Radialspannung
radiation damping	Abstrahldämpfung
rail	Bahn
raise-boring	Aufbruchbohrung
advance step	Abschlagslänge, -tiefe
reaming	Aufweitung
recycling	Wiederverwendung
reinforcement	Bewehrung
relative displacement	Relativverschiebung
removable wing	absetzbarer Ladetisch
resultant	resultierend
retaining wall	Stützwand
rigid block	Starrkörper
rigid carriageway	feste Fahrbahn
ring closure	Sohlschluß
road construction	Straßenbau
road traffic	Straßenverkehr
roadheader	Teilschnittmaschine
rock class	Gebirgsklasse
rock mass	Felsverband
rock measurement techniques	Felsmeßtechnik
rock pressure	Gebirgsdruck
rock rating	Gebirgsklassifikation
rockburst	Bergschlag
rock	Fels, Festgestein
rod anchor	Stabanker
rod extensometer	Stangenextensometer

rotating cutter	Schneidrad
rotation	Drehung
roughly	ungefähr
rule	Vorschrift
safety	Sicherheit
safe	sicher
saturated	gesättigt
scalar product	Skalarprodukt
scale effect	Maßstabseffekt
scaling	Felsabräumung
scatter	Streubreite, streuen, Streuung
scratching class	Schrämmklasse
scree	Hangschutt
screw conveyor	Förderschnecke
gasket	Dichtung
sealing	Abdichtung, Versiegelung
segment	Tübbing
semicross ventilation	Halbquerlüftung
settlement trough	Setzungsmulde
settlement	Setzung
shaft	Schacht
shallow	seicht
shear band	Scherfuge
shear strength	Scherfestigkeit
shear	Schub
shield	Schild
shoulder	Kämpfer
shutter	Schalwagen
side gallery	Ulmenstollen
side	Ulme
silo	Silo
slickenside	Harnischfläche
sliding wedge	Gleitkeil
slope	Steigung, Böschung
slurry shield	Thixschild
softening	Entfestigung
soften	entfestigen
soil layer	Bodenschicht
soil-structure	Boden-Bauwerk
soil	Lockergestein
solubility	Lösbarkeit
solution	Lösung
spalling	Abplatzung
spatial direction	Raumrichtung
special case	Sonderfall
spherical symmetry	Kugessymmetrie
spherical	kugelförmig
spoil	Ausbruchsmaterial
sprayed concrete	Spritzbeton

spring	Feder
squeezing	druckhaft
stand-up time	Standzeit
standpipe piezometer	Grundwasserpegel
standpipe	Standrohr
state variable	Zustandsvariable
steel rib	Ausbaubogen
step	Schritt
stiffness	Steifigkeit
strain	Dehnung
strength	Festigkeit
stress deviator	Spannungsdeviator
stress distribution	Spannungsverteilung
stress field	Spannungsfeld
stress path	Spannungspfad
stress state	Spannungszustand
stress-free	spannungsfrei
stress	Spannung
stroke	Hub
subgrade reaction approach	Bettungsansatz
subgrade reaction modulus	Bettungsmodul
subgrade reaction	Bettung
substep	Teilschritt
superficial velocity	Filtergeschwindigkeit
support core	Brustkeil
support measure	Sicherungsmittel
support pressure	Ausbauwiderstand
supporting force	Stützkraft
support	Ausbau, Sicherung
surface settlement	Oberflächensetzung
survey	meßtechnische Überwachung
swell	schwellen
tail gap	Ringspalt
tender	Ausschreibung
tension	Zug
thermal conductivity	Wärmeleitzahl
threedimensional	dreidimensional
thrust	Vorschubkraft
top heading	Kalottenvortrieb
torque	Drehmoment
traffic lane	Fahrstreifen
transducer	Aufnehmer
transformation	Umformung
transition condition	Übergangsbedingung
trapdoor	Falltür
trimming	Nachprofilierung
tube	Röhre
tunnel axis	Tunnelachse
tunneler	Tunnelbauer

tunnelling,	Tunnelvortrieb, Tunnelbau
turning point	Umkehrnische
twist	Verdrehung
twodimensional	zweidimensional
uncertainty of measurement	Meßunsicherheit
uncoupled	entkoppelt
undercutting	Hinterschneidetechnik
undercut	Unterfahrung
underground	unter Tage
underpining	Unterfangung
undrained	undräniert
unit vector	Einheitsvektor
unjointed	ungeklüftet
unpreloaded	unvorbelastet
unsaturated	ungesättigt
unstable	nachbrüchig
vault	Gewölbe
vector	Vektor
vehicular tunnel	Straßentunnel
vertical stress	Vertikalspannung
viscosity	Viskosität
volume force	Volumenkraft
volume loss	Volumenverlust
walkway	Gehweg
wall friction	Wandreibung
water inrush	Wassereinbruch
waterproofing	Abdichtung, Versiegelung
watertight concrete	WU-Beton
weak zone	Störzone
weak	gebräch
weathered	verwittert
weathering	Verwitterung
weightless	gewichtslos
wheel loader	Radlader
wire mesh	Baustahlgitter
wooden grid	Holzrost
wooden plank	Holzbohle
Young´s modulus	Elastizitätsmodul

9.2 German - English

Abbaukammer	working chamber
absetzbarer Ladetisch	removable wing
Abdichtung	sealing, waterproofing
Abluft	exhaust
Abplatzung	spalling
Abraummaterial	muck, spoil
Abpressversuch	pump-in-test
Abschlagslänge, -tiefe	rate of advance, advance step
Abschottung	bulkhead
Abstandhalter	bar spacer
Abstrahldämpfung	radiation damping
Abziehbohle	levelling bar
Achse	axe, axis
Aggressivität	aggressiveness
Altlast	contaminated material
Anhaltenische	emergency lay-by
Ankerlänge	anchor length
Aufbruchbohrung	raise-boring
auffahren	to drive
Auffangbecken	catchpit
Auflager	abutment
Auftrieb	buoyancy
Auftragnehmer	contractor
auffüllen	fill
Auffahrung	excavation
Aufnehmer	transducer
Aufwcitung	reaming
Ausbau	support
Ausbaubogen	steel rib
Ausbauwiderstand	support pressure
Ausbläser	blowout
Ausbreitungsgeschwindigkeit	propagation velocity
Ausbruch	excavation
Ausbruchradius	excavation radius
Ausbruchsart	method of excavation
Ausbruchsmaterial	spoil
Ausführungsplanung	construction design
Ausfall	failure
Ausheben	excavation
Auskleidung	lining
Ausklingen	fading
Auskragung	cantilever
Ausschreibung	tender
Auswertung	evaluation
Bagger	excavator
Baggerschaufel	digging bucket
Bahn	rail

Bauherr	owner
Baulos	construction lot
Baumaßnahme	construction
Baustahlgitter	wire mesh
Baustelle	construction site
Bauwerksfuge	building splice
Bauzustand	construction level
Beanspruchung	load
Beiwert	coefficient
Belastungsprozeß	loading process
Beleuchtung	lighting
Bentonit	bentonite
Beobachtungsmethode	observation method
Bergmann	mining worker
Bergschlag	rockburst
Bergwasser	groundwater
Bettung	subgrade reaction
Bettungsansatz	subgrade reaction approach
Bettungsmodul	subgrade reaction modulus
Bettungszahlverfahren	method of subgrade reaction
beulen	buckle
Bewehrung	reinforcement
Beweissicherung	perpetuation of evidence
bindig	cohesive
Bitumen	bitumen
Boden-Bauwerk	soil-structure
Bodenschicht	soil layer
Bogenlänge	arc-length
Bohrgut	muck spoil
Bohrkern	core
Bohrloch	borehole
Bruch	failure, collapse
Bruchlast	collapse load
Bruchmechanismus	collapse mechanism
Bruchmuster	pattern of failure
Brustkeil	support core
Bruststützung	face support
Chemismus	chemical composition
Dämpfung	damping
Dehnung	strain
Dehnungsfuge	expansion joint
Dichtigkeit	impermeability
Dichtung	seal, gasket
Dilatanz	dilatancy
Dilatanzwinkel	angle of dilatancy
Diskenrolle	disk cutter
diskontinuierlich	discontinuous
Dissipationsleistung	power of dissipation
dräniert	drained

Drehmoment	torque
Drehung	rotation
dreidimensional	threedimensional
Drittel	gang
Druckentlastung	pressure relief
druckhaft	squeezing
Druckkissen	flat jack
Druckverteilung	pressure distribution
Druckwand	pressure bulkhead
duktil	ductile
durchörtern	penetrate
Durchbruch	break through
Durchlässigkeit	permeability
Durchstoßpunkt	breakthrough point
eben	plane
Ebene	plane
einachsial, einaxial	uniaxial
eindeutig	unique
eindimensional	one-dimensional
einfachheitshalber	for sake of simplicity
eingeschwommener Tunnel	immersed tunnel
Einheitsvektor	unit vector
Einlage	inlet
Einrichtung	installation
Einwirkung	action
elastisch	elastic
Elastizitätsmodul	Young´s modulus
Endkonfiguration	end configuration
entfestigen	soften
Entfestigung	softening
entkoppelt	uncoupled
entrüsten	outrage
erfahrungsmäßig	by experience
Erfassen	investigation
explizit	explicite
Extensionsversuch	extension test
Fahrbahn	carriageway
Fahrstreifen	traffic lane
fallen	decrease
Falltür	trapdoor
Feder	spring
Felsabräumung	scaling
Felsdecke	cavity roof
Felsmeßtechnik	rock measurement techniques
Felsverband	rock mass
Fensterstollen	adit
Fernfeld	far field
Fernzug	long distance train
feste Fahrbahn	rigid carriageway

Festgestein	rock
Festigkeit	strength
Festigkeitsgrenze	strength limit
Filtergeschwindigkeit	superficial velocity
Filterkuchen	cake
Findling	boulder
Firste	crown
Firstsenkung	crown displacement
Firstsetzung	crown settlement
Fixpunktiteration	functional iteration
flächengleich	of equal area
Frontbereich	heading
Förderschnecke	screw conveyor
funktionale Ausschreibung	design-build contract
günstig	favorable
Gasaustritt	leaking
Gebirge	ground
Gebirgsdruck	rock pressure
Gebirgskennlinie	ground reaction line
Gebirgsklasse	rock class
Gebirgsklassifikation	rock rating
Gebirgsklassifizierung	rock rating
gebräch	weak
Gehweg	walkway
geklüftet	jointed
Geländeoberfläche	ground surface
Geländeoberkante	ground surface
Gelenk	articulation
genau	precise
geneigt	inclined
geostatisch	geostatic
Geotechnik	geotechnical engineering
geotechnisch	geotechnical
gesättigt	saturated
gewichtslos	weightless
Gewölbe	vault
Gitternetz	lattice girder
Gleichgewichtsbedingung	equation of equilibrium
Gleichgewichtsgleichung	equation of equilibrium
Gleitkeil	sliding wedge
Glg.	equ.
Größe	quantity
Grenzbedingung	limit condition
Grenze	boundary
Grenzzustand	limit state
Grundwasserabsenkung	lowering of groundwater table
Grundwasserpegel	standpipe piezometer
Höchstgeschwindigkeit	maximum velocity
Haftlänge	interlocked length

Halbkugel	hemisphere
Halbquerlüftung	semicross ventilation
Halbraum	halfspace
Hangschutt	scree
Harnischfläche	slickenside
Hauptspannung	principal stress
Hereinbrechen	cave-in
Herstellung	construction
Hinterschneidetechnik	undercutting, back-cutting
Holzbohle	wooden plank
Holzrost	wooden grid
Hub	stroke
Hydraulikzylinder	hydraulic piston
hydrogeologisch	hydrogeological
Impulserhaltungssatz	momentum balance
Injektion	grouting
Injektionskörper	grouted ground
Injizierbarkeit	groutability
irreversibel	irreversibel
Iterationsschema	iteration scheme
Kabelkanal	cable duct
Kalotte	crown
Kalottenvortrieb	top heading
Kämpfer	shoulder
Kennlinie	line
Kern	core
Kernbohrung	core drilling
kinematisch	kinematic
klüftig	jointed
Kluft	joint
Kluftabstand	rock spacing
Knotenkraft	nodal force
Knotenlast	nodal load
Knotenpunkt	nodal point
Koeffizient	coefficient
Kohäsion	cohesion
kohäsionslos	non-cohesive
Konsistenz	consistency
Konsolidierung	consolidation
Kontinuumsmechanik	continuum mechanics
Kontraktanz	contractancy
Konvergenz	convergence
Korngerüst	grain skeleton
Krümmungsradius	curvature radius
Kraft	force
kraftschlüssig	locked
Kraftwerk	power station
Kreisquerschnitt	circular cross section
kugelförmig	spherical

9 Glossary

Kugessymmetrie	spherical symmetry
Lafette	boom
Längenänderung	extension
Längenschnitt	longitudinal section
Längsschnitt	longitudinal section
Lösbarkeit	solubility
Lösung	solution
Lager	bearing
Lagesicherheitsnachweis	proof of operability
Lastannahme	design load
Lastplattenversuch	plate load test
Leitfähigkeit	conductivity
lichte Breite	clearance
Lochrand	cavity wall
Lockergestein	soil
Maßstabseffekt	scale effect
Massenkraft	body force
Massenkraft	mass force
Matrix	matrix
Maulprofil	mouth profile
Meßbolzen	pin
meßtechnische Überwachung	survey
Meßunsicherheit	uncertainty of measurement
Mehrausbruch	overexcavation
Meissel	chisel
Mergel	marl
Messung	measurement
mobilisiert	mobilized
Mulde	basin, recess
NÖT	NATM
Nachprofilierung	trimming
Näherung	approximation
Näherungsformel	approximative solution
nachbrüchig	unstable
Nachbrechen	cave in
Nachläufer	back-up
Nachweis	proof, assessment
Nagel	nail
Neigungswinkel	inclination angle
Neubau	construction
niederbringen	undertake
Niederbruch	cave in
normalkonsolidiert	normal consolidated
numerisch	numerical
Oberflächensetzung	surface settlement
Ortbeton	cast concrete
Ortsbrust	excavation face
Pannenbucht	breakdown bay
Pfahl	pile

PFC	PFC
Pilotstollen	pilot bore
planen	design
plastifiziert	plastified
Plastifizierung	plastification
plastisch	plastic
Plastizität	plasticity
Polarkoordinate	polar coordinate
Porenwasserdruck	pore pressure
Potenzgesetz	power law
Pressiometer	pressuremeter
primär	primary
Primärspannung	primary stress
Polier	foreman
Probestollen	exploration gallery
quadratisch	quadratic
quasistatisch	quasi-static
Querschlag	crossover
Röhre	tube
Rückschlag	drawback
Radialspannung	radial stress
Radius	radius
Radlader	wheel loader
Rand	boundary
Randbedingung	boundary condition
Randwertproblem	boundary value problem
Raumrichtung	spatial direction
Reibungswinkel	friction angle
Reibungsmaterial	frictional material
Relativverschiebung	relative displacement
resultierend	resultant
Riß	fissure
Ringspalt	tail gap
Riss	fissure
Fels	rock
Rohrschirm	pipe arch
Rohrschirmdecke	pipe screen cover
Rohrspieß	pipe spile
rollig	non-cohesive
Schüttgut	granulate
Schacht	shaft
Schale	lining
Schalwagen	shutter
Scherfestigkeit	shear strength
Scherfuge	shear band
Schild	shield
schlaff	stress-free
Schleuse	lock
Schlitzwand	diaphragm wall

Schneidrad	rotating cutter
Schrämmklasse	scratching class
Schrämmeissel	dray bit
Schranke	bound
Schrankentheorem	bound theorem
Schritt	step
Schub	shear
Schuttern	mucking
Schutterung	mucking
Schutzhelm	hard hut
schwellen	swell
seicht	shallow
Seitendruck	lateral stress
Seitendruckbeiwert	lateral stress coefficient
Setzung	settlement
Setzungsmulde	settlement trough
sicher	safe
Sicherheit	safety
Sicherung	support
Sicherungsmittel	support measure
Silo	silo
Skalarprodukt	scalar product
SN-Anker	cement grouted rock dowel
Sofortsetzung	initial settlement
sog.	so-called
Sohle	invert
Sohlschluß	ring closure
Sonderfall	special case
Spannung	stress
Spannungsdeviator	stress deviator
Spannungsfeld	stress field
Spannungspfad	stress path
Spannungsverteilung	stress distribution
Spannungszustand	stress state
Spezialfall	special case
spröde	brittle
Spritzbeton	sprayed concrete
Stangenextensometer	rod extensometer
Steinbrecher	crusher
Störzone	weak zone
Stützkraft	supporting force
Stützwand	retaining wall
Stabanker	rod anchor
Standrohr	standpipe
Standzeit	stand-up time
Starrkörper	rigid block
Station	chainage
Steifigkeit	stiffness
Steigung	slope

Stoffkonstante	material constant
Stollen	gallery
Strömung	flow
Straßenbau	road construction
Straßentunnel	vehicular tunnel
Straßenverkehr	road traffic
Streubreite	scatter
streuen	scatter
Streuung	scatter
Strosse	bench
Systemankerung	systematic anchoring
Tagbruch	daylight collapse
Tauchwand	diving plate
technisch	from technical point of view
Teilausbruch	partial face advance
Teilschnittmaschine	roadheader
Teilschritt	substep
Temperaturleitzahl	diffusivity
Thixschild	slurry shield
Ton	clay
Träger	beam
Trägheitskraft	inertia force
tragen	bear
Tragsystem	bearing construction
Tübbing	segment
Tunnelachse	tunnel axis
Tunnelbau	tunnelling
Tunnelbauer	tunneler
Tunnelvortrieb	tunnelling, heading
Überdeckung	overburden, cover
Übergangsbedingung	transition condition
Überlagerungshöhe	overburden
überkonsolidiert	overconsolidated
Ulme	side
Ulmenstollen	side gallery
Umfang	circumference
Umformung	transformation
Umhüllende	envelope
Umkehrnische	turning point
undräniert	undrained
ungeklüftet	unjointed
ungesättigt	unsaturated
unter Tage	underground
Unterfahrung	undercut
Unterfangung	underpining
unvorbelastet	unpreloaded
Vektor	vector
Verband	mass
Verbruch	cave-in

Verdrehung	twist
verformbar	deformable
Verformbarkeit	deformability
Verlauf	plot
verlaufen	run
Vernachlässigung	neglection
verrohrt	cased
Verrohrung	casing
Versagen	collapse
Verschiebung	displacement
Verschiebungsinkrement	increment of displacement
Versiegelung	sealing
Versuch	test
Vertikalspannung	vertical stress
Verwerfung	fault
verwittert	weathered
Verwitterung	weathering
Verzerrung	distortion
Verzweigung	bifurcation
Viskosität	viscosity
Vollausbruch	full face advance
Volumenkraft	volume force
Volumenverlust	volume loss
Vorausinjektion	advance grouting
vorbelastet	preloaded
Vordimensionierung	preliminary design
Voreinschnitt	entry cut
Vorerkundung	advance investigation
Vorschrift	rule
Vorschubkraft	thrust
Vortrieb	excavation
Vortriebsverfahren	method of excavation
Wärmeleitzahl	thermal conductivity
Wandreibung	wall friction
Wassereinbruch	water inrush
Wechselwirkung	interaction
weitmaschig	loose
Wendepunkt	inflection point
Wiederverwendung	recycling
WU-Beton	watertight concrete
zeitabhängig	time dependent
Zeitintegration	time integration
Zementmörtel	cement mortar
Zug	tension
Zulassung	admission
Zustandsvariable	state variable
zweidimensional	twodimensional
Zyklon	hydrocyclone
Zylindersymmetrie	axial symmetry

Index

A

abrasion 120, 124
accelerators 134
acoustic emission 251
advance grouting 168
aerodynamic air pressure rise 17
AESCHBACH, M. 219
air cushion 104
air gap membrane 186
air lock 103, 104, 197
air opacity 41
air pollution 38
air velocity..................... 41
alignment...................... 18
ANAGNOSTOU, G. 316
analytical solutions 271
anchors 138, 305
ANDREAE...................... 171
ANFO 86
ANHEUSER, L................... 92
anisotropy 252
arching.................. 131, 273
ARZ, P. 208
asbestos 219
ATKINSON, J.H. 328, 333, 340

B

BABENDERERDE, S. 100
back grouting 100
backup....................... 90
BARTON, N. 70, 71, 152
base tunnel 19
BAUMANN, L. 124
BAWDEN, W.F................. 249

BECKER, E.................... 388
Belgian tunnelling method 144
bench 6, 76
BETTELINI, M. 48
BIENIAWSKI, Z.T. 3, 66, 69–71, 143, 151
BINGHAM-fluid................. 385
BJARNASON, B.................. 360
blastholes.................. 85, 116
BLIEM, C. 303
blow-out 103, 201
BOCK, H. 353
bolt 138, 305
BONALA, M.V.S. 258
boom cutter.................... 82
borehole expansion test 270
boreholes 116
bound theorems 242
boundary condition............. 367
box-jacking 109
BRAMESHUBER, W.............. 156
Brazilian test.................. 244
brittle 244, 246
BRUX, G. 32
buckling...................... 349
BÜRGER, W. 388
BULITSCHEW, N.S. 352
BURLAND, J.B................. 340
BUSBY, J. 205

C

cables....................... 222
caisson 206
CAQUOT, A. 332
carbon dioxide................. 218

CARRANZA-TORRES, C. 263
cartridges 86
chainage 6
CHEATHAM, J.B. 120
chipping 121
claims 57
CLAYTON, C.R.I. 358
climbing lane 16
clogging 186
CO-concentration 41
cohesion 240
collapse 77, 174, 226
collapse theorems 242
compensation grouting 161
compensation methods 362
composite materials 399
compressed air 103, 197
conditioning 107, 161
conflagration 42
constitutive equations . . . 235, 365, 370
construction experts 29
construction supervisors 29
contour drillholes 87
contracting 25
contractor 26
contracts 27
controlling 29
controls 223
convergence 354
conveyors 99, 128
core discing 248
core recovery 65
core-heading 75
CORKUM, B. 263
CORNET, F.H. 278
CORY, W.T.W. 34
cost management 28
costs 24, 114
cover 99
cover and cut 198
COX, B.G. 244
cracks 346
creep 254
critical mass 380
crown 6
crown arch 76
CUNDALL, P.A. 260
cutterhead 82

D

DAHL, J. 46
damping 370
DAVIS, E.H. 327
DAY, J. 42
decompression 199
DEERE 65, 70
deflectometer 355, 357
DERMATAS, D. 265
design as you go 26
design checkers 29
design fire 43
design-bid-build 27
design-build 26
designers 29
desk study 58
detonating cord 86
detonation 379
detonators 86
DIEDERICHS, M.S. 247
diesel combustion products (DCP) . 220
dilatometer 270
dip 65
DIRKSMEIER, R.A. 48
disc cutter 95, 113, 118
discrete element method 260
dispersivity 117
Dispute Review Boards 28
Disturbance Factor D 263
diving tables 199
DOMKE 352
doorstopper 363
double packer 159
double shield 102
dowel 138
downreaming 211
drainage 52, 183
drill & blast 34, 84
drill & split 125
drillability 113
drilling 116
drive-in operation 112
drive-out operation 112
dry mix 133
dry pack 184
dual drainage system 53
ductile 244, 246, 263
ductings 34

DUSSEAULT, M.B. ... 64, 116, 137, 143, 179, 363
dust 37, 82, 218
dynamic relaxation 370

E

earth-pressure-balance shield 105
earthquake 335
effective stress 179
EICHORN, B. 255
EINSTEIN, H.H. 265
elasticity 237
electrical installations in tunnelling 221
elephant feet 76
EMCH, J.P. 46
emergency calls 32
emergency plan 225
emergency recess 48
equilibrium 272
evolution equation 366
EWERT, F.K. 181
excavation 80, 94
excavation face 329
excavation platform 212
experts 29, 30
exploration 57
exploration drilling 62
explosion fumes 219
explosives 88
extensometer 355
extinguishers 46

F

face support 95
faceplate 141
FAIRHURST, CH. 382
fans 38
faults 67
FEDER, J. 159
FELLIN, W. 226
FENNER-PACHER-curve 286
field tests 267
filter cake 385
fire
    combat 44
    detectors 46
    extinguisher 48
    protection 42
fire resistant concrete 45

firefighting 47
FLANAGAN, R.F. 90
flash-over 43
flat jacks 268
flickering 50
FLIEGNER, E. 203
foliation 68
forepoling 147
formwork 153
fractal dimension 257
fractals 257
fragmentation 122
FRANKLIN, J.A. 64, 116, 137, 143, 179, 363
friction 240
friction of joints 251
FRIEBEL, W.D. 156
frost heave 169
frost propagation 387
full face excavation 75
fume 86

G

GARG, S.K. 179
GARSHOL, K.F. 165
gas intrusion 217
gasket 99
Geological Strength Index GSI 70, 263
geophysical exploration 64
geospacers 183
geostatic primary stress 276
geosynthetics 195
Geotechnical Baseline Record 73
Geotechnical Data Record 73
geotechnical investigation 57
German heading method 76
GERTSCH, R.E. 123
GIRMSCHEID, G. 88, 186
glass fibre reinforced concrete 46
GLOSSOP, R. 58, 197
GÖTZ, H.-P. 353
GOODMAN, R.E. 178
Goodman jack 270
GOTTSTEIN, H. 53
GRIFFITH 250
gripper 83
gripper shield 102
ground loss 341
ground reaction line 285

Grout Intensity Number GIN 167
grouting 159
grouts 163
GRÜTER, R. 17
GUDEHUS, G. 256
guidance 101
GUNN, M.J. 340
GURKAN, E. 5

H

HÄRLE, B. 368
HALBACH, G. 41
hammer 80
hard clearance 16
HART, R.D. 260
HASHASH, Y.M.A. 335
haulage 127
heading 75
headrace tunnel 20
headroom 16
health hazards 217
health problems 199
Heathrow collapse 229
HEIJBOR, J. 46, 97, 200
HEIM 171
HERLE, I. 371
HERRENKNECHT, M. 92
high level tunnel 19
HOEK, E. 25, 29, 66, 249, 263, 264, 296
HOEK-BROWN criterion 263
HOLZHÄUSER, J. 315
HOONAARD, J. VAN DER 200
HSE review 173
HUDER-AMBERG test 265
HUGHES, J.M.O. 288
hydrant 47
hydration heat 194
hydraulic conductivity 179
hydraulic fracturing 360
hydrogen sulphide 218
hydrostatic primary stress 279

I

ignition 86, 382
ignition pattern 87
illumination 49
illumination during construction .. 223
inclinometer 355, 357
indentation 121

initial conditions 366
installations for traffic control 31
interference 383
invert 6

J

jacks 92
JAKY 314
JANCSECZ, S. 344
JANNSEN, H.A. 313
JESSBERGER 169
jet grouting 163
JOHANSEN, J. 87
joints 64
JOOSTEN 165

K

KAISER, P.K. . 227, 247, 249, 263, 349, 363
KAISER-effect 251
kakirite 67
karst 164, 177
kataklasite 67
kerb 16
KERRY ROWE, R. 343
KÉZDI, A. 169, 215
KHARITON, YU. 380
KIRSCHKE, D. 194
KLEIN, J. 211
KNIGHTS, M. 90
KNOLL, E. 17
KOLYMBAS, D. 260
KORDINA, K. 42, 43
KOVÁRI, K. 41, 77, 92, 172, 175
KRAMER, S.L. 336
KRETSCHMER, M. 203
KROHN, C.E. 258
KUTUSOW, B.N. 379
KUTZNER, C. 163, 165

L

L-R-mechanism 291
LAABMAYR, F. 172
LAMÉ 279, 280
LAMÉ's solution 279
lamination 69, 252
LAPLACE 387
lattice girders 146
LAUFFER, H. 70, 173
LAVRIKOV, S.V. 291

lay-by 16
LECA, E. 332
LEHMANN, B. 64
levelling 354
LIJON, B. 360
LINDE, F.W.J. VAN DE 200
LINGENFELSER, H. 206
lining 154, 155
lining segments 96
LOMBARDI, G. 167
longitudinal drains 53
longitudinal slope 19
loosening 131, 248
lower bound of the support pressure 327
LUGEON-test 181
LUNARDI, P. 149
LUONG, M.P. 245
LUONG-test 244
LUTZ, H. 173

M
MACKLIN, S.R. 342
MÄHR, M. 101, 344
MAIDL, B. 92
MAIR, R.J. 342
MALVERN, L. 279
MASHIMO, H. 6
MAUERHOFER, S. 124
maximum allowable concentrations
  (MAC) 33
measurements 60
MÉLIX, P. 324
methane 217
MEYER-OTTENS, R. 43
micropiles 76
microtunnels 110
MINDLIN, R.D. 276
mixed system 52
mixshield 106
MÖBIUS transformation 393
MOGI, K. 174
MOMBER, A.W. 118
monitoring 353
monocoque lining 156
MORGAN, S. 149
mouth profile 7
mucking 99, 127
MÜLLER-SALZBURG, L. . . 154, 171, 344
MUIR-WOOD, A. 3, 27, 172

multiphase media 399
mylonite 67

N
NÖT 171
nail 138, 305
NATM 171
natural longitudinal ventilation 38
New Austrian Tunnelling Method . 171
NIKOLAI 352
nitrose gases 219
noise 218
non-linearity 368
NOVA, R. 245
nozzles 48
numerical analysis 365
numerical simulation 365

O
OBERGUGGENBERGER, M. 226
observational method 174
Old Austrian Tunnelling method . . . 76
overbreak 126
owner 26
owner-design 27

P
PACHER 171
packer test 181
PALIGA, K. 45
panel of experts 29
PANET, M. 267
PAPAMICHOS, E. 318
parallel cut 87
partial face excavation 75
passive design 266
PATERSON, M.S. 246, 251
pattern bolting 143, 305, 311
PECK, R.B. 340
penstock 20
Perforex-method 148
permanent lining 152
permeability of rock 179
perpetuation of evidence 191
PERZYNA 256
piezometer 178
pipe roof 148
pipe-jacking 109
plane deformation 275

planning 25
plasticity 239
plastification 281
point load test 249
POLUBARINOVA-KOCHINA, P.YA. . . 188
polymer foames 105
polypropylene fibres 45
polyurethane 166
pore pressure 178
porosity of rock 177
porous concrete 186
portal 20
post-peak deformation 248
PRANDTL, L. 122, 254
pre-stressing 20
prediction 365
pressiometer 270
pressure cell 357
pressure chamber 268
pressure rise 397
pressure shafts 20
pressuremeter 288
principle of *design and construct* . . 173
principle of effective stress 242
PROCTOR, R.J. 335
profiling 126
PROMMERSBERGER, G. 184
PUCHER, K. 43
punching 122

R

RAABE, E.W. 162
RABCEWICZ, VON 171, 172
radial press 268
radio communication 32
radon 218
ramp 77
rate dependence 253
rating 55
RAYMER, J.H. 189
rebound 134
recommendations 22
refuge 48
reinforcement 137, 154, 305
relaxation 254, 294
reliability 226
reports 72
required excavation and support sheet 224

rescue concept 225
rigid block deformation mechanism 290
rigid-design 266
risk
   allocation 26
   control 224
   management 224
risk of building damage 344
road headers 82
ROBBINS, R.J. 211
ROBINSON, R.A. 57, 72
robust design 230
rock classification 69
rock grouting 166
Rock Mass Rating-System 70
rock mass strength 261
Rock Quality Designation 70
rock rating 69
rock reinforcement 137
Rock Structure Rating 70
rockbursts 349
RODIONOV, V.N. 258
RÖH, P. 90
ROGGE, A. 311
rolling formwork 153
ROZSYPAL, A. 149
RUINA, A. 255
RŽIHA 171

S

safety 55, 217
   provisions 89
   quantification 226
safety and rescue plan 44
SAGASETA, C. 343
SAVILLE, G. 244
SCAVIA, C. 258
SCHIELE, W. 174
schistosity 68, 301
SCHMIDT, H.G. 208
SCHNEIDER, U. 42
SCHREYER, J. 156
SCHUCK, W. 353
SCHWEIGER, H.F. 366
sciage 148
screw conveyor 99, 107
sealing 149, 191
SEEBER, G. 268
SEEDSMAN, R.W. 178

seepage force 189
seismic waves 403
SEITZ, J. 208
self-boring anchor 141
self-boring pressuremeter 270
self-drilling anchor 141
self-rescue phase 44
self-similar 257
semi-transverse ventilation 38
SEMPRICH, S. 198, 208
sensors 46
separate system 53
service telephones 32
servo-controlling 249
SERWAS, H.D. 217
settlements 339
shaft 211
shear test 267
shield heading 92
shockwave 379
shotcrete 132, 201
shotcrete lining 373
shrinkage 194
side 6
side galleries 77
sidewall drift 77
silica fume 134
silo equation 313
single-shell lining 156
SINTZEL, M. 3
site investigation 57
size effect 256
  in rock 256
  in soil 258
slake durability test 118
sleeve pipe 159
slickenside 68
sliding micrometer 356
slurry shield 104
SMITH, R.J. 28
smoke 42
smooth blasting 88
smouldering 43
snapper 63
soft data 73
softening 131, 301
soil fracturing 161
soil freezing 168
solubility 178

SORNETTE, D. 174
spalling 45
SPANG, J. 172
speed of advance 111
spiles 148
spot bolting 143
sprayed concrete 132
sprinklers 48
squeezing 293
squeezing in anisotropic rock 301
squeezing rock 113
stability problems 349
stand-up time 333
standards 22
standpipe test 180
start shaft 112
steel fibre reinforced concrete 136
steel fibres 97
steel meshes 137
STEINERT, C. 43
STEPHANSSON, O. 360
STEYRER, P. 191
stoking 125
strain tensor 271
strain-controlled 248
strength 240, 243
strength index 249
stress-controlled 248
strike 65
Subaqueous tunnels 203
subgrade reaction 371
support 131, 150
  arches 146
  measures 151
  pressure 319
  reaction line 289
surging 21
Swellex 139
swelling 265, 405
syneresis 166
SZÉCHY, K. 3, 215, 315

T

tail gap grouting 100
tail void closure 100
tamping 86
TAN, T.L. 44
TANNANT, D.D. 150
TANSENG, P. 277

target plates 354
target shaft 112
TATSUOKA, F. 254
TBM performance prediction 118
TEICHERT, P. 132
TEICHMANN, G. 37
temperature.................... 41
temporary invert 76
temporary lining 152
tendon 141
tensile strength 244
tension stiffening 306
TERZAGHI, K. 70, 317
testing machine 249
third-party impact 25
Thixshield 105, 107
THOMAS, T. 44
THOMPSON, A.G. 305
timbering 144
time management 28
TIMOSHENKO 352
tomography 64
top heading 76
towing and lowering method 204
traffic capacity in tunnels 15
traffic lane 16
transformers 221
transition section 49
transverse ventilation 39
triaxial test 244
tube à manchette 159
tunnel
    boring machine 82
    cross section 7
    heading machines 102
    history 3
    lining 325
tunnel stopper 44
TURCOTTE, D.L. 258
turn-around points 16

U

undercutting 124
underground explosions 381
underground water conduits 20
upper bound of the support pressure 328
upreaming 211

V

V-cut 87
vacuum erectors 97
VAUGHAN, P.R. 333
velocity index 65
ventilation 32, 90
ventilator 35
VERMEER, P.A. ... 315, 316, 318, 330
VERRUIJT, A. 340
VERSPOHL, J. 90
vertical clearance............... 16
vibrations 80, 90
video control 47
viscohypoplasticity 256
viscosity....................... 253
volume loss 341
VUILLEUMIER, F. 48

W

WAGNER, A. 211
walkways 16
wall friction 94
WALLIS, S. 128
WALZ, B. 215
water
    inflow 393
    ingress 187
    inrush..................... 181
    intake 22
    mist 48
water-tight concrete 204
waterjet drilling............... 118
waterproofing 191
watertight concrete 194
weathering 69
wedge......................... 139
WEINDL, K. 84
wet mix 133
WHITTLE, R.W. 288
WICKHAM 66, 70
WINDLE, D. 288
WINDSOR, C.R. 305
WROTH, C.P. 288
WU, W........................ 76
WYSS, R. 217

Y

yielding support 299
YOUNG's modulus 235

Z

ZELDOVICH, IA.B. 380
ZHANG, L. 189
ZIEGLER, M. 157
ZILCH, K. 311

Printing: Krips bv, Meppel
Binding: Stürtz, Würzburg

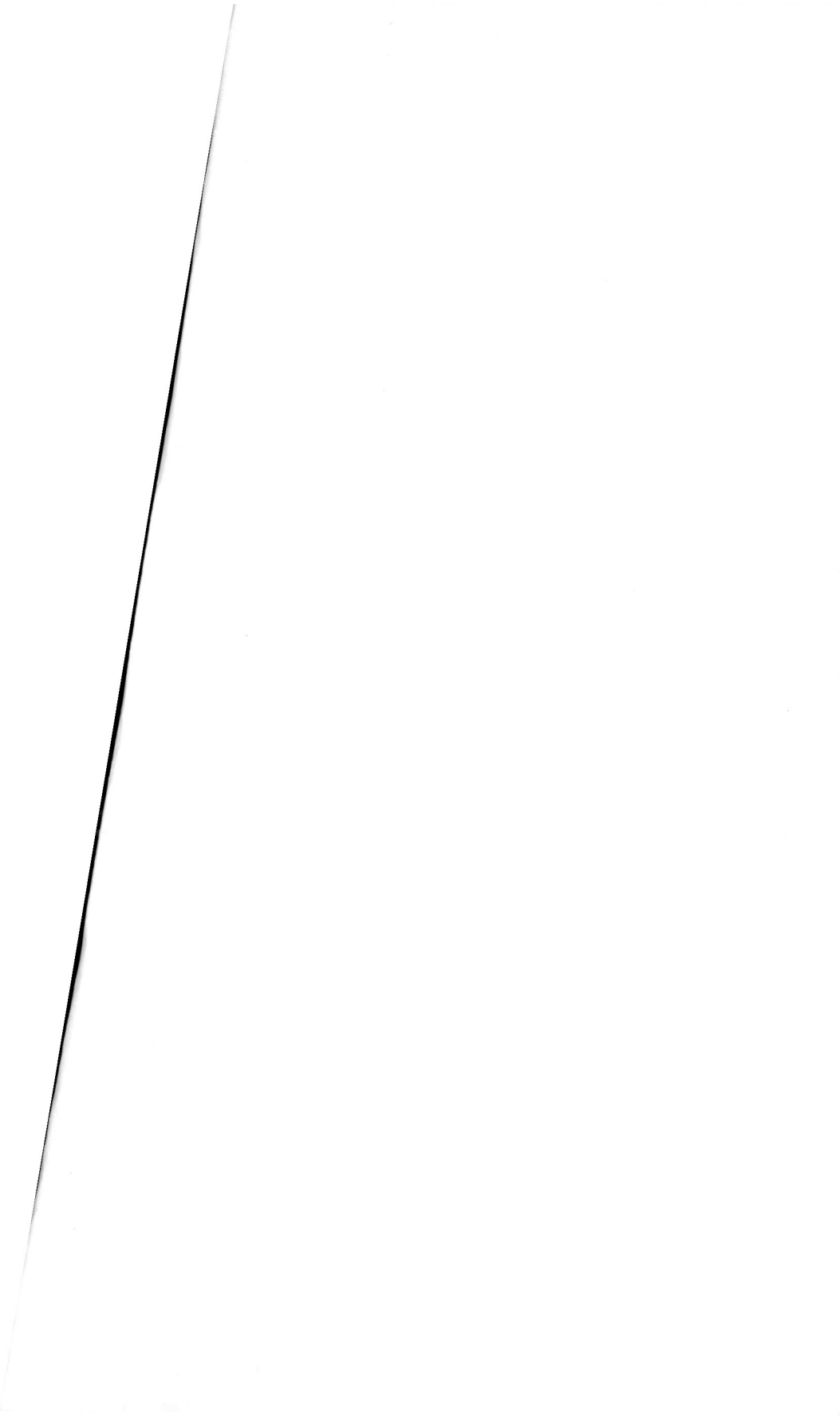